Lecture Notes in Networks and Systems 656

The series "Lecture Notes in Networks and Systems" publishes the latest developments in Networks and Systems—quickly, informally and with high quality. Original research reported in proceedings and post-proceedings represents the core of LNNS.

Volumes published in LNNS embrace all aspects and subfields of, as well as new challenges in, Networks and Systems.

The series contains proceedings and edited volumes in systems and networks, spanning the areas of Cyber-Physical Systems, Autonomous Systems, Sensor Networks, Control Systems, Energy Systems, Automotive Systems, Biological Systems, Vehicular Networking and Connected Vehicles, Aerospace Systems, Automation, Manufacturing, Smart Grids, Nonlinear Systems, Power Systems, Robotics, Social Systems, Economic Systems and other. Of particular value to both the contributors and the readership are the short publication timeframe and the world-wide distribution and exposure which enable both a wide and rapid dissemination of research output.

The series covers the theory, applications, and perspectives on the state of the art and future developments relevant to systems and networks, decision making, control, complex processes and related areas, as embedded in the fields of interdisciplinary and applied sciences, engineering, computer science, physics, economics, social, and life sciences, as well as the paradigms and methodologies behind them.

Indexed by SCOPUS, INSPEC, WTI Frankfurt eG, zbMATH, SCImago.

All books published in the series are submitted for consideration in Web of Science.

For proposals from Asia please contact Aninda Bose (aninda.bose@springer.com).

Noureddine Aboutabit · Mohamed Lazaar ·
Imad Hafidi
Editors

Advances in Machine Intelligence and Computer Science Applications

Proceedings of the International Conference ICMICSA'2022

Editors
Noureddine Aboutabit
National School of Applied Sciences
of Khouribga
Sultan Moulay Slimane University
Khouribga, Morocco

Imad Hafidi
National School of Applied Sciences
of Khouribga
Sultan Moulay Slimane University
Khouribga, Morocco

Mohamed Lazaar
ENSIAS
Mohammed V University
Rabat, Morocco

ISSN 2367-3370 ISSN 2367-3389 (electronic)
Lecture Notes in Networks and Systems
ISBN 978-3-031-28845-6 ISBN 978-3-031-29313-9 (eBook)
https://doi.org/10.1007/978-3-031-29313-9

This Springer imprint is published by the registered company Springer Nature Switzerland AG
The registered company address is: Gewerbestrasse 11, 6330 Cham, Switzerland

Preface

In computer science and in the field of computers more generally, the word "artificial intelligence" has played and plays a very important role. Lately, this term has gained popularity due to recent advances in the field of artificial intelligence and machine learning. Machine learning is the sphere of artificial intelligence in which machines are assigned to perform everyday tasks and are supposed to be more intelligent than humans.

This book is one of the important series that encloses the latest and advanced researches on artificial intelligence and its applications in computer science. It is an interesting manuscript that aims to help students, researchers, industrialists and policy-makers to understand, promote and synthesize innovative solutions and think of new ideas with the application of artificial intelligence concepts. It will also allow to know the existing scientific works and contributions in the literature.

This book identifies original research in new directions and advances focused on multidisciplinary areas and closely related to the use of artificial intelligence in applications of computer science, communication and technology.

The present book contains selected and extended papers of the first international conference on Machine Intelligence and Computer Science applications (ICMICSA'2022). It is the result of a reviewed, evaluated and presented work in ICMICSA'2022 held on November 28–29, 2022, in Khouribga-Morocco.

We thank all authors from across the globe for choosing ICMICSA'2022 to submit their manuscripts.

A sincere gratitude to all keynotes speakers for offering their valuable time and sharing their knowledge with the conference attendees. Special thanks are addressed to all organizing committee members, to local chairs and local committee in National School of Applied Science of Khouribga, to all program committee members, to all chairs of sessions for their efforts and the time spent in order to make this event a success. Many thanks to the Springer staff for their support and guidance.

Noureddine Aboutabit
Mohamed Lazaar
Imad Hafidi

Contents

Artificial Intelligence

Machine Vision

Intelligent Systems

Artificial Intelligence

A New Adaptation Mechanism of the ALNS Algorithm Using Reinforcement Learning

Hajar Boualamia(✉), Abdelmoutalib Metrane, Imad Hafidi, and Oumaima Mellouli

Laboratory of Process Engineering Computer Science and Mathematics (LIPIM), University Sultan Moulay Slimane, Beni Mellal, Morocco
{hajar.boualamia,oumaima.mellouli}@usms.ac.ma,
{a.metrane,i.hafidi}@usms.ma

Abstract. Adaptive Large Neighborhood Search (ALNS) is used to solve NP-hard practical problems. Selecting operators and changing parameters to match a specific purpose is a difficult aspect of metaheuristic design. Our proposal concerns ALNS operator selection. Classical ALNS uses "roulette wheel selection" (RWS) to pick operators during the search phase. Choosing operators with RWS is a big challenge because an operator will almost always take the best spot in the roulette, whereas evolutionary algorithms require a balance between exploration and exploitation. We provide an improved ALNS metaheuristic for the capacitated vehicle routing problem (CVRP) that balances exploration and exploitation. The suggested strategy favors the most successful operators using reinforcement learning, notably the Q-learning algorithm. The experimental study shows that the suggested approach works well and is comparable to the classic ALNS.

Keywords: Reinforcement Learning · Adaptive Large Neighborhood Search · Capacitated Vehicle Routing Problem

1 Introduction

Hyperheuristic is a novel optimization paradigm called "heuristic to pick heuristics." It includes search and learning approaches for generating and selecting heuristics in order to solve optimization problems. The primary difference between hyperheuristics and metaheuristics is that hyperheuristics explore a space of heuristics [1], while metaheuristics directly search a problem space. According to the research on combinatorial optimization problems [2], a classical heuristic can produce the best results for only a subset of instances and perform badly for the rest. Hyperheuristics, on the other hand, may produce significantly better solutions than traditional heuristics or metaheuristics. This is because hyperheuristics can discover the best properties of specific low-level heuristics (move operators, neighborhoods, etc.). Since this is the case, it stands to reason that combining

N. Aboutabit et al. (Eds.): ICMICSA 2022, LNNS 656, pp. 3–14, 2023.
https://doi.org/10.1007/978-3-031-29313-9_1

many heuristics in an effective manner may produce better solutions compared to using them individually. Hyperheuristics are not designed to deal with state-of-the-art problem-specific approaches but rather to give a general framework that may yield high-quality solutions to many different optimization problems. Cowling introduced the term "hyperheuristic" for the first time in [3]. They characterized a hyperheuristic as a technique that works at a higher level than metaheuristics and handles the selection of which low-level heuristic method should be performed at any given moment, based on the features of the area of the solution space being explored. In other words, the hyperheuristic chooses the most promising simple, low-level heuristic (or combination of heuristics) at each stage of the solution in the hopes of improving the solution. If no improvement is discovered or a locally optimal solution is found, the hyperheuristic shifts the search to another region of the solution space by picking relevant heuristics from the specified set. It is not necessary to be familiar with the inner-workings of each lower-level heuristic or the specifics of the objective function of the issue in order to use hyperheuristics. It is enough to know how the optimization process will go (whether it will maximize or minimize), and it will then analyze the value of one or more objective functions that the low-level heuristic returns to it after its call. There are two primary categories of hyperheuristic approaches: heuristic generation and heuristic selection. When selecting, we have a range of atomically predetermined algorithms to choose from. When it comes to generation, we are dealing with a set of heuristic elements.

This chapter will focus on selection heuristics. It starts with a solution that has already been made and then tries to make it better by using local searches and the right neighborhood structure. The most difficult part of hyper-heuristics is determining the optimal sequence of heuristics for a specific scenario instead of solving the problem explicitly. Recently, many algorithms equipped with selection processes have been developed. One of these is the Adaptive Large Neighborhood Search (ALNS) created by Ropke and Pisinger [4]. In this sense, the ALNS framework enables us to apply multiple neighborhoods within the same search procedure in an adaptable manner. This is done by keeping track of how each neighborhood does and changing the choice of methods based on the neighborhood's performance. Since the dynamic selection method stops inefficient operators from running, the adaptive selection of neighborhoods gives the designers more freedom to add more operators. At each step in the ALNS framework, a roulette wheel selection (RWS) is used to choose a destroy operator to destroy the current solution and a repair operator to repair it. The RWS picks the most successful operators based on a probability calculation.

According to the literature, the most difficult aspect of selection heuristics is maintaining a balance between exploration and exploitation in order to avoid the local optimum. In this way, choosing operators with RWS is a big challenge because an operator will take the best spot in the roulette and will almost certainly be chosen if it has a better solution than the other operators. As a result, the algorithm can produce less variety and converge on suboptimal solutions. The results of this study [5] show that only one paper by [6] looked at an adaptive way to choose which destroy and repair operators to send out, using a stochastic universal sampling method made by [7]. This shows a hole in the ALNS hyper-heuristic's ability to come up with new ways that keep exploration and exploitation in balance.

In the same line of thinking, with its demonstrated success in data modeling and predictive analytics, reinforcement learning (RL) has begun to receive immense interest in terms of its applicability across all industries. Reinforcement learning makes the algorithm adaptive by reducing the drawbacks of parameterized techniques. Reinforcement learning has several appealing features. Because it doesn't need the time-consuming meta-optimization process, the RL technique has clear advantages when it comes to guiding the selection of operators in a smart way. It's not necessary to have a complete model of the problem at hand, and RL approaches let the model learn from experience.

In this chapter, we solve the capacitive vehicle routing problem (CVRP) by developing an adaptive large neighborhood search. We utilize the Q-Learning approach as the selection method to choose the destroy and repair operators throughout the search process, which is one of the most widely used RL strategies. This will allow it to self-adapt to the search space and prevent it from being stuck at a local minimum. The suggested method strikes a balance between exploration and exploitation, letting the most productive operators keep coming in during the search phase.

The rest of the chapter is structured as follows. In Sect. 2, we present a set of concepts used in our suggested method. Section 3 explains the suggested approach in detail, including a detailed discussion of the methodologies employed. Section 4 contains several computer tests that show that the suggested method outperforms the classical ALNS. The paper is concluded in Sect. 5.

2 Background Concepts

2.1 Problem Definition and Mathematical Formulation

The Capacitated Vehicle Routing Problem, also known as the CVRP, is a well-known NP-hard (non-deterministic polynomial time) combinatorial optimization problem. The objective of this research was to find a solution to the CVRP. In 1959, Dantzig and Ramser [8] presented the first mathematical formulation and algorithm for solving the CVRP. Five years later, in 1964, Clarke and Wright [9] put forward the first heuristic for resolving the problem. There have been many CVRP solution methods written about and published up to this point.

The goal of the CVRP is to find a set of routes with the lowest total cost for a fleet of vehicles with limited capacity that are based at a single depot and serve a set of clients. There are three restrictions on the routes. In the first case, each client can only be seen once. In the second constraint, the total number of people who want to use each route shouldn't be more than what each vehicle can hold. The last constraint takes into account the fact that each tour begins and ends at the depot. We look at a complete directed graph $G = (V, \text{A})$, where V is the set of nodes labeled i and j, and A is the set of arcs, with node 0 representing the depot. A collection of p vehicles of the same size Q. Every customer $i \in V - \{0\}$

whose value ranges from 0 to V has a positive demand, $d_i \leq Q$. Each arc (i, j) has a trip cost c_{ij} that is not negative. When the vehicle moves in an arc, the value of the binary decision variable x_{kij} is 1. The mathematical model for the CVRP is shown below.

$$Minimize \sum_{k=1}^{p} \sum_{i=0}^{n} \sum_{j=0, i \neq j}^{n} c_{ij} x_{kij} \tag{1}$$

Subject to

$$\sum_{k=1}^{p} \sum_{i=0, i \neq j}^{n} x_{kij} = 1, \forall j \in \{1, \dots, n\}, \tag{2}$$

$$\sum_{j=1}^{n} x_{k0j} = 1, \forall k \in \{1, \dots, p\}, \tag{3}$$

$$\sum_{i=0, i \neq j}^{n} x_{kij} = \sum_{i=0}^{n} x_{kji}, \forall j \in \{0, \dots, n\}, k \in \{1, \dots, p\}, \tag{4}$$

$$\sum_{i=0}^{n} \sum_{j=1, i \neq j}^{n} d_j x_{kij} \leq Q, \forall k \in \{1, \dots, p\}, \tag{5}$$

$$\sum_{k=1}^{p} \sum_{i \in S} \sum_{j \in S, i \neq j} x_{kij} \leq |S| - 1, \forall S \subseteq \{1, \dots, n\}, \tag{6}$$

$$x_{kij} \in \{0, 1\}, \forall k \in \{1, \dots, p\}, i, j \in \{0, \dots, n\}, i \neq j. \tag{7}$$

The goal of Eq. (1) is to lower the total cost of travel. The first restriction (2) in the model makes sure that only one vehicle goes to each customer. The flow constraints (3) and (4) make sure that each vehicle can only leave the depot once and that the number of vehicles arriving at each customer and entering the depot equals the number of vehicles leaving. Constraints (5) list the limits on the vehicle's capacity. This assures that the total number of requests from customers along a route is less than or equal to the vehicle's capacity. The subtour elimination restrictions (6) ensure that in the solution, no cycles get too far from the depot. The remaining essential limitations (7) define the variable definition domains.

2.2 Adaptive Large Neighborhood Search

This section talks about the adaptive large neighborhood search (ALNS) metaheuristic, which was produced by Ropke and Pisinger in 2006 [4] and is used to solve a wide range of hard optimization problems. Adaptive large neighborhood search (ALNS) has been used in a number of routing applications [10–16].

Given the benefits of the ALNS framework, multiple studies [25,26] are turning to ALNS to address vehicle routing problems. Despite the fact that the Pickup and Delivery Problem with Time Windows (PDPTW) and the class of problems in Vehicle Routing Problems were the initial applications of ALNS, multiple studies have proven the significant efficacy of ALNS and the adaptability of this approach in multiple domains.

The ALNS framework is an extension of the Large Neighborhood Search, which was first described in [17]. ALNS improves the proposed solution by repeatedly applying the destroy and repair operator. The conventional ALNS heuristic structure is shown in Algorithm 1.

Algorithm 1. Adaptive Large Neighborhood Search

1: Input : a feasible solution x ; $x^b = x$
2: **repeat**
3: Select a destroy operator and a repair operator using a roulette wheel mechanism based on their past performance $\{\pi_j\}$
4: $x^t = r(d(x))$
5: **if** x^t can be accepted **then**
6: $x = x^t$
7: **end if**
8: **if** $c\left(x^t\right) < c\left(x^b\right)$ **then**
9: $x^b = x^t$
10: **end if**
11: Update scores π_j
12: **until** stop criterion is met
13: Return x^b

The ALNS comprises a set of destroy operators and a set of repair operators, represented by $\Omega^- = \{\Omega_1^-, ..., \Omega_{|\Omega-|}^-\}$ and $\Omega^+ = \{\Omega_1^+, ..., \Omega_{|\Omega+|}^+\}$ respectively, in contrast to its predecessor, which only takes into account one destroy operator and one repair operator. The ALNS heuristic was used by [4] to solve the PDPTW. In the initial design of ALNS, three destroy operators and two repair operators were created. Each destroy or repair operator uses specific criteria to remove or insert requests from the solution under consideration since a PDPTW solution is made up of a series of pick-up and delivery requests. A roulette wheel selection mechanism is used in every iteration of ALNS to choose one destroy and one repair operator from the pool of possible operators. The probability of selecting each operator from the collection of available operators is represented on the roulette wheel by the symbols $p^- = \{p_1^-, ..., p_{|\Omega-|}^-\}$ for the destroy operators and $p^- = \{p_1^-, ..., p_{|\Omega-|}^-\}$ for the repair operators. These probabilities are determined using the appropriate weights w_i^- and w_i^+, which indicate how each operator has performed historically. Every w_i^- and every w_i^+ start off with equal values, which are subsequently modified every s iterations.

In this respect, $p_i^- = \frac{w_i^-}{\Sigma_j^{|\Omega^-|} w_j^-}$ and $p_i^+ = \frac{w_i^+}{\Sigma_j^{|\Omega^+|} w_j^+}$ may be used to mathematically describe how probabilities p_i^- and p_i^+ are calculated. These equations suggest that the operator with the greatest contribution has a high probability of being chosen. The ALNS of [4] uses Metropolis criteria rather than the traditional acceptance criterion used in [17] to minimize the possibility of being stuck at a local optimal point. If $f(S') \langle f(S^*)$, the ALNS accepts the S' as S^*, the ALNS assigns a probability of acceptance of $e^{-(f(s')-f(s))/T}$. In this case, T stands for the current temperature. The starting temperature, T_0, is set as T. T is steadily reduced in each iteration by multiplying it by the cooling rate α parameter. The probability of accepting a worse solution decreases as the T value decreases. The goal of using these acceptance criteria is to allow the ALNS to focus on finding excellent solutions during later iterations while exhaustively exploring the solution space during the first rounds. An ALNS structure typically includes at least four significant components, as shown by Algorithm 1. These elements include the algorithm's stopping criteria (a), the acceptance criterion for a recently discovered solution (b), the architecture of the destroy and repair operators (c), and the adaptive method for choosing the used operators (d). We will concentrate on the ALNS's adaptive mechanism in this work and how it selects its operators over time.

2.3 Operator Selection and Adaptive Mechanism

Adaptive large neighborhood search has already shown that it works well in a variety of situations. This success is due, in large part, to the fact that ALNS can handle mixed or unstructured search spaces. For ALNS algorithms to work well on a given problem, two decisions about operator variation must be made. The first is to decide which operator to use to make a new solution, and the second is to decide how often each of the chosen operators should be used. So, one of the functions that ALNS needs is operator selection.

The main reason for operator selection is that some operators are more useful at different stages of the search process than others. The standard version of the ALNS uses the roulette wheel selection mechanism to select the successful operator at every stage of the space. According to the research [5], only one paper by [6] looked at an adaptive way to choose which destroy and repair operators to use. This was a stochastic universal sampling technique developed by [7]. As with the roulette wheel, random probabilities are given to groups of neighborhood operators (in this case, the sets of destroy and repair operators) in order to pick some of those groups. In contrast, stochastic universal sampling utilizes several N evenly spaced pointers rather than a single pointer, resulting in a decrease in bias in the selection of operators.

Research conducted by [18] also showed the value and significance of the ALNS's adaptive mechanism. Based on 134 investigations and 25 various ALNS implementations, this analysis carried out a thorough investigation of the adaptive layer of ALNS. To evaluate whether the adaptive mechanism in ALNS makes a major contribution or not, they looked at the "average improvement of the goal function produced by the adaptive layer." Their research showed that adding an adaptive mechanism increased the objective value by 0.14% and should be done in a number of situations.

To make an adaptive operator selection method, we have to do two things: assign credit and choose the operator. The Credit Assignment (CA) scheme gives an operator credit based on the effect of their most recent application on the current search optimization process. Various CA techniques have been described in the literature, such as learning automata-based CA [19], extreme value-based CA [20], and Q-learning-based CA [21]. Soft-Max Selection [22], Probability Matching Selection [23], and Epsilon Greedy Selection [24] are the three best-known and most promising algorithms.

3 Proposed Method

The ALNS metaheuristic makes it possible to locate solutions of high quality. The use of the roulette wheel as a method for operator selection enables the selection of either destruction or construction operators on an individual level for each iteration. In this scenario, selecting the same operator each time is likely to occur if a single operator produces solutions of satisfactory quality throughout the research phase. As a consequence of this, the algorithm can stop exploring new possibilities and instead converge on solutions that are less than optimal. When it comes to metaheuristics, it is essential to keep in mind the importance of maintaining a good balance between exploring and exploiting potential solutions. As shown in the previous section, only one article from [6] looked into an adaptive way to choose which destroy and repair operators to use. This shows that there is a big gap in the literature in the development of new operator selection mechanisms for ALNS. The exploration-exploitation trade-off affects accuracy and convergence speed. Choosing the right neighborhood structure during a search or after a change depends on many things and affects how well the structure works.

So, we try to use smart methods, such as reinforcement learning, particularly the Q-Learning algorithm, to choose operators (destroy, repair) within the ALNS. Q-learning is a method for reinforcement learning that uses trial and error rather than a pre-existing model. In Q-learning, the value of the cumulative reward for each state-action combination is instantly calculated using a Q-function on the Eq. (8).

$$Q(s_t, a_t) = (1 - \alpha) \cdot Q(s_t, a_t) + \alpha \cdot [r_t + \gamma \cdot \max Q(s_{t+1}, a_{t+1})] \tag{8}$$

In this equation, $Q(s_t, a_t)$ represents the total discounted benefit of acting in state s_t at time t. The learning rate, denoted by α, must be an integer between 0 and 1, and the reward or punishment received when acting in state s_t at time t is denoted by r_t. The highest $Q-$value of the next state, and the best action the agent can take in the next state, is denoted by $\max Q(s_{t+1}, a_{t+1})$, and γ is the discount factor which also must have a value between 0 and 1.

The steps of this program are shown in Algorithm 2.

Algorithm 2. Q-Learning

1: Initialize $Q(s, a)$ (for each episode)
2: Initialize s
3: **repeat**
4: Choose a from s using policy (e.g., epsilon-greedy)
5: Take action a, observe r and s'
6: $Q(s_t, a_t) = (1 - \alpha) \cdot Q(s_t, a_t) + \alpha \cdot [r_t + \gamma \cdot \max Q(s_{t+1}, a_{t+1})]$
7: **until** s is terminal

Combining ALNS with RL is an innovative idea. Instead of selecting the destruction and construction operators independently at each iteration, the reinforcement agent uses the Q-Table values as a reference for the search agents of the ALNS metaheuristic and selects a couple of destruction and construction operators. A system of rewards and punishments is used to maintain this table.

The suggested method is presented in (see Fig. 1).

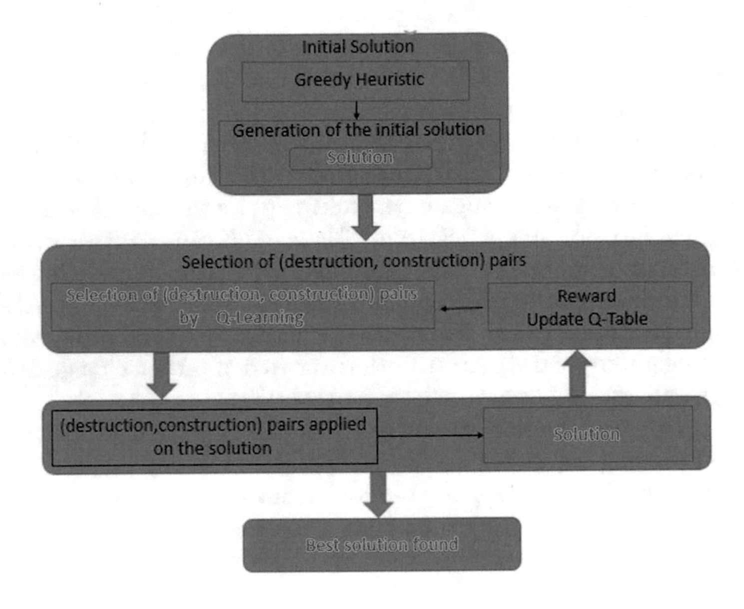

Fig. 1. The proposed method

To construct the first solution, we employ a straightforward greedy approach. It decides what to do based on what seems to be the best choice at the moment. The greedy method tries out different configurations to find the best place for each node that hasn't been inserted based on where it will cost the least.

Then, we use two subclasses of operators-destroy and repair-because the ALNS architecture enables us to use several neighborhoods within the same searching process. We set up the following destroy operators (random removal, worst removal, related removal, and cluster removal) to destroy the current solution. The repair operators (basic greedy heuristic and regret heuristic) are then used to repair it. The ALNS generates a new solution by removing and inserting elements from the existing solution at each iteration. During the search process, we should choose the operator that performs the best.

Using a Q-Learning algorithm, we automatically pick the destruction-construction operator pairs in our method. After the greedy heuristic constructs the first solution, as seen in (see Fig. 1), it picks a pair of destruction-construction operators from those that are presently accessible. When the state's destruction-construction pairs reach a local minimum, a new pair is chosen as an action. The algorithm selects the action with the greatest Q-value. However, in order to explore other options, a random component is employed. The Q-value is updated based on the execution performance of the specified action. The executed action transforms into a state. The state is then updated, and the best global solution is recorded.

The research suggests that one of the most challenging tasks in this situation is to design a selection procedure that may achieve a balance between taking use of behaviors that have been shown to be rewarded and exploring novel behaviors. In this study, the behavior policy is epsilon-greedy whereas the goal policy of Q-Learning is greedy. The agent selects a random action with probability ϵ and then selects the action with the greatest Q-value with probability $1 - \epsilon$ in the epsilon-greedy search. Using an epsilon-greedy strategy with an exponential decay, we begin with a value of $\epsilon = 0.6$ and gradually lower it. This method restricts exploration while gradually increasing exploitation. It chooses where to achieve a balance between exploration and exploitation.

4 Computational Experiments

The objective is to compare the performance of traditional ALNS with the newly presented approach before discussing the computational results. All of the algorithms were performed in the trials with the same number of iterations and the same configurations for the common adjustable parameters. To evaluate how well the recommended strategy worked, we did a number of computer experiments. Four datasets from the CVRP library (http://vrp.atd-lab.inf.puc-rio.br) were used to test the algorithms once they were developed in Python. Since the algorithms are non-deterministic, they were run 20 times on each instance for a predetermined period of time.

Table 1 shows the average of 20 runs with 10 CVRP instances (value of the objective function, processing time in seconds).

Table 1. Performance of traditional ALNS and the proposed ALNS.

Instances	Optimal CVRP	Initial Solution	Traditional ALNS		Proposed ALNS		gap (%)	
	Obj	Obj	Obj	Time	Obj	Time	Obj	Time
B-n45-k6	678	933,63	709,5	218,55	691,05	140,36	2.60	35.77
B-n64-k9	861	1136.47	894,15	365,41	874,13	213,41	2.23	41.59
B-n78-k10	1221	1871.65	1279,19	203,26	1242,97	195,51	2.83	3.81
tai75a	1618,36	2439	1664,25	238,46	1632,18	186,32	1.92	21.86
tai75d	1365,42	2199.79	1674,23	118,72	1384,77	102,62	17.28	13.56
Golden1	5623,47	6305,45	5716,61	390,4	5667,16	370,02	0.86	5.22
Li21	16212,83	20509,28	20147,82	825,36	18906,44	700,8	6.16	15.09
Li25	16665,7	18141,13	18091,13	896,32	17564,67	703,6	2.91	21.50
Li26	23977,73	30417,96	30005,96	985,7	28123,45	809,6	6.27	17.86
Li32	37159.41	55317,16	49123.17	1080	41036.42	920.40	16.46	14.77

According to the results shown in Table 1, the suggested technique with RL component outperforms the traditional ALNS in terms of mean results and reduces the amount of time required for processing.

5 Conclusion

This work develops the adaptive large neighborhood search (ALNS) approach to solve the CVRP. Learning has been incorporated into the operator selection process for construction and destruction. More precisely, Q-learning has been developed to select the appropriate operators at each iteration. The computer results shown in this paper show that the proposed method is better than the traditional ALNS in terms of both the quality of the solution and the time it takes to run. For future works, it is fascinating to perform the computational tests with other combinatorial optimization problems that are available in the literature. Depending on the context of the hyperheuristics, it is important to keep the balance between exploration and exploitation. In this situation, a way to find new research might be to come up with many ways to choose actions in order to find the best one.

References

1. Mlejnek, J., Kubalik, J.: Evolutionary hyperheuristic for capacitated vehicle routing problem. In: Proceedings Of The 15th Annual Conference Companion on Genetic and Evolutionary Computation, pp. 219-220 (2013)
2. Sanchez, M., Cruz-Duarte, J., Ortız-Bayliss, J., Ceballos, H., Terashima-Marin, H., Amaya, I.: A systematic review of hyper-heuristics on combinatorial optimization problems. IEEE Access **8**, 128068–128095 (2020)
3. Cowling, P., Kendall, G., Soubeiga, E.: A hyperheuristic approach to scheduling a sales summit. In: International Conference on the Practice and Theory of Automated Timetabling, pp. 176-190 (2000)

4. Ropke, S., Pisinger, D.: An adaptive large neighborhood search heuristic for the pickup and delivery problem with time windows. Transp. Sci. **40**, 455–472 (2006)
5. Mara, S., Norcahyo, R., Jodiawan, P., Lusiantoro, L., Rifai, A.: A survey of adaptive large neighborhood search algorithms and applications. Computers & Operations Research, pp. 105903 (2022)
6. Chowdhury, S., Marufuzzaman, M., Tunc, H., Bian, L., Bullington, W.: A modified Ant Colony Optimization algorithm to solve a dynamic traveling salesman problem: a case study with drones for wildlife surveillance. J. Comput. Design Eng. **6**, 368–386 (2019)
7. Baker, J., et al.: Reducing bias and inefficiency in the selection algorithm. In: Proceedings of the Second International Conference On Genetic Algorithms, vol. 206, pp. 14–21 (1987)
8. Dantzig, G., Ramser, J.: The truck dispatching problem. Manage. Sci. **6**, 80–91 (1959)
9. Clarke, G., Wright, J.: Scheduling of vehicles from a central depot to a number of delivery points. Oper. Res. **12**, 568–581 (1964)
10. Arda, Y., Crama, Y., François, V.: An adaptive large neighborhood search for a vehicle routing problem with multiple trips and driver shifts (2013)
11. Chen, S., Chen, R., Wang, G., Gao, J., Sangaiah, A.: An adaptive large neighborhood search heuristic for dynamic vehicle routing problems. Comput. Electr. Eng. **67**, 596–607 (2018)
12. Hof, J., Schneider, M.: An adaptive large neighborhood search with path relinking for a class of vehicle-routing problems with simultaneous pickup and delivery. Networks **74**, 207–250 (2019)
13. Li, Y., Chen, H., Prins, C.: Adaptive large neighborhood search for the pickup and delivery problem with time windows, profits, and reserved requests. Eur. J. Oper. Res. **252**, 27–38 (2016)
14. He, L., Liu, X., Laporte, G., Chen, Y., Chen, Y.: An improved adaptive large neighborhood search algorithm for multiple agile satellites scheduling. Comput. Oper. Res. **100**, 12–25 (2018)
15. Sacramento, D., Pisinger, D., Ropke, S.: An adaptive large neighborhood search metaheuristic for the vehicle routing problem with drones. Transp. Res. Part C: Emerg. Technol. **102**, 289–315 (2019)
16. Shirokikh, V., Zakharov, V.: Dynamic adaptive large neighborhood search for inventory routing problem, pp. 231–241. Modelling, Computation And Optimization In Information Systems And Management Sciences (2015)
17. Shaw, P.: Using constraint programming and local search methods to solve vehicle routing problems. In: International Conference on Principles and Practice of Constraint Programming, pp. 417-431 (1998)
18. Turkeš, R., Sörensen, K., Hvattum, L.: Meta-analysis of metaheuristics: quantifying the effect of adaptiveness in adaptive large neighborhood search. Eur. J. Oper. Res. **292**, 423–442 (2021)
19. Gunawan, A., Lau, H., Lu, K.: ADOPT: combining parameter tuning and adaptive operator ordering for solving a class of orienteering problems. Comput. Ind. Eng. **121**, 82–96 (2018)
20. Silvestre Fialho, Á.: Adaptive operator selection for optimization. (Paris 11, 2010)
21. Wauters, T., Verbeeck, K., Causmaecker, P., Berghe, G.: Boosting metaheuristic search using reinforcement learning. In: Hybrid Metaheuristics, pp. 433-452 (2013)
22. Gretsista, A., Burke, E.: An iterated local search framework with adaptive operator selection for nurse rostering. In: International Conference On Learning And Intelligent Optimization, pp. 93-108 (2017)

23. Fialho, Á., Costa, L., Schoenauer, M., Sebag, M.: Extreme value based adaptive operator selection. In: International Conference on Parallel Problem Solving From Nature, pp. 175-184 (2008)
24. Santos, J., Melo, J., Neto, A., Aloise, D.: Reactive search strategies using reinforcement learning, local search algorithms and variable neighborhood search. Expert Syst. Appl. **41**, 4939–4949 (2014)
25. Mehdi, N., Abdelmoutalib, M., Imad, H.: A modified ALNS algorithm for vehicle routing problems with time windows. J. Phys. Conf. Ser. **1743**, 012029 (2021)
26. Nasri, M., Hafidi, I., Metrane, A.: Multithreading parallel robust approach for the VRPTW with uncertain service and travel times. Symmetry. **13**, 36 (2020)

A Solution Based on Faster R-CNN for Augmented Reality Markers' Detection: Drawing Courses Case Study

Hamada El Kabtane[1(✉)], Fatima Zohra Ennaji[2], and Youssef Mourdi[3]

[1] SMARTE Systems and Applications (SSA), National School of Applied Science - UCA, Marrakesh, Morocco
h.elkabtane@uca.ma

[2] Laboratory of Process Engineering, Computer Science and Mathematics (LIPIM), National School of Applied Science - USMS, Khouribga, Morocco
f.ennaji@usms.ma

[3] Modeling and combinatorial laboratory, Polydisciplinary Faculty of Safi - UCA, Safi, Morocco
y.mourdi@uca.ac.ma

Abstract. This work aims to prove the relevance of Augmented Reality (AR) in distant Practical Activities. To do so, several practical activities have been proposed to two groups (AR-Grp and N-Grp). The participants of N-Grp use only the 2D printed models during the period of the experimentation while the participants of AR-Grp use the proposed solution where the models are presented in three dimensions using AR. In this work, customs markers have been proposed and was detected in the scene using the Faster R-CNN so the 3D object will be placed in the appropriate position. Finally, a quantitative and qualitative mertrics were used to explore the impact the integration of AR on the drawing courses.

Keywords: Augmented reality · Faster R-CNN · Virtual environments · E-learning

1 Introduction

The significant development of digital technologies and computer tools, in addition of their adoption in all our daily activities, actualize the need for their use in the educational process [4].

E-learning refers to the use of electronic technologies to access educational programs. Following an automated process, these technologies are good tools for making available to learners' courses with defined materials and automatically marked tests.

Now that affordable e-learning solutions exist, all it takes is a good e-learning tool to make learning easy from any place. The technology has evolved so much that the geographic divide is being bridged by the use of tools that give the impression of being in the classroom.

N. Aboutabit et al. (Eds.): ICMICSA 2022, LNNS 656, pp. 15–25, 2023.
https://doi.org/10.1007/978-3-031-29313-9_2

Besides all the strengths proposed by e-learning platforms, it did not prevent the feeling of isolation and lack of assistance and support towards students, and a feeling of speaking in a vacuum because of the absence immediate feedback from students towards instructors.

Students from various disciplines (such as Mechanical and industrial disciplines) may employ real 3D models during the learning process (Engineering Graphics and Descriptive Geometry, Engineering and Computer Graphics etc.). The use of these 3D real objects helps the students to understand metric and positional problems in descriptive geometry and look at solutions from different perspectives [12].

Unfortunately, this solution suffers from several problems like the high cost of these real objects, over and above the limited access to these objects since they belong to universities and educational institutions.

These last years, digital technology has made a huge leap in the development and expansion of areas of use. Augmented reality (AR) is a technology that combined real world with virtual objects in real time. In the beginning, technology was used mainly in the industry, military and computer games, but now AR enters almost all spheres of human social activity: education, architecture, advertising, medicine, economics, etc. [11].

To solve the listed problems, we propose to use Augmented Reality (AR). The creation of virtual version for the real objects used during the classroom session, can solve the problem of the expensive cost needed to create the objects and also there is no need to think about the storage problem so the learners will have the possibility to access to the objects and manipulate them anytime and anywhere.

2 Related Works

2.1 Augmented Reality (AR)

Augmented Reality (AR) was first used by Caudell and Mizell [2] to reference the act of adding and superimposing virtual information (2D or 3D computer generated 3D objects) on the real world. We can also say that AR is the combination of the real physical world captured by a camera and virtual objects generated by the computer.

Augmented reality must integrate three characteristics [7]:

- Combination of the real world with the virtual world
- Feasibility in three dimensions
- Real-time interaction

The general process in an AR system is composed of seven steps. This process begins with capturing the real world using a camera, then the two worlds (real and virtual) must align in order to adjust their overlap afterwards. As a third step, the detection and the tracking of the objects is necessary in order to ensure the positioning location of virtual objects in the real world that is the fourth step.

The fifth step is the generation of the virtual scene taking into consideration the new calculations from the third step before performing the sixth step where the overlay of the virtual scene on the captured real-world video is performed. The last step is to display the result on the display device.

Researchers advocate the AR technology and find it efficient since it is applied in several fields: military, educational, medical, advertising, architectural, industrial, artistic, etc.

2.2 AR in E-learning

One of the missing aspects in e-learning systems is user interaction. Users are compelled to learn without any form of interaction when learning with previous online learning systems. As a result, they risk losing interest quickly, and the content produced can be easily forgotten after being learned. The e-learning system can be extended to include intuitive interaction using AR technology. Virtual information overlaid on actual content can help AR users improve their perception of the real world and help them better understand real objects. This is one of the advantages offered by the integration of the AR in the learning process.

AR technology offers students an interactive interface to learn and explore in different thematic environments in a more engaging and motivating way. The teachers are aware that the use of 3D images and any visualization technique for the introduction of content helps and reinforces learning. Indeed, a large body of published research about the application of AR in educational settings, for a wide variety of learning domains, shows that the implementation of AR in the classroom helps improve the learning process, increases student motivation and facilitates the teacher's job [6]. According to Wang and al [12], and Akçayır [1], research in this area should continue and should be approached to discover the affordances and characteristics of AR in education.

As the summary in Table 1 shows, there are many areas where AR technology is suitable and applied for teaching - learning. Most of the research studies have demonstrated positive feedback from participants about the RA system under study.

3 The Proposed Solution

3.1 The Purpose of the Study

The global objective of this article is to test the relevance of AR technologies in drawing lessons on students' academic level and their spatial imagination. In addition, the answers to the following questions will be sought in the following section:

- Is there an effect on the integration of AR in Practical Activities?
- What is the opinion of the participants (students of the 2 groups and teachers) on this technology?

Table 1. A summary of learning areas using AR

Study	Area	Purpose of using AR
Jorge Martín-Gutiérrez and al. (2015) [8]	Electrical engineering	Provide interactive and autonomous training and offer collaborative performance in the virtual lab between students without the help of a teacher.
Chang and al. (2013) [13]	Medical education (Surgical training)	Provide surgical training, plan and guide medical interventions
Xiao and al. (2016) [14]	Astronomy	watching the solar system video and recognizing and observing planets' features
Serrano Vergel (2020) [10]	Medical education (anatomy)	Teach and test knowledge of anatomy
Cerqueira and Kirner (2012) [3]	Mathematics	Teach geometry using 3D geometric concepts
Fleck and Simon (2013) [5]	Astronomy	Show augmented views of celestial bodies and support learning using visual spatial guides and views of an Earth observer

3.2 Participants

The group, constituting the subject of our study, were first-year students of Engineering's department. The total number of the attendees were 73 students, where 58.9% of them were males. All the participants were divided, randomly, into two groups; 36 students in the first group, entitled 'AR-Grp' (used AR in PAs) and 37 students in the normal group 'N-Grp' (used static images and papers for the same PA as AR-Grp). Both groups attended the same course and therefore received the same theoretical foundations.

3.3 Research Method

To test the efficiency of the integration of AR in drawing courses, a group of instructors proposed six practical activities. Those last are PAs of 'perspective-drawing', the learners had to draw objects from an angles.

The proposed PA were delivered to the designers' staff to create its 3D version. Those 3D objects were used to create the Augmented Reality environment. We distributed a survey to the learners; this last contains quantitative and qualitative questions. To evaluate the understanding level, we distributed a quiz to the participants at the end of each course.

The process used in our solution is presented in Fig. 1. This process describes how we proceed to deliver the PA to the learners. Firstly, the instructor starts by the defining the canvas of the PA, in addition of the objects that have to be drawn. Then, the creation and design of the 3D models is performed. To do so, we used SketchUp, 3ds Max and Cinema4d. The listed softwares are paid and require appropriate knowledge and skills. Therefore, we used the services of a designer.

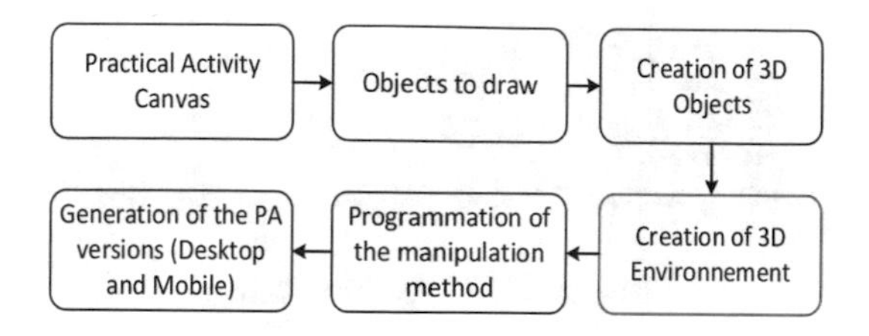

Fig. 1. The process of the proposed solution

The next step is the development of the solution. To manipulate the 3D objects used in practical activities, we have proposed the use of markers.

The machine that generate the AR content focuses on theses markers to know and to determinate the orientation and the position where to put the AR object. Also these markers, which are placed in the real world, can used to ensure the tracking phase (Fig. 2).

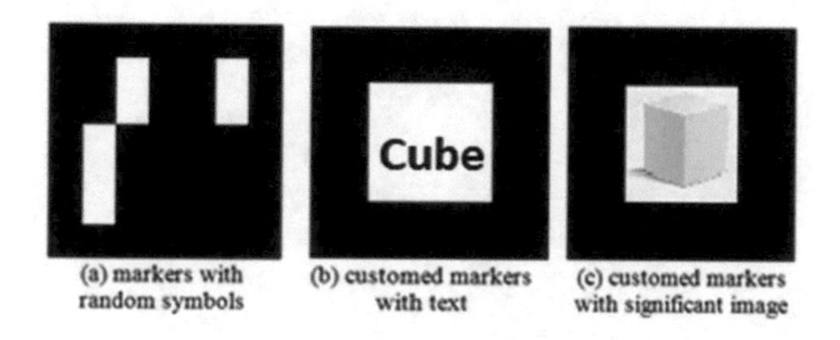

Fig. 2. Some examples of markers

In our approach, a hybrid type of markers have been proposed based on the use of a black frame including an image inside the white square. This choice will ensure easy and precise detection of markers, an optimal processing time and customizable and meaningful markers.

The detection of markers is carried out using "Faster R-CNN" [9]. This last has been proposed in 2015 and it is considered as one of the famous object detection architectures. It has been proved that The Faster R-CNN produces better region proposals compared to generic methods like Selective Search and EdgeBoxes [9]. Faster R-CNN is composed in 3 steps (Figs. 3 and 4):

- Convolution layers: it aims to train filters in order to extract the features from the training set.

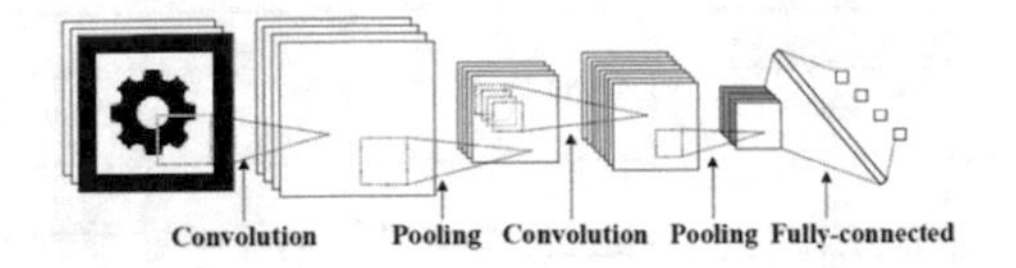

Fig. 3. CNN Architecture

- Region Proposel Network: RPN is a small neural network sliding over the last feature map of the convolution layers. It aims to predict the existence of an object and it's bounding box.
  - The fully connected layer "cls" is a binary classifier (2D) that generates the objectness score for each region. If the first element is 0 and the second is 1 then the region represents a marker, otherwise it represents a background.
  - "REG" is a fully connected layer that returns a 4D vector defining the bounding box of the region.

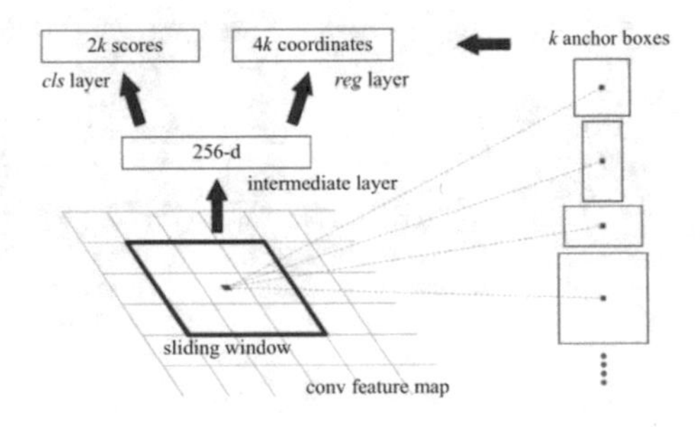

Fig. 4. The Region Proposel Network

- Classes and Bounding Boxes prediction: At this point, a fully connected neural network is used which takes as input the regions proposed by the RPN and predicts the class of objects.This step will help to figure out the location of the marker in order to place the appropriate 3D object in the appropriate location.

It is obligatory to test the display of the PAs on screens, the operability of the movement and the rotation of the virtual objects in addition of the manipulation method. To do so, we have tested the new PA version in several computers like:

- HP EliteBook 840 G3 Processor: Intel Core i5-6200U frequency: 2.30 GHz, 2400 MHz MHz, 2 cores, 4 logical processors, RAM 8 Go DDR4.
- Lenovo 80MX Processor: Intel Core i5-6200U frequency: 2.30 GHz, 2401 MHz MHz, 2 cores, 4 logical processors, RAM 8 Go DDR3.
- Asus VivoBooK 14 s, processor Intel Core i3-8145U frequency: 2.10 GHz RAM 8 GB DDR4.

According to testing results, we can conclude that the solution works correctly on computer with Windows OS regardless of processor type, screen matrix and RAM size.

4 Findings

The 73 learners were divided randomly into two groups: N-Grp and AR-Grp. The content of the courses was prepared and presented to the N-Grp and AR-Grp in a similar way so we can evaluate the benefit of integrating virtual manipulations in drawing courses. The AR-Grp's participants had the possibility to perform some AR manipulations using markers.

To assess the quality of the proposal, we were concerned with the satisfaction regarding the setting of goals and objectives, we can quote: increasing the levels of satisfaction and understanding of participants, and evaluating the ease of manipulation of the virtual objects created.

A survey was distributed, at the end of the experimentation, on the students of the two groups to see the quality of the course content, the clarity of the PAs' statements and the degree of satisfaction of the students.

The result of this survey (in Fig. 5, Fig. 6 and Fig. 7) shows that the learners in the AR-Grp were more satisfied (94,44% for the AR-Grp vs 89,19% for the N-Grp) and they found the content of the courses more interesting (94,44%) then the learners of the N-Grp (83,78%).

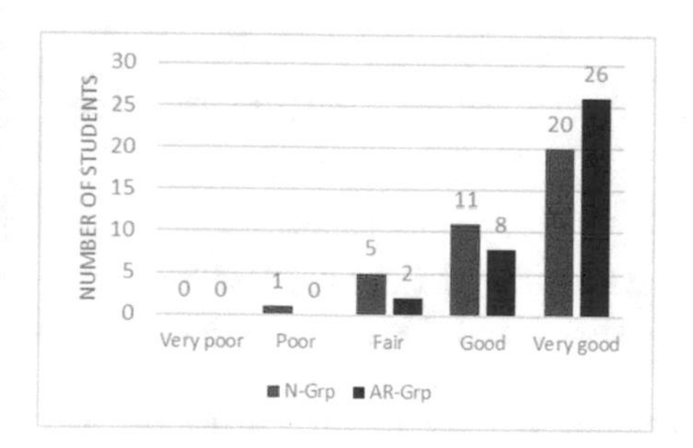

Fig. 5. The quality of the course content

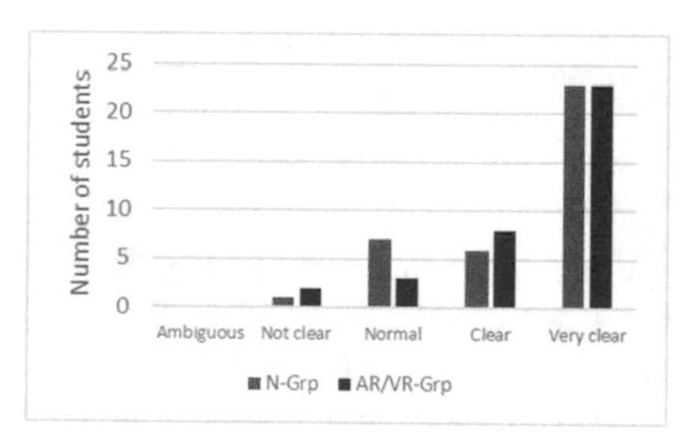

Fig. 6. The clarity of the PA's statements

To test the effectiveness of our proposition the integration of the AR in the drawing courses, all the 36 students (the AR-Grp) were invited to answer a survey that contains several questions about the integration of AR in the drawing class and comparing it to the traditional way.

The quantitative questions were structured as a five-level Likert item, like the first survey, where the first two choices are negatives, the last two answers are positives and the middle answer is neutral.

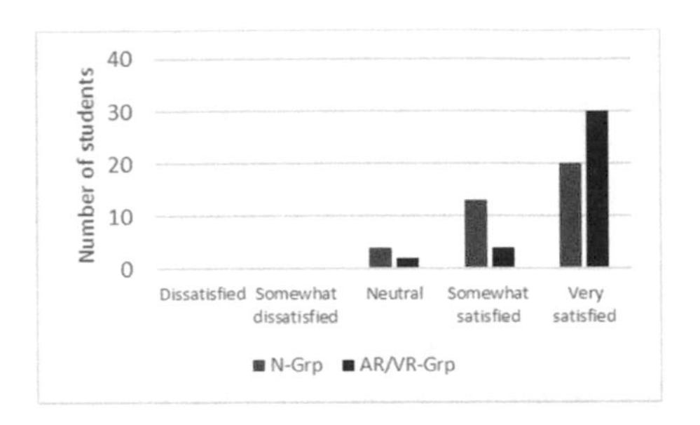

Fig. 7. Are you satisfied?

The survey was composed from four questions, the first one, in Fig. 8, aims to test the ease of the use of our proposition, 84,93% of the participants found that the PA are easy to use and to manipulate.

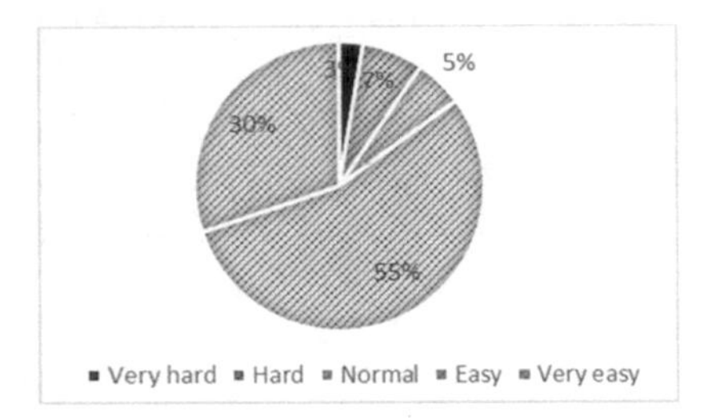

Fig. 8. The ease of the use of the PA?

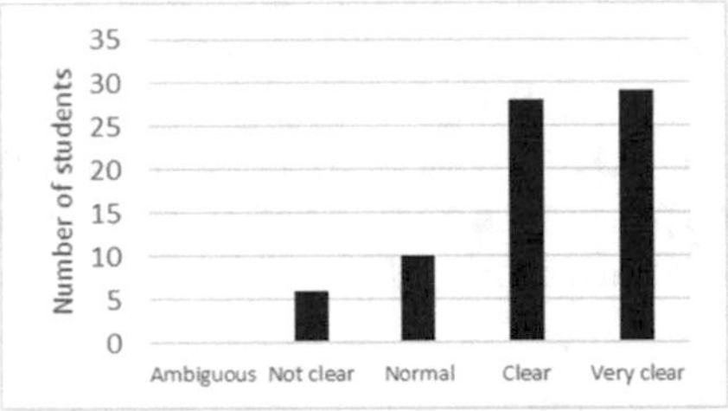

Fig. 9. How clear was the practical activities' descriptions?

The second question, in Fig. 9, shows the clarity of the practical activities' descriptions, 78,08% of the students answer that the descriptions was clear and easy to read and to understand.

The third question (Fig. 10), focus on the added value of integrating AR into drawing courses, 90,41% of the learners answered "Very interesting" on the fact of adding AR in drawing courses.

The Fig. 11 focuses on adding AR to drawing lessons, does that help? It has been noticed that 78,08% of the students answered that it's "helpful" and "very helpful" to integrate AR in drawing course. For the students who rated this question higher than the rate "Neutral", another question was asked. "Where does the added value of the integration of AR appear in drawing lessons?" The majority answered that these technologies help them to imagine the appearance of the object from a precise view, also the passage from one view to another is very easy and does not require effort.

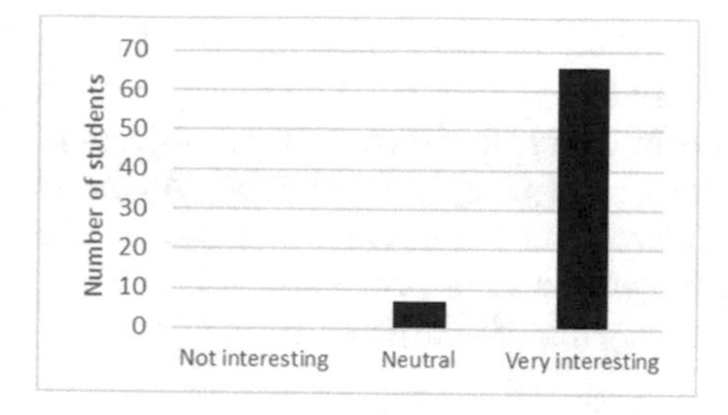

Fig. 10. Something interesting captured your attention?

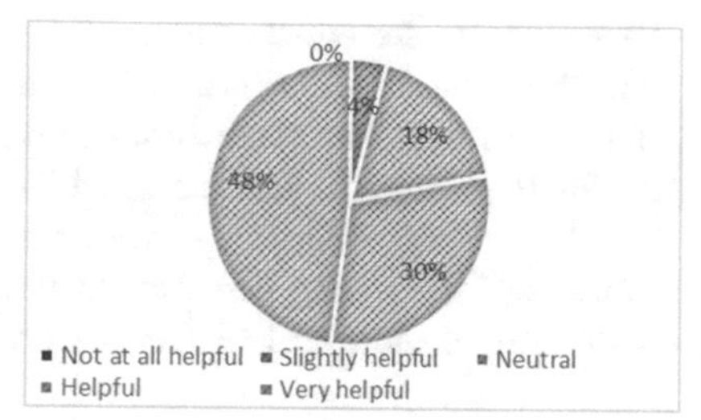

Fig. 11. Did you find the integration of virtual practical activities based on AR in drawing courses helpful?

At the end of each PA, a quiz was proposed to the learners in both groups. The purpose is to evaluate the impact of integrating the AR in the practical activities on the imagination and the understanding level of the students.

The majority of the students in AR-Grp (81%) successfully passed the quizzes, of which 48,65% scored above 90%. However, only 48,65% of the N-Grp were above the average (Fig. 12). Moreover, the evolution rate of the participants in AR-Grp is developing significantly compared to the N-Grp.

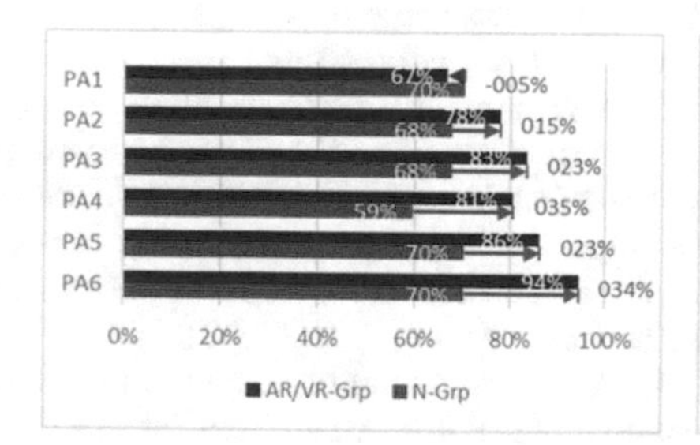

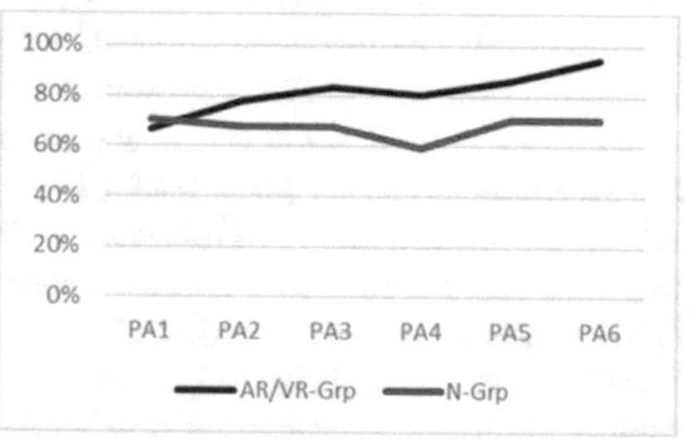

Fig. 12. The understanding-level based on the quizzes in each PA results

5 Discussion and Conclusion

In the drawing courses, the learners have to draw virtual objects, these last can be in their real form, in front of the learners or just printed on paper and the learners have to imagine them. The first solution, in most of the time, can be expensive, the instructor must provide objects, one or more models, for each activity or exercise to the learners and that requires additional costs.

In this paper, we proposed the integration of virtual Activities in the drawing courses. To do so, customized markers have been used and detected using Faster R-CNN in order to place the created 3D object in the right place.

To attest to the effectiveness of the solution, 73 students took part in this experience, those last were divided randomly to two groups. The first one is an ordinary class that use traditional method in course and just printed objects in papers for the practical activities, while the second group uses Virtual PAs based on AR. The collected findings from the quizzes and the survey have proved the usefulness of integrating the virtual manipulations to improve and enhance the learners' imagination and to facilitate the drawing of complex objects.

References

1. Akçayır, M., Akçayır, G.: Advantages and challenges associated with augmented reality for education: a systematic review of the literature. Educ. Res. Rev. (2017). https://doi.org/10.1016/j.edurev.2016.11.002
2. Caudell, T., Mizell, D.: Augmented reality: an application of heads-up display technology to manual manufacturing processes. In: Proceedings of the Twenty-Fifth Hawaii International Conference on System Sciences **ii**, vol. 2, pp.659–669 (1992). https://doi.org/10.1109/HICSS.1992.183317. http://ieeexplore.ieee.org/lpdocs/epic03/wrapper.htm?arnumber=183317
3. Cerqueira, C., Kirner, C.: Developing educational applications with a non-programming augmented reality authoring tool. In: EdMedia: World Conference on Educational Media and Technology, pp. 2816–2825 (2012)
4. Fedorenko, E.H., Velychko, V.Y., Stopkin, A.V., Chorna, A.V., Soloviev, V.N.: Informatization of education as a pledge of the existence and development of a modern higher education. In: CEUR Workshop Proceedings (2019)
5. Fleck, S., Simon, G.: An augmented reality environment for astronomy learning in elementary grades: an exploratory study. In: Proceedings of the 25ième conférence francophone on IHM, pp. 14–22 (2013). https://doi.org/10.1145/2534903.2534907
6. Khan, T., Johnston, K., Ophoff, J.: The impact of an augmented reality application on learning motivation of students. Advances in Human-Computer Interaction (2019). https://doi.org/10.1155/2019/7208494
7. Krüger, J.M., Buchholz, A., Bodemer, D.: Augmented reality in education: three unique characteristics from a user's perspective. In: ICCE 2019 - 27th International Conference on Computers in Education, Proceedings (2019)
8. Martin-Gutierrez, J., Fabiani, P., Benesova, W., Meneses, M.D., Mora, C.E.: Augmented reality to promote collaborative and autonomous learning in higher education. Comput. Human Behav. **51**, 752–761 (2015). https://doi.org/10.1016/j.chb.2014.11.093
9. Ren, S., He, K., Girshick, R.B., Sun, J.: Faster R-CNN: towards real-time object detection with region proposal networks. CoRR abs/1506.01497 (2015). http://arxiv.org/abs/1506.01497
10. Serrano Vergel, R., Morillo Tena, P., Casas Yrurzum, S., Cruz-Neira, C.: A comparative evaluation of a virtual reality table and a hololens-based augmented reality system for anatomy training. IEEE Trans. Hum.-Mach. Syst. (2020). https://doi.org/10.1109/THMS.2020.2984746
11. Syrovatskyi, O.V., Semerikov, S.O., Modlo, Y.O., Yechkalo, Y.V., Zelinska, S.O.: Augmented reality software design for educational purposes. In: CEUR Workshop Proceedings (2018)

12. Wang, P., Wu, P., Wang, J., Chi, H.L., Wang, X.: A critical review of the use of virtual reality in construction engineering education and training (2018). https://doi.org/10.3390/ijerph15061204
13. Wu, H.K., Lee, S.W.Y., Chang, H.Y., Liang, J.C.: Current status, opportunities and challenges of augmented reality in education. Comput. Educ. **62**, 41–49 (2013). https://doi.org/10.1016/j.compedu.2012.10.024
14. Xiao, J., Xu, Z., Yu, Y., Cai, S., Hansen, P.: The design of augmented reality-based learning system applied in U-learning environment. In: El Rhalibi, A., Tian, F., Pan, Z., Liu, B. (eds.) Edutainment 2016. LNCS, vol. 9654, pp. 27–36. Springer, Cham (2016). https://doi.org/10.1007/978-3-319-40259-8_3

Long-Term Average Temperature Forecast Using Machine Learning and Deep Learning in the Region of Beni Mellal

Hamza Jdi(✉) and Noureddine Falih

LIMATI Laboratory, Polydisciplinary Faculty, Sultan Moulay Slimane University, Beni Mellal, Morocco
hamzajdi@gmail.com

Abstract. The weather has an immense impact on agriculture, hence the need for a smart model that accurately predicts and forecasts weather based on available meteorological data. The region of Beni Mellal-Khenifra in particular, located in Morocco, relies heavily on the agricultural sector, making it the primary if not the only considerable source of income for the region. This stresses the importance and the benefits of advanced and early weather prediction in the region. In this paper, we are interested in forecasting the weather in the region of Beni Mellal. Two types of Artificial Intelligence were used in forecasting average temperature: the first is a machine learning algorithm called AutoRegressive Integrated Moving Average (ARIMA); the second is a deep learning algorithm named Gated Recurrent Unit (GRU). The results suggest that GRU model outperformed ARIMA model by 1.51% in respect to Mean of Absolute Error.

Keywords: Temperature forecast · Machine learning · Deep learning · Times series · Big data analytics

1 Introduction

Agriculture is one of the main impacted sectors by abnormally high or low temperatures. This could result in catastrophic crop failures. Accurate weather forecast [1,2] is essential to prevent the harm caused by thermal disasters, as severe changes in temperature negatively impact agriculture [3]. Temperature predicting methods are various. On one hand, we find the Numerical Weather Prediction (NWP) [4], a set of procedures that predict future atmospheric conditions by resolving equations in dynamics and physics that describe how the atmosphere moves and changes. The NWP approaches confront a number of difficulties, including the need for powerful computational resources, the lack of a thorough understanding of physical causes, and the difficulty in extracting knowledge from the vast amount of observational data [5]. Machine learning on the other hand is

N. Aboutabit et al. (Eds.): ICMICSA 2022, LNNS 656, pp. 26–34, 2023.
https://doi.org/10.1007/978-3-031-29313-9_3

data driven, meaning it is based on the idea that machines can discover patterns in data, learn from it, and draw conclusions. We are interested in time series modeling which is a data driven approach, also known as the Box-Jenkins model [6]. In statistical signal processing science, it refers to a model frequently used to measure time-based data. A stochastic process is thought of as one realization of an observed time series. The most basic model for simulating time series is made up of a process where events occur at different times and at regular intervals, with each event being independent of other values [7]. In this study, we choose the data driven approach. This method is simple and affordable because gathering local weather information has become easily accessible and publicly available online. As a result, it makes it easy for farmers to predict temperature. This work comes as a continuation of the previous work [8] where big data analytic tools, namely Hadoop and MapReduce, were used to predict minimum and maximum temperature as well as wind speed. We come to the conclusion that the accuracy noticeably changes depending on the month. To solve this issue, we would like to explore more advanced machine learning and deep learning algorithms such as ARIMA and GRU using time series to forecast temperature. We shall use the collected 8369 records of day-to-day average temperature in Beni Mellal to predict temperature using two algorithms implemented in Python with the open-source library Keras. We aim at showing the different methodologies adopted by each of the two models, produce reliable temperature prediction, and compare the two chosen algorithms using MAE.

2 Related Work

In this part, we will provide an overview of studies conducted on temperature forecasting using big data analytics, machine learning and deep learning.

This study [9] makes use of the ECWMF ERA-5 dataset to study and compare GRU and ANFIS techniques in order to gain a better understanding of their ability to anticipate the maximum temperature. ANFIS and GRU performed well since their CC values are more than 0.95 and their RMSE and MAPE are less than 2.

The authors [10] present a novel technique incorporating ARIMA and Artificial Neural Network (ANN). ARIMA is good for linear prediction, whereas ANN is good for nonlinear prediction. This research also looks at how to properly model short-term air conditioning load time series using a novel approach that predicts the ANN weights and ARMA model parameters. The experimental findings show that the hybrid air conditioning load forecasting model may be used to increase forecasting accuracy.

In this work [11], the authors focus on six climatic elements that impact crop growth: temperature, humidity, lighting, carbon dioxide concentration, soil temperature, and soil humidity. They propose a GCP lstm model for greenhouse climate prediction. Because greenhouse climate change is nonlinear, They characterize the link between historical climate data using a long short-term memory (LSTM) model. Furthermore, short-term climate has a greater impact on the long-term tendency of greenhouse climate change.

The authors [12] propose through a new approach that increases the accuracy of predictions by utilizing Bidirectional Gated Recurrent Units (Bi-GRU) and the Sparrow Search Algorithm (SSA). The Bi-GRU may utilise both past and future data inside the production sequence and associate characteristics. The Bi-GRU model's hyperparameters are tuned via SSA. In regards to effectiveness, the observations reveal that the Bi-GRU outperforms the others.

The researchers [13] offer a deep learning strategy based on the Hybrid GRU - long short-term memory (LSTM) model. GRU and LSTM are excellent at predicting time series as the model skips the time series decomposition procedure by embedding a time layer to provide efficient predictions and a deep understanding of time to produce trustworthy performance. The prediction is measured using root mean squared error (RMSE), mean absolute error (MAE), and R–Squared (R2). The best RMSE value is 0.07499. The best MAE result is 0.0578.

In order to choose the pertinent input meteorological variables, the authors [14] used Pearson correlation coefficients; as a result, eight different input combinations were built. In order to predict the minimum temperature based on the eight different input combinations, three generalized machine learning models and two deep learning models were used: Random Forest, SVM, and Multiple Linear Regression (for ML); Long-Short Term Memory and GRU (for DL). In comparison to generalized machine learning and deep learning models, the results demonstrated that the RF and GRU models had the best prediction performance.

The authors [15] presented a comparative analysis using simplified rainfall estimation models based on traditional machine learning algorithms and deep learning architectures, using climate data from 2000 to 2020 from five major cities in the United Kingdom. In order to estimate hourly rainfall volumes using time-series data, models based on LSTM, Stacked-LSTM, Bidirectional-LSTM Networks, XGBoost, and an ensemble of Gradient Boosting Regressor, Linear Support Vector Regression, and an Extra-trees Regressor were tested. The performance of the models was assessed using the following assessment metrics: Loss, RMSE, MAE, and RMSL. The Stacked-LSTM Network with two hidden layers and the Bidirectional-LSTM Network outperformed all other examined models.

3 Methodology

In this study, we collected data published by the National Climatic Data Center. As it has some missing parts, this data is completed; the missing parts are obtained from the Oum Er-Rbia Hydraulic Basin Agency, located in Beni Mellal, Morocco. The result is more than 22 years' worth of day-to-day average temperature from January 1, 2000, to November 30, 2022, hence a total of 8369 records. We choose to invest 87%, the equivalent of 7305 records to train the two models, and 13%, 1064 records, for testing. We use GRU and ARIMA to compute this important amount of collected records. The Python programming language is used to implement both algorithms. We can already see a huge difference between the two algorithms. On one hand, GRU needs no further preprocessing. On the other hand, ARIMA needs the data to be stationary, i.e., there should not be any trend or seasonality. In another words, constant mean and variance of the studied data. Data stationarity is confirmed using the Augmented Dickey-Fuller test after applying a first order differentiating. After the prediction stage, which yields two distinct readings, we proceed to the comparison stage. It is difficult to compare different forecasting systems because there is not a single measurement system for evaluating the accuracy of a given prediction. This is primarily caused by the multiple time scales, scales of the estimated data, and the unpredictability of the meteorological data. To compare the predicted and observed data, some metrics have been developed such as the Mean Absolute Error [16], Mean Absolute Percentage Error [17], and R-Squared [18]. The MAE is the metric used in this study to assess and compare the machine learning and the deep learning algorithm.

4 Study Area

Beni Mellal lies at the foot of the Middle and High Atlas Mountains. It is the principal city and capital of the Beni Mellal-Khenifra region. Beni Mellal benefits from its administrative capital status, agricultural land availability, and its new status as a university town. The climate of Beni Mellal is considered mild and temperate, with an average temperature of 62° Fahrenheit. The climate is classified as Csa by the Köppen-Geiger climate classification system. Summer rainfall in Beni Mellal is less common than winter rainfall.

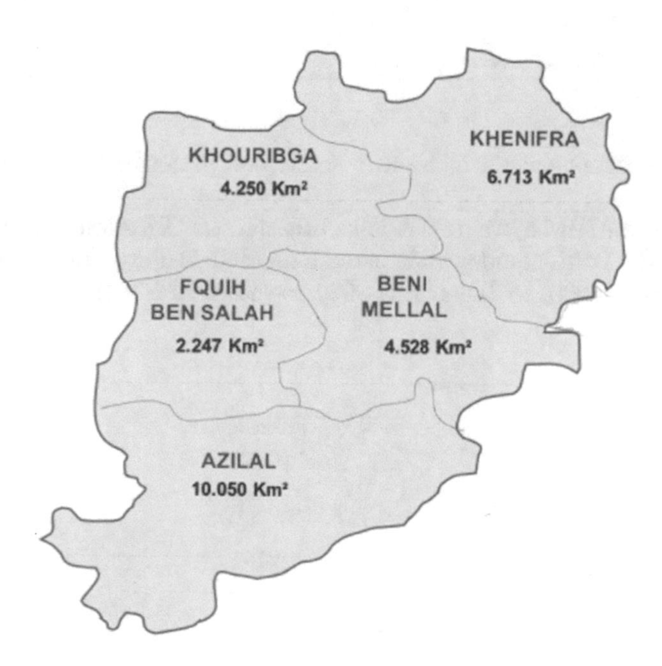

Fig. 1. The Beni Mellal-Khenifra region.

5 Results and Discussion

In this research, we study a dataset containing daily average temperature of Beni Mellal. This temperature is the average for a day in Fahrenheit. The dataset contains 8369 daily observations; we divided the dataset into two sections: training and testing. The training set has 7305 observations, which accounts for around 87% of the raw dataset, while the testing set contains 1064 observations, which accounts for roughly 13% of the raw dataset. The Mean Absolute Error is the primary prediction accuracy evaluation criterion, Table 1 shows the parameters used per algorithm.

Table 1. The parameters used per algorithm.

Algorithm	Parameters used
GRU	Number of units of a the GRU=256 Embedding length =8 Number of neurons in the dense layer followed by the RNN layer =64 Learning rate =0.0000000001 Epochs = 500
ARIMA	p (AutoRegressive) = 2 d (Integrated) = 1 q (Moving Average) = 3

The parameters with which each algorithm performed best are listed in Table 1.

In regard to ARIMA, we used Akaike Information Criterion (AIC) [19] to find the best fitted ARIMA model while using Beni Mella's daily average temperature from January 1, 2000, to January 1, 2020 as shown in Fig. 2

Out[11]:

	(p, d, q)	AIC
1	(2, 1, 3)	43450.200788
2	(8, 1, 7)	43453.500899
3	(3, 1, 6)	43454.497694
4	(5, 1, 9)	43454.978985
...	...	...
95	(0, 1, 2)	43846.874130
96	(2, 1, 0)	43943.485971
97	(0, 1, 1)	44048.504087
98	(1, 1, 0)	44057.747016
99	(0, 1, 0)	44082.002561

Fig. 2. AIC best ARIMA fitted models classified in an ascending order.

In respect to GRU, the amount of training epochs to utilize might be a challenge while developing neural networks. 500 epochs are chosen because the models stopped improving. Figure 3 shows MAE (LOSS) at a given epoch. The GRU layers model is shown in Table 2.

Table 2. Proposed GRU model architecture.

Layer (type)	Output Shape	Parameters
gru_1 (GRU)	(None, 256)	204288
dense_2 (Dense)	(None, 64)	16448
dense_3 (Dense)	(None, 1)	64

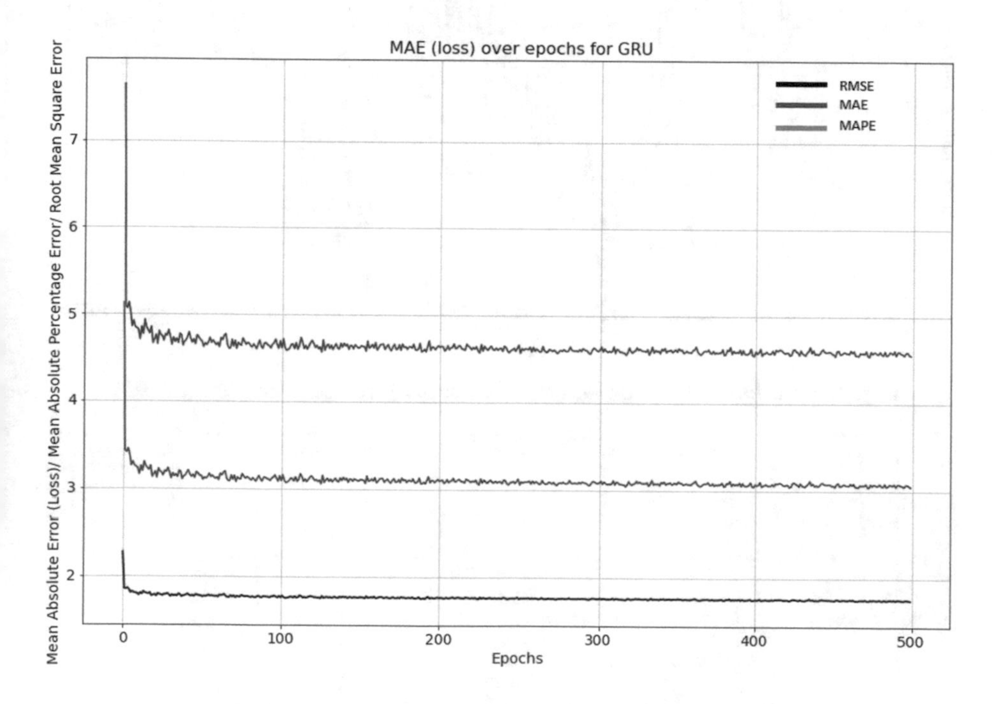

Fig. 3. GRU MAE, MAPE and RMSE values over epochs.

According to Table 3, the MAE of the GRU model is the smallest. When compared to the ARIMA model, the MAE of the GRU model is better by 1.5108%. The experimental results demonstrate that the GRU model has the optimal temperature forecasting prediction ability of the two. Let us show the results of the GRU and ARIMA models when attempting to predict temperature for the period spanning from September 1, 2020, to November 31, 2022.

Table 3. Results of the various forecasting models classified in an ascending order.

	MAE	MAPE
GRU	3.0446	4.5547
ARIMA	3.0906	-

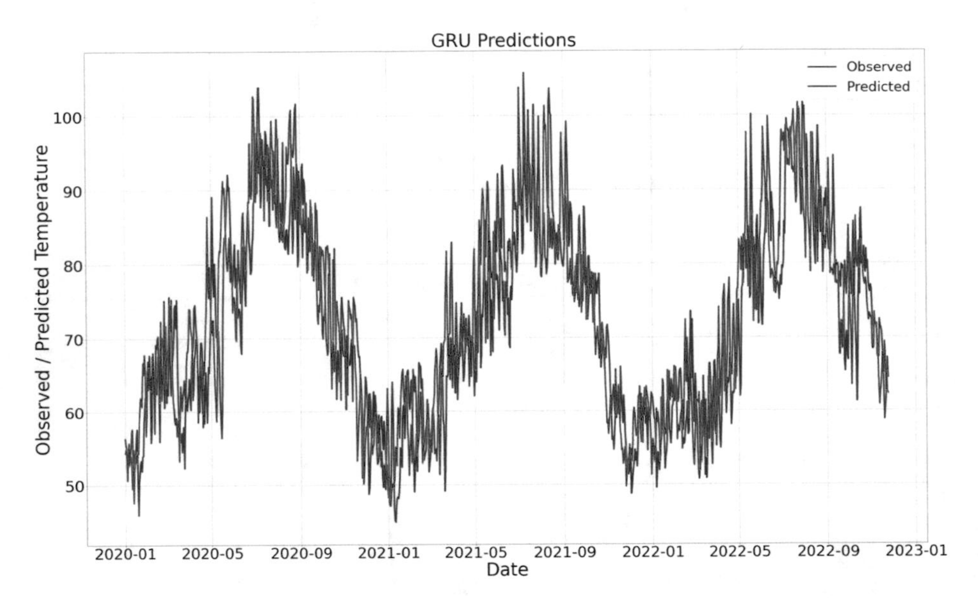

Fig. 4. Beni Mellal's actual and forecasted daily temperature using GRU.

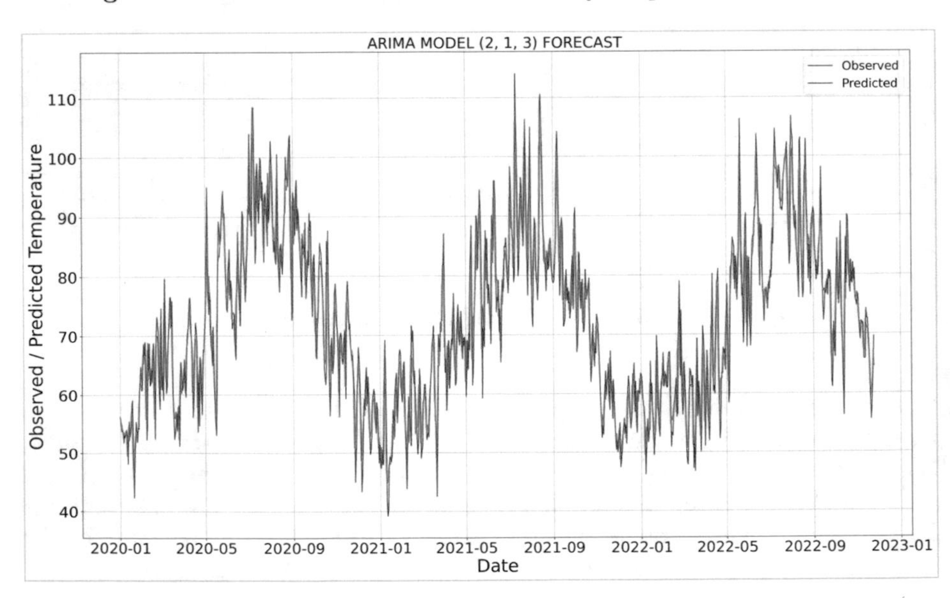

Fig. 5. Beni Mellal's actual and forecasted daily temperature using ARIMA.

The deep learning model presented trough GRU and the machine learning model ARIMA both did well when predicting the temperature in the region of Beni Mellal, with GRU realizing an overall better performance.

6 Conclusion

In this paper, two types of Artificial intelligence were presented: ARIMA, a Machine learning algorithm, GRU, a deep learning algorithm. We set to fulfill three objectives: first, to show the different methodologies of each type of AI in regard to prediction; second, to compare the chosen algorithms; third, to forecast long-term average temperature in Beni Mellal, Morocco.

The results demonstrate that GRU model outperformed the ARIMA model by 1.51% in respect to MAE. In future work, we will study Beni Mellal's precipitation with concepts of Artificial Neural network and machine learning using Python.

References

1. Singh, N., Chaturvedi, S., Akhter, S.: Weather forecasting using machine learning algorithm. In: 2019 International Conference on Signal Processing and Communication (ICSC), NOIDA, India, Mar. 2019, pp. 171–174. https://doi.org/10.1109/ICSC45622.2019.8938211
2. Ukhurebor, K.E., et al.: Precision agriculture: weather forecasting for future farming. In: AI, Edge and IoT–based Smart Agriculture, Elsevier, 2022, pp. 101–121. https://doi.org/10.1016/B978--0--12--823694--9.00008--6
3. Malhi, G.S., Kaur, M., Kaushik, P.: Impact of climate change on agriculture and its mitigation strategies: a review. Sustainability **13**(3), 1318 (2021). https://doi.org/10.3390/su13031318
4. Eyre, J.R., et al.: Assimilation of satellite data in numerical weather prediction. Part II: recent years. Q. J. R. Meteorol. Soc. **148**(743), 521–556 (2022). https://doi.org/10.1002/qj.4228
5. Ren, X., et al.: Deep learning-based weather prediction: a survey. Big Data Res. **23**, 100178 (2021). https://doi.org/10.1016/j.bdr.2020.100178
6. Monika, P., Ruchjana, B.N., Abdullah, A.S.: The implementation of the ARIMA-ARCH model using data mining for forecasting rainfall in Bandung city. Int. J. Data Netw. Sci. **6**(4), 1309–1318 (2022). https://doi.org/10.5267/j.ijdns.2022.6.004
7. Aghelpour, P., Mohammadi, B., Biazar, S.M.: Long-term monthly average temperature forecasting in some climate types of Iran, using the models SARIMA, SVR, and SVR-FA. Theor. Appl. Climatol. **138**(3–4), 1471–1480 (2019). https://doi.org/10.1007/s00704-019-02905-w
8. Jdi, H., Falih, N.: Weather forecast using sliding window algorithm based on hadoop and MapReduce. In: Maleh, Y., Alazab, M., Gherabi, N., Tawalbeh, L., Abd El-Latif, A.A. (eds.) ICI2C 2021. LNNS, vol. 357, pp. 122–132. Springer, Cham (2022). https://doi.org/10.1007/978-3-030-91738-8_12

9. Zamelina, A.J.F., Adytia, D., Ramadhan, A.W.: Forecasting of maximum temperature by using ANFIS and GRU algorithms: case study in Jakarta, Indonesia. In: 2022 10th International Conference on Information and Communication Technology (ICoICT), Bandung, Indonesia, Aug. 2022, pp. 222–227. https://doi.org/10.1109/ICoICT55009.2022.9914885.
10. Xuemei, L., Lixing, D., Ming, S., Gang, X., Jibin, L.: A novel air–conditioning load prediction based on ARIMA and BPNN Model. In: 2009 Asia–Pacific Conference on Information Processing, Shenzhen, China, Jul. 2009, pp. 51–54. https://doi.org/10.1109/APCIP.2009.21
11. Liu, Y., et al.: A long short-term memory-based model for greenhouse climate prediction. Int. J. Intell. Syst. **37**(1), 135–151 (2022). https://doi.org/10.1002/int.22620
12. Li, X., Ma, X., Xiao, F., Xiao, C., Wang, F., Zhang, S.: Time-series production forecasting method based on the integration of Bidirectional Gated Recurrent Unit (Bi-GRU) network and Sparrow Search Algorithm (SSA). J. Pet. Sci. Eng. **208**, 109309 (2022). https://doi.org/10.1016/j.petrol.2021.109309
13. Sari, Y., Arifin, Y.F., Novitasari, N., Faisal, M.R.: Deep learning approach using the GRU-LSTM hybrid model for air temperature prediction on daily basis. Int. J. Intell. Syst. Appl. Eng. **10**(3), 430–436 (2022)
14. He, Z., et al.: Gated recurrent unit models outperform other Machine learning models in prediction of minimum temperature in greenhouse Based on local weather data. Comput. Electron. Agric. **202**, 107416 (2022). https://doi.org/10.1016/j.compag.2022.107416
15. Barrera-Animas, A.Y., Oyedele, L.O., Bilal, M., Akinosho, T.D., Delgado, J.M.D., Akanbi, L.A.: Rainfall prediction: a comparative analysis of modern machine learning algorithms for time-series forecasting. Mach. Learn. Appl. **7**, 100204 (2022). https://doi.org/10.1016/j.mlwa.2021.100204
16. Karunasingha, D.S.K.: Root mean square error or mean absolute error? use their ratio as well. Inf. Sci. **585**, 609–629 (2022). https://doi.org/10.1016/j.ins.2021.11.036
17. Rubi, M.A., Chowdhury, S., Abdul Rahman, A.A., Meero, A., Zayed, N.M., Islam, K.M.A.: Fitting Multi–layer feed forward neural network and autoregressive integrated moving average for dhaka stock exchange price predicting. Emerg. Sci. J. **6**(5), 1046–1061, Aug. 2022. https://doi.org/10.28991/ESJ--2022--06--05--09
18. Zhang, X., Hedeker, D.: Defining R-squared measures for mixed-effects location scale models. Stat. Med. **41**(22), 4467–4483 (2022). https://doi.org/10.1002/sim.9521
19. Kanageswari, S., Gladis, D.: Predicting air pollutants SO2, NO2 and PM10 in chennai using autoregressive integrated moving average model,' presented at the PROCEEDINGS OF THE INTERNATIONAL CONFERENCE ON RESEARCH ADVANCES IN ENGINEERING AND TECHNOLOGY - ITechCET,: Kerala. India **2022**, 030015 (2021). https://doi.org/10.1063/5.0103378

Unsupervising Denoising Model Based Generative Adversarial Networks

H. Hafsi(✉), A. Ghazdali, H. Khalfi, and N. Lamghari

LIPIM, ENSA Khouribga, USMS, Khouribga, Morocco
hind.hafsi@usms.ac.ma

Abstract. This paper addresses a novel method, based on generative adversarial network (GAN) for denoising images especially in an unsupervised setting from corrupted and unpaired datasets of images. The only source of information is offered by the observations and the measurement process statistics. The process aims to find the maximum a posteriori (MAP) estimate of the distribution given each measurement, optimizing the noise generating distribution represented analytically defined noise as the likelihood and the GAN as the prior. In which the generator takes a corrupted observation as input to generate realistic reconstructions, and then adds a penalty term tying the reconstruction to the related observation. We test our approach on a variety of common datasets with varied sizes and levels of corruption. The proposed method offers an alternative to unsupervised denoising and achieves results that are comparable to the state-of-the-art in generative models noise removal.

Keywords: Generative adversarial networks · Denoising image · Inverse problems · Unsupervided learning

1 Introduction

Image denoising is a very big challenge for image processing. The noise is a random variation of information in images caused usually by different intrinsic and extrinsic situations. Removing noise from images has been a very active area of research in the recent 5 decades [1]. The goal of image denoising is to obtain clean image from a noisy image with any kind of distribution noise. This rises various applications from image restoration, image registration and image segmentation to image classification and visual tracking.

Many approaches are available to reduce noise in images and the solutions has fallen mainly in the statistical area working with modified Gaussian-like sliding filters in design, producing a smoother image in terms of pixels but leaves the effect of blurring and softness. Recent deep learning techniques in the field of image denoising have received much attention. Usually, denoising algorithms imply a dataset of pairs images noisy and clean where the process is based on a comparison between the reconstructed clean images of the noisy and the clean data. However neural network technology has been providing new technologies that consider the noise as a signal and remove it from the data instances.

N. Aboutabit et al. (Eds.): ICMICSA 2022, LNNS 656, pp. 35–46, 2023.
https://doi.org/10.1007/978-3-031-29313-9_4

A substantial amount of works in image denoising has been established in a supervised setting. In contrast, modeling the denoising problem in an unsupervised context has proved challenging. The latest advancement successes in unsupervised learning rely on generative modeling in particular generative adversarial network (GAN) in order to recover clean images using knowledge acquired from a set of noisy images corrupted by a given noise distribution without access to ground truths.

In this paper we present how neural networks technologies have shown problems scenarios in solving a big challenge in image processing which is how to recover clean image from a noisy image and how generative adversarial networks show more advancements in this area of research and we will discuss our methods which is based on generative adversarial network with only access to a single signal of the noise data without paired instances and it show good results similar to the state of art and provide new future opportunities.

2 Related Work

Over the years, image denoising has been a hot topic in machine learning and image processing area of research but each released method and technique has its own assumptions, benefits, and drawbacks. Review of several important works on picture denoising is provided in this part.

A traditional way to reduce noise from image data is to employ filters pixel wise in the form of spacial functions. These filters can be further categorized into two classes non-linear and linear filters.

Non-linear filters minimize noise without directly identifying it. If the noise is in the high-frequency range, sets of pixels can be filtered using a lower pass filter. These filters typically reduce noise to some extent, but they also have the undesirable effect of blurring [7,18].

In terms of mean square error, a mean filter for linear filters is best for Gaussian noise. Furthermore, linear filters have a tendency to lose characteristics like lines and sharp edges and perform very poorly for noise that depends on the signal. The Wiener filter approach [11], only functions well if the underlying signal is smooth and requires knowledge of the noise and spectrum of the original signal. The study [5] suggested the wavelet denoising technique as a way to solve the drawbacks of Wiener filtering.

Image denoising remains a fundamental problem in the field of image processing. Wavelets perform well in image denoising due to properties such as sparsity and multiresolution structure. With the popularity of wavelet transforms over the last two decades, various algorithms have been introduced for denoising in the wavelet domain. The focus has shifted from the spatial and Fourier domains to the wavelet transform domain. Wavelet denoising can be as simple as multiresolution thresholding of wavelet coefficients [20], or more complex statistical analysis of wavelets [14].

Before delving into denoising in the deep learning field, it is necessary to first define deep learning. Deep learning is a type of machine learning method that employs numerous layers to extract higher-level features from raw input [4]. The background methods section of this paper contains further information on the fundamental mechanisms and math behind deep neural networks, convolutional neural networks, and generative adversarial networks.

Deep learning has transformed many fields, including natural language processing, computer vision, and signal denoising. Deep learning and denoising were not involved problems until 2015, when deep autoencoders were employed quite well [12]. Both autoencoders and variational autoencoders have proven to be effective denoising methods throughout time. A downside of autoencoders and variational autoencoders, similar to the kernel filtering technique, is that their objective function produces blurred output [10].

The generative adversarial network (GAN), developed by Ian Goodfellow in 2014 [6], is an adversarial technique for training deep generative models. It offered a novel method for recognizing and sampling the distributions of images. In recent years, there has been a shift of attention toward GAN-based solutions in denoising applications, as state-of-the-art results have been achieved utilizing this architecture. Some early experiments used GANs to duplicate a noise distribution in order to augment a dataset. In a supervised learning scenario, other approaches rely on the GAN to remove the noise from a collection of clean and noisy paired images [17].

In other research [2,16], generative models are used to try to address ill-posed inverse problems. The general strategy entails first training a generative model on the signal distribution that has not been distorted. The signal is then inverted using a measurement from which we want to reconstruct it, by minimizing the mean square error between the corrupted reconstruction and the measurement, and by identifying the latent input code that produced the uncorrupted image [15]. For each image, this involve the solution of an optimization problem, which takes several minutes on a GPU and require random restarts to prevent sliding into a terrible local minima. The environment is again entirely supervised.

3 Methods

Deep learning techniques have rapidly advanced during the last ten years, having a considerable impact on signal and information processing. The research on artificial neural networks served as the foundation for the idea of deep learning (ANNs) [9]. Deep learning techniques were introduced by Hinton's layer-wise-greedylearning training method, which he suggested in 2006 [8].

3.1 Convolutional Neural Networks

A convolutional neural network is a series of convolutional and pooling layers that allow for the extraction of the primary features from images that best respond to the ultimate objective.

In general, an image can be mathematically represented as a tensor with the dimensions: (n_H, n_W, n_C).

where: n_H is the size of the Height, n_W is the size of the Width and n_C is the number of channels. The convolutional product, with a given image I and filter K, we have:

$$s(I,K)_{x,y} = \sum_{i=1}^{n_H}\sum_{j=1}^{n_W}\sum_{k=1}^{n_C} K_{i,j,k} I_{x+i-1,y+j-1,k} \tag{1}$$

Convolutional Layer. As previously demonstrated, at the convolutional layer, we apply convolutional products to the input, this time utilizing multiple filters, preceded by an activation function ψ. More importantly, at thel^{th} layer, we indicate: input: $a^{[l-1]}$ with size $(n_H^{[l-1]}, n_W^{[l-1]}, n_C^{[l-1]})$, $a^{[0]}$ being the image in the input, padding: $p^{[l]}$, stride: $s^{[l]}$, number of filters: $n_C^{[l]}$ where each $K^{(n)}$ has the dimension: $(f^{[l]}, f^{[l]}, n_C^{[l-1]})$, Bias: of the nth convolution: $b^{[l]_n}$, Activation function: $\psi^{[l]}$, pooling function: $\phi^{[l]}$ and Output: $a^{[l]}$ with size $(n_H^{[l]}, n_W^{[l]}, n_C^{[l]})$.

And we have:

$\forall n \in [1, 2, ..., n_C^{[l]}]$

$$s(a^{[l-1]}, K^{(n)})_{x,y} = \psi^{[l]}(\sum_{i=1}^{n_H^{[l-1]}}\sum_{j=1}^{n_W^{[l-1]}}\sum_{k=1}^{n_C^{[l-1]}} K_{i,j,k}^{(n)} a_{x+i-1,y+j-1,k}^{[l-1]} + b_n^{[l]}) \tag{2}$$

Thus:

$$a^{[l]} = [\psi^{[l]}(s(a^{[l-1]}, K^{(1)})), \psi^{[l]}(s(a^{[l-1]}, K^{(2)})), ..., \psi^{[l]} \quad (s(a^{[l-1]}, K^{(n_C^{[l]})}))] \tag{3}$$

Pooling Layer. The pooling layer seeks to downsample the features of the input without reducing the number of channels. The pooling layer seeks to downsample the features of the input without reducing the number of channels. We take into account the following representation:

We can assert that:

$$p(a^{[l-1]})_{x,y,z} = \phi^{[l]}((a_{x+i-1,y+j-1,z}^{[l-1]})_{(i,j)\in[1,2,...,f^{[l]}]^2}) \tag{4}$$

We commonly use average pooling, in which we average the items on the filter, or max pooling, in which we return the maximum given all of the components in the filter.

Fully Connected Layer. A fully connected layer is a limited number of neurons that takes a vector $a^{[i-1]}$ as input and outputs a vector $a^{[i]}$ as output.

Taking into account the jth node of the ith layer as a whole, we have the following formula:

$$z_j^{[i]} = \sum_{l=1}^{n_{i-1}} w_{j,l}^{[i]} a_l^{[i-1]} + b^{[i]} \rightarrow a_j^{[i]} = \psi^{[i]}(z_j^{[i]}) \tag{5}$$

Input $a^{[i-1]}$ could be the output of a convolution or pooling layer with dimensions $(n_H^{[i-1]}, n_W^{[i-1]}, n_C^{[i-1]})$.

In order to be able to plug it into the fully connected layer we flatten the tensor to a 1D vector having the dimension: $(n_H^{[i-1]} \times n_W^{[i-1]} \times n_C^{[i-1]})$, thus, to plug it into the fully connected layer, we flattening the tensor to a 1D vector.

3.2 Generative Adversarial Networks

Generative adversarial networks structure was defined by [6], who used sophisticated deep generative networks to generate high-quality synthetic data for a wide range of applications [3].

So, GANs is another architecture to generate new data following the same distribution of the input data. The Adversarial part comes because in this architecture two neural networks contesting with each other in a minimax game where one network gain is the other network loss.

As feed, generator models create a sample in the domain from a fixed-length random vector. This vector is selected at random from a Gaussian distribution and serves as the starting point for the generating process. Points in this multidimensional vector space correspond to points in the task field after training, resulting in a reduced representation of the data distribution.

In more precise terms, the generator network may be represented as a differentiable function that collects random noise from the latent space Z with the distribution $p_z(z)$ and data from the exact actual data space with the same distribution $p_{data}(x)$.

$$G : Z \rightarrow \mathbb{R}^n \tag{6}$$

The discriminator model requires an example in the domain as input (real or generated) and guesses a binary class label for real or fake. Actual instances originate from the testing set. The created instance is the generator model's output.

Here Z is the latent space and n is the data space dimension. D is a simple neural classifier network that may be mathematically expressed as a function that maps from the data distribution to the probability $p \in [0, 1]$that expresses how probable the input data vector is actual.

$$D : \mathbb{R}^n \rightarrow [0, 1] \tag{7}$$

Generative adversarial networks is based on a game theory scenario called the minimax game. In this scenario, discriminator D and generator G compete with each other. The discriminator attempts to detect whether the data is real (from the training set) or faked by the generator network, which produces it using stochastic noise (from the generator network).

It is crucial that the generating network only trains from the discriminator and is not given access to the real data.

The minimax game is represented by the optimal solution:

$$\min_{G} \max_{D} V(D, G) \tag{8}$$

Following their definition, the two objective functions are jointly learnt via alternating gradient descent. Up until the generator provides excellent images, we train the two networks in opposite directions. Where:

$$V(D, G) = \mathbb{E}_{x \sim p_d ata(x)}[\log(D(x))] - \mathbb{E}_{z \sim p_z(z)}[1 - \log(D(G(x)))] \tag{9}$$

4 Our Approach

Suppose we have $X \sim p_X$ as the signal that we want to acquire, and assuming that the signal is only accessible through out a incorrect measurements $Y \sim p_Y$.

The MAP estimation a maximum a posteriori estimate, is the foundation of the reconstruction. The necessity to recreate some data X from a finite set of observations Y gave rise to the concept of image reconstruction. For instance, where a linear sampling operation $A \in R^{M \times N}$ maps from a finite signal $X \in R^N$ to

$$Y = AX + e \tag{10}$$

with $Y \in R^M$, while $e \in R^M$ models the impact noise has on the measured signal Y. A models the sampling of X to the observed signal Y and is often called the system matrix.

Let the measurement process modeled with a stochastic function F that map signals X to their associated observations Y. To put it another way, we can state that $Y = F(x)|x \in X$ and that we want to construct $F(.)$ for every $y \in Y$ considering a set of noisy images $\hat{Y}$ and the measurement function $\hat{x}$ s.t. $y = F(\hat{x})$.

The goal is to reconstruct x based on the measurements y which is an inverse problem. Treating this inverse problem as a MAP estimate is how we plan to solve it. The objective of the MAP estimate is to identify the uncorrupted image $\hat{x}$ that maximizes its probability given the lossy measurement y by treating x and y as random variables with independent distributions.

This is defined as

$$\begin{aligned}
\hat{x} &= \arg\max_{X} p_{X|Y}(x|y) \\
&= \arg\max_{X} \frac{p_{Y|X}(y|x)p_X(x)}{p_Y(y)} \\
&= \arg\max_{X} p_{Y|X}(y|x)p_X(x)
\end{aligned}$$

While $p_X(x)$ is referred to as the prior, $p_{Y|X}(y|x)p_X(x)$ is referred to as the posterior, and $p_{Y|X}(y|x)$ is known as the likelihood. Finding the Maximum A

Posteriori (MAP) estimate is the process of identifying the x that maximizes the posterior. The prior represents the certainty that x is a component of the real image distribution, whereas the likelihood represents the certainty that y is the outcome of a corruption measurement on x.

In order to find the correct image, our method uses a Generative Adversarial Network. To do this, we devised the challenge of finding a generator $G : Y \leftarrow X$ such that the generated output is each input's corresponding MAP estimate $\hat{x}$ for each input y.

We can then describe our objective as finding:

$$\hat{G} = \arg\max_{G} \mathbb{E}_{p_Y} \{\log p_{Y|X}(y|G(y)) + \log p_X(G(y))\} \tag{11}$$

The prior is $\log p_X(G(y))$ and the likelihood is also now $\log p_{Y|X}(y|G(y))$. We applied the equation's log and changed x by G(y). In the general case, evaluating the likelihood $\log p_{Y|X}(y|G(y))$ requires marginalizing on the unobserved noise variable, which involves computing an intractable integral. Most probabilistic model for image denoising make assumptions on the structure of the measurement operator F and on the distribution in order to obtain an analytic form for the expectation.

Here, we consider more general measurement operators which do not necessarily lead to such a simplification and therefore proceed in a different way. Maximizing w.r.t. the prior term $p_X(G(y))$ is similar to learning a mapping G, such that the distribution induced by G(y), $\mathbb{E}_{p_Y} p_X(G(y))$is close to the distribution p_X.

The prior p_X being unknown, the only sources of information are the lossy measurements y and the known prior on the measurement process. In order to learn an approximation of the true prior p_X, we will use a form of generative adversarial learning, and build on an idea introduced in the AmbientGAN. Given an observation y, one wants to reconstruct a latent signal approximation $\hat{x} = G(y)$ so that a corrupted version of this signal $\hat{y} = F(\hat{x})$ will have a distribution indistinguishable from the one of the observations y.

The generator G and a discriminator D are trained on observations y and generated samples $\hat{y}$. The corresponding loss is the following:

$$\hat{D} = \arg\max_{D} \mathbb{E}_{p_Y} \{\log D(y) + \log(1 - D(F(G(y))))\} \tag{12}$$

We showed that an expression likelihood that can be found and optimized. Similar to this, we showed that even with access to only noisy data, it is still possible to get generator to produce examples from a clean distribution.

Our overall objective function can finally be represented as follows when both penalties are combined:

$$loss_{total}(G) = \lambda . loss_{likelihood}(G) + loss_{prior}(G) \tag{13}$$

The algorithm of training procedure is described below.

Algorithm 1. Training Process algorithm

Require: Initialize parameters of the generator G and the discriminator D.
for `number of training iterations` **do**
  for `number of batches` **do**
    for `k steps` **do**

$Y_{real} \leftarrow sample\_batch(Y)$
$Y_{fake} \leftarrow F(G(sample\_batch(Y)))$
Update D by ascending its stochastic gradient:

$$\nabla\theta_d \frac{1}{n} \sum_{i=1}^{n} \{\log D(y) + \log(1 - D(F(G(y))))\} \tag{14}$$

    end for
    $Y_{fake} \leftarrow sample - batch(Y)$
    Update G by descending its stochastic gradient:

$$\nabla\theta_g \frac{1}{n} \sum_{i=1}^{n} \lambda \{\|y - G(y)\|^2\} + \{1 - \hat{D}(F(G(y)))\} \tag{15}$$

  end for
end for

5 Experiments

5.1 Model Architecture and Collections

The outcomes of our experiments are presented in this section. For the datasets, we use mainly CelebA image datasets to test our method, CelebA: Celebrity data set with roughly 200 000 samples [13].

The GAN design in [19] serves as a model for our network architectures. For the reconstruction network G, we suggest an image-to-image variation of their latent-to-image generator using the same discriminator.

The size of each image has been changed to 64 × 64. Every image has only ever been corrupted once, i.e., there are never numerous instances of the same image corrupted with various corruption settings, in order to put ourselves in the most realistic situation possible.

5.2 Corruptions

Let us present the different measurement processes F used in the experiments, also named corruptions:

- Remove_Pixel: This measurement process randomly samples a fraction p of pixels uniformly and sets the associated channel values to 0. All the corresponding channel values are set to 0.

- Remove_Pixel Channel: Instead of setting to 0 a pixel for all channels as in Remove-Pixel, one samples a pixel coordinate and a channel, and sets the corresponding value to 0.
- Convolve_Noise: Here $F(x;\theta) := k * x + \theta$, where $*$is the convolution operator and k is a mean filter of size l. For each pixel, noise θ sampled from a zero-mean Gaussian of variance added to the previous result.
- Patch Band: A horizontal band of height h whose vertical position in the image is uniformly sampled from the set of possible positions. For each pixel falling inside the band, its associated value is set to 0. The resulting measurement for pixel at column i and row j can be summarized as:

$$F(x;\theta)_{i,j} = \begin{cases} 0, & \text{if} \quad j \in \theta, ..., \theta + h \\ x_{i,j}, & \text{otherwise} \end{cases}$$

where θ is uniformly sampled from $\{1, ..., H - h\}$, and H is the image height.

5.3 Baselines

- Unpaired version: This is a model version in which we have access to samples of the signal distribution p_X. This indicates that, while we do not have paired samples from $p_{X,Y}$, we do have access to unpaired samples from p_X and p_Y.
- Paired version: This is a model version in which we have access to signal measurement pairs (y, x) from the joint distribution $p_{Y,X}$. Regressing y to the related signal x yields the reconstruction given an input measurement y.

5.4 Results

This section present our results. First, we compare our model quantitatively on CelebA. Then, using samples from our model and these baselines, we offer qualitative results.

Quantitative Results. We contrast our model with the baselines presented in the section before. We present mean square error (MSE) values between the true signal x utilized to create the input y and the reconstructed $\hat{x}$ (Table 1).

Table 1. Average mean square error of our model and its versions on datasets

	Our model	Unpaired version	Paired version
Remove_Pixel	0.0537	0.049	0.0506
Remove_Pixel Channel	0.0532	0.0459	0.0524
Patch Band	0.0288	0.046	0.027
Convolve_Noise	0.0211	0.0226	0.0096

In terms of numbers, our model works nicely. In regards of MSE, all of the approaches are pretty comparable. Our unsupervised model performs as well as versions of it that were trained with more supervision. Additionally, we observe that outcomes are equivalent - and occasionally even better - when the aligned signal-observation pairs are not employed (as in our Unpaired version) than when these pairings are used directly (as in our Paired version). This shows that the reconstruction can be conditional on the input signal using just our likelihood term.

Qualitative Results. Therefore, quantitative results gives us only insufficient information. We now examine the quality of our reconstruction on two different datasets. These figures illustrates reconstructions obtained from several models on the CelebA dataset. Visually, the quality of our model's reconstructions are congruent with the quantitative results: they are equivalent to their paired and unpaired counterparts. Our model is able to produce images with good visual quality while being cohesive with the underlying uncorrupted images (Fig. 1).

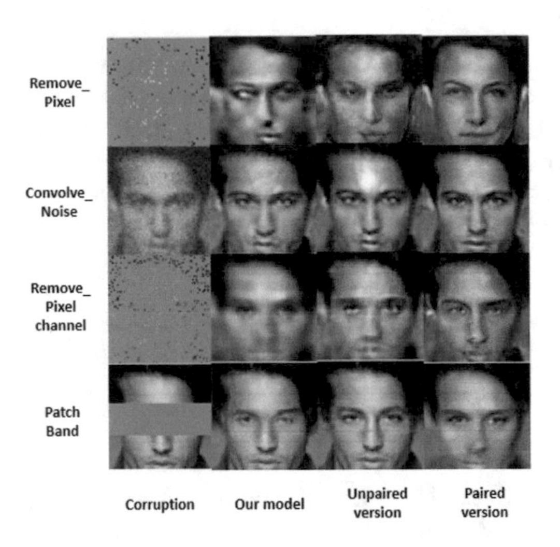

Fig. 1. On CelebA, model reconstructions for various corruption types. Each column represents a different model, and each row represents a particular corruption type.

6 Conclusion

In this paper, we have presented a general framework to recover an uncorrupted image from lossy measurements using a neural network, without having access to uncorrupted image data. For all of the observations in the training set, the task has been phrased as determining the highest a posteriori estimate of the picture given its observation. As a result, we are provided with a natural objective for our neural network that is made up of a linear combination of an adversarial loss for recovering realistic images and a reconstruction loss to link the reconstruction to its related observation. Results from our strategy are comparable to baselines that have access to more demanding kinds of supervision.

Multiple directions for future study and in-depth inquiry are present within the immediate scope of this paper. The first opportunity is to keep modifying the deep network architecture that is used in our model, as well as adding new testing, training, and learning rate schedulers. Applying our framework to various corruption processes and assessing our model's efficiency in real-world scenarios are also excellent areas for future research.

References

1. Bernstein, R.: Adaptive nonlinear filters for simultaneous removal of different kinds of noise in images. IEEE Trans. Circuits Syst. **34**(11), 1275–1291 (1987)
2. Bora, A., Jalal, A., Price, E., Dimakis, A.G.: Compressed Sensing using Generative Models. arXiv e-prints, p. 1703.03208 (2017)
3. Choi, E., Biswal, S., Malin, B., Duke, J., Stewart, W.F., Sun, J.: Generating multi-label discrete patient records using generative adversarial networks. In: Proceedings of the 2nd Machine Learning for Healthcare Conference, vol. 68, pp. 286–305, 18–19 Aug 2017
4. Deng, L., Yu, D.: Deep learning: methods and applications. Found. Trends Signal Process. **7**(4), 197–387 (2014)
5. Donoho, D.L., Johnstone, I.M.: Ideal spatial adaptation by wavelet shrinkage. Biometrika **81**(3), 425–455 (1994)
6. Goodfellow, I.J., et al.: Generative Adversarial Networks. arXiv e-prints p. 1406.2661 (2014)
7. Hamza, A., Luque-Escamilla, P., Martínez-Aroza, J., Roman Roldan, R.: Removing noise and preserving details with relaxed median filters. J. Math. Imaging Vision **11**, 161–177 (1999)
8. Hinton, G., Salakhutdinov, R.: Reducing the dimensionality of data with neural networks. Science (New York, N.Y.) **313**, 504–7 (2006)
9. Hinton, G., Osindero, S., Teh, Y.W.: A fast learning algorithm for deep belief nets. Neural Comput. **18**, 1527–54 (2006)
10. Jabbar, A., Li, X., Omar, B.: A Survey on Generative Adversarial Networks: Variants, Applications, and Training. arXiv e-prints p. 2006.05132 (2020)
11. Jain K. A.: Fundamentals of digital image processing. Prentice-Hall (1989)
12. Liang, J., Liu, R.: Stacked denoising autoencoder and dropout together to prevent overfitting in deep neural network. In: 2015 8th International Congress on Image and Signal Processing (CISP), pp. 697–701 (2015)
13. Liu, Z., Luo, P., Wang, X., Tang, X.: Deep learning face attributes in the wild. CoRR abs/1411.7766 (2014)
14. Portilla, J., Strela, V., Wainwright, M., Simoncelli, E.: Image denoising using scale mixtures of gaussians in the wavelet domain. IEEE Trans. Image Process. **12**(11), 1338–1351 (2003)
15. Ulyanov, D., Vedaldi, A., Lempitsky, V.S.: Deep image prior. CoRR abs/1711.10925 (2017)
16. Van Veen, D., Jalal, A., Soltanolkotabi, M., Price, E., Vishwanath, S., Dimakis, A.G.: Compressed Sensing with Deep Image Prior and Learned Regularization. arXiv e-prints p. 1806.06438 (2018)

17. Yang, Q., Yan, P., Zhang, Y., Yu, H., Shi, Y., Mou, X., Kalra, M.K., Zhang, Y., Sun, L., Wang, G.: Low-dose ct image denoising using a generative adversarial network with wasserstein distance and perceptual loss. IEEE Trans. Med. Imaging **37**(6), 1348–1357 (2018)
18. Yang, R., Yin, L., Gabbouj, M., Astola, J., Neuvo, Y.: Optimal weighted median filtering under structural constraints. IEEE Trans. Signal Process. **43**(3), 591–604 (1995)
19. Zhang, H., Goodfellow, I., Metaxas, D., Odena, A.: Self-Attention Generative Adversarial Networks. arXiv e-prints p. 1805.08318 (2018)
20. Zhang, M., Gunturk, B.K.: Multiresolution bilateral filtering for image denoising. IEEE Trans. Image Process. **17**(12), 2324–2333 (2008)

Meta-heuristic Algorithms for Text Feature Selection Problems

Issam Lakouam[1](✉), Imad Hafidi[1], and Mourad Nachaoui[2]

[1] National School of Applied Sciences Khouribga, Sultan Moulay Slimane University, Beni Mellal, Morocco
issam.lakouam.info@gmail.com, i.hafidi@usms.ma
[2] Faculty of Sciences and Technologies, Sultan Moulay Slimane University, Beni-Mellal, Morocco
m.nachaoui@usms.ma

Abstract. The increasing unstructured amount of text information on the Internet has become a significant obstacle to user needs. The text clustering approach is used in text mining to divide a group of texts into a predetermined number of clusters. The clustering algorithms suffer from the increasing amount of non-informative words in the corpus. Non-informative text features are removed from each document to improve clustering approach performance and computation. Here, several particle swarm optimization variants, such as inertia weight and constriction factor, are compared to improve the particle exploration experience. The PSO method is compared with the other commonly used metaheuristics (i.e., the Genetic algorithm and the Harmony search algorithm). Also, various exploration and initialization characteristics are integrated with the PSO to improves its performance. The experiments were carried out on four standard datasets: Reuters-21578, 20Newsgroups, Classic4, and WebKB. The experimental results show that PSO outperforms the other competing approaches in terms of clustering Accuracy, Precision, Recall, and F-measure. Moreover, adjusting PSO parameters improves its performance.

Keywords: Text clustering · Unsupervised feature selection · Particle swarm optimization

1 Introduction

In recent years, the amount of high-dimensional digital text documents has grown widespread in many fields, affecting the text mining process. The rapid growth in data volume increased the significance of text clustering in organizing digital text documents. The primary objective of a text clustering technique is to divide massive volumes of text documents into a predetermined number of clusters. When applied to high-dimensional data, the presence of noisy, irrelevant, and redundant features reduces the effectiveness and increases the computing costs of the text clustering process [1].

N. Aboutabit et al. (Eds.): ICMICSA 2022, LNNS 656, pp. 47–58, 2023.
https://doi.org/10.1007/978-3-031-29313-9_5

Digital text documents contain both informative and uninformative features, with uninformative features including noise, irrelevant, and redundant features. Both feature extraction and feature selection are two of the most powerful tools for dealing with these challenges. During Feature extraction, the original feature space is converted, either linearly or nonlinearly, into a new low-dimensionality feature space. On the other hand, Feature selection is an optimization problem that selects a subset of useful features for each document while removing uninformative features [2,3]. They are both viewed as an efficient preprocessing stage. However, if the data contains comprehensible features to the underlying technique, the generated set of new features in the features extraction process retains no physical meanings of the original features, posing a new feature analysis challenge. As a result, feature selection is often used in text mining applications such as text clustering, anomaly detection, classification, and text retrieval.

Real-world applications contain a large number of irrelevant, redundant, and noisy data features; selecting a new subset of relevant features from these digital text documents by feature selection will achieve several objectives: (1) reduces memory storage, (2) increases computational efficiency, and (3) improves the text clustering algorithm's effectiveness. Based on various selection strategies, feature selection methods are categorized as filter, wrapper, and embedded approaches. Wrapper approaches depend on the learning algorithm to assess the quality of each selected subset of features; these methods are effective but computationally expensive. Filter methods rely on statistical aspects of text data to determine feature subset relevance, independent of any learning algorithms. Therefore, they are computationally efficient and are commonly employed to solve the feature selection problem and increase clustering accuracy. However, the selected feature subset may not be optimal for the underlying algorithm compared to wrapper methods. Embedded approaches involve interactions with the learning algorithm by embedding feature selection into model learning. Thus, they are more effective than wrapper methods [1,5].

The remainder of the paper is organized as follows. Section 2 describes the text preprocessing steps and the VSM representation for text mining. Section 3 explains the Mathematical model and categorization of traditional feature selection algorithms. Section 4 illustrates some Meta-heuristic algorithms for text feature selection problems. Section 5 displays empirical results. Finally, the conclusion and recommendations for future directions are shown in Sect. 6.

2 Text Clustering and Preprocessing Steps for Feature Selection Problems

The growing unorganized volume of text content on the internet has posed a significant challenge to user needs. As a result, the importance of text clustering techniques has increased in recent years to help reorganize the massive amount of information. Text clustering is an appropriate unsupervised learning method that aims to construct a predetermined number of clusters from a set of digital text documents, thereby simplifying access to the vast amount of information available on the internet. The text clustering search for the best solution based on both the intrinsic characteristics of the documents and the evaluation criteria. However, before clustering, documents are preprocessed and represented numerically in a standard format. The vector space model (VSM) is the most commonly used model for representing documents. It considers document features as a vector of term weights, with each cell of each vector containing a weighted value TF-IDF derived for the corresponding term within the document [6].

Standard preprocessing steps include tokenization, stop word removal, stemming, term weighting, and vector space model representation. These procedures are used to convert the contents of a digital text document into a suitable format. The following subsections provide a brief overview of the preprocessing steps and the text clustering.

2.1 Preprocessing

The first step is to divide the digital text document stream into smaller tokens (i.e., single term, character, or sequence of words). This process is known as tokenization. Second, a stop word removal phase is applied, in which a list of frequently occurring words throughout text documents, such as articles and pronouns, are deleted due to their high frequency in the corpus and non-utility to the clustering process, this step is commonly used in text-mining applications as a preprocessing step. Finally, before representing documents numerically, they are stemmed to reduce terms to their stem or root form. As a result, the feature space size decreases, and the clustering technique's effectiveness increases.

2.2 Vector Space Model

The vast majority of text clustering approaches consider each unique term a feature [7]. The most commonly used weighting scheme in the literature is the TF-IDF, also known as the term frequency-inverse document frequency [6]. The TF-IDF scheme converts digital text documents to a numerical vector representation of terms by assigning each term a weighted value w_{ij} based on its frequency of occurrence in documents. If a term frequently appears in a few documents, it is considered an informative feature for the text clustering process to distinguish between the document's text content [8]. Otherwise, the term is treated as a non-informative text feature. The term weighting value for word j in the document i is represented mathematically (see Eq. 1) in a standard model known as the vector space model.

$$w_{ij} = tf_{ij} \times \log(\frac{n}{df_j}). \tag{1}$$

where w_{ij} is the weight of feature j in the document i, tf_{ij} represents the frequency of the term j in the ith document, df_j is the document frequency of feature j, which means the number of documents containing the word j and n represents the number of documents in the collection. The vector space model

(VSM) represents the documents as vectors of weights, with each vector representing a text document and each weight indicating the importance of its term in the clustering process. The vector space model standard format is shown in the following expression:

$$VSM = \begin{bmatrix} w_{1,1} & \cdots & w_{1,j} & \cdots & w_{1,t} \\ \vdots & & \vdots & & \vdots \\ w_{i,1} & \cdots & w_{i,j} & \cdots & w_{i,t} \\ \vdots & & \vdots & & \vdots \\ w_{n,1} & \cdots & w_{n,j} & \cdots & w_{n,t} \end{bmatrix} \quad (2)$$

2.3 Text Clustering

The text clustering approach is employed at this stage to divide a set of documents into a preset number of clusters. The clustering techniques will fail due to the increasing amount of non-informative terms in the corpus, and the irrelevant, redundant, and noisy features will slow the text clustering algorithm and reduce its effectiveness. Thus, feature selection is critical as a preprocessing step to improve clustering approach performance and computation. Following the generation of a new subset of relevant features, the text clustering algorithm partitions D a large set of text documents $D = d_1, d_2, d_3, ..., d_n$ into K a set of coherent clusters to evaluate the accuracy of the feature selection methods and demonstrate their importance in text mining applications. Where d_i is the vector of weights for the document number i as shown in (see Eq. 3).

$$d_i = w_{i1}, w_{i2}, ..., w_{ij}, ..., w_{it} \quad (3)$$

where w_{ij} is the weight of the jth term in the document number i, n represents the number of documents in the corpus, and t is the total number of features.

3 Text Feature Selection Problem

The feature selection problem has been resolved by removing non-informative text features from each document in the corpus [3]. This stage involves mathematically modeling the feature selection problem as an optimization problem with a fitness function to assess the significance of the selected subsets of terms to the text clustering process. One application is to execute the features selection procedure at the document level, then combine the generated subsets of informative features from the documents to create a new dataset collection comprised entirely of relevant terms [2,9]. The primary aim of feature selection is to reduce the massive amount of noisy, irrelevant, and redundant features to prepare comprehensible, clean, and understandable text data for text mining processes. As a result, the computing efficiency and effectiveness of the text clustering technique increase.

Feature selection is an NP-hard problem. A Brute Force approach to solving the problem is impractical and computationally complex due to the exponentially required time to find the optimal solution. Several selection strategies have been proposed in the literature, such as Wrapper, Filter, Embedded, and Hybrid approaches. They are distinguished by whether and how to use the predictive learning method. However, even in the case of low dimensionality, feature selection improves the accuracy and performance of the underlying algorithm by retaining the most informative features and removing the irrelevant ones.

3.1 Wrapper Methods

Wrapper techniques employ a search strategy to produce a new subset of text features for each iteration and rely on the underlying algorithm's predictive performance to measure the effectiveness of the selected feature subset. Traditional wrapper methods are divided into two stages: searching for a subset of informative text features, then evaluating the quality of this selected subset of terms based on the performance of the text clustering algorithm, and repeating the two steps until satisfying some stopping criteria, such as achieving the highest clustering effectiveness or generating the desired size of feature subset. Thus, The selected features for the clustering stage are the optimal subset of features in terms of accuracy.

However, because of the extensive use of the clustering process, approaches in this category outperform other feature selection methods in terms of clustering accuracy while being computationally expensive [1,8]. As a result, wrapper methods are commonly applied in practice. Many wrapper search techniques are proposed to improve text clustering, including sequential search [3], branch-and-bound search [21], and best-first search [22,23].

3.2 Filter Methods

Filter selection methods evaluate the importance of a subset of features based on statistical characteristics of the data features to solve the feature selection problem without interacting with the text clustering algorithm [4]. A conventional filter method has two stages. First, the relevance of the words is assessed and ranked individually or in a batch based on specific evaluation criteria. Individually ranked terms are evaluated independently of one another, whereas batch ranking considers the relationship between words in the feature importance evaluation process. Then, in the second stage of a technique in this category, low-ranked features are removed as noisy, irrelevant, and redundant terms to improve the performance of the text clustering technique.

Filter methods have less computational cost than wrapper methods since they do not use a learning algorithm throughout the evaluation process. Thus, in the feature selection problem, these methods are commonly used for the case of high dimensional feature space. However, for the same reasons, filter approaches perform worst than wrapper methods in terms of text clustering accuracy. Various filter approaches evaluation criteria have been proposed in the literature, including mutual information [24], information gain [25], feature correlation [3], mean median [10], document frequency (DF) [17], term strength [26], and mean absolute difference [27].

3.3 Embedded Methods

Embedded techniques generate and evaluate a subset of features by embedding the feature selection process into the learning technique. They interact with the learning algorithm without assessing the subset of features iteratively. As a result, the effectiveness and computational efficiency of the embedded approaches are a trade-off between the wrapper and filter methods [11]. The most commonly used embedded technique is the regularization model, in which feature selection and learning model fitting take place simultaneously [5]. Then, the optimal features subset and the regularization model are returned as the final results.

3.4 Hybrid Methods

From the selection strategy perspective, The hybrid feature selection approach is the fourth and last category of feature selection methods [12,13]. This category's key objective is to overcome the perturbation problem of many existing feature selection approaches. Hence, increasing the text clustering technique's performance. Hybrid methods combine the benefits of some well-known techniques while minimizing the disadvantages of others by integrating several feature selection search strategies (e.g., wrapper, filter, and embedded) and aggregating their selected feature subsets [8]. The resulting set of features is more credible and robust than competing methods. Thus, This technique receives more attention for solving the text feature selection problem [10,28,29].

3.5 Mathematical Model

The features selection problem is represented as an optimization problem (see Eq. 4) to find the optimal subset of features, where the mean absolute difference (MAD) is a common objective function for the text feature selection problem. Given F a set of features, defined as a vector of unique terms $F = f_1, f_2, ..., f_i, ..., f_t$, where t denotes the total number of distinct features. The applied features selection technique generates a new subset of informative text features for each text document, $SF_i = s_{i1}, s_{i2}, ..., s_{ij}, ..., s_{it}$, where t is the total number of unique text features, i is the document number. For each document number i and term number j, we have $s_{ij} \in \{0, 1\}$, $i = 1, ..., n$ and $j = 1, ..., t$ where n is the total number of documents in the corpus. If $s_{ij} = 1$, it signifies that the jth term in document number i has been selected as a useful text feature for the clustering process. Otherwise, $s_{ij} = 0$ indicates that the jth text feature is missing from the document or has been eliminated as a non-informative feature [10].

$$\begin{array}{ll} Max & MAD_i \\ s.t. & s_{i,j} \in \{0,1\} \\ & \forall i = 1..n, \forall j = 1..t \end{array} \quad (4)$$

The vector space model (VSM) is a standard traditional model defined as vectors of term weights, expressed as $VSM_i = w_{i1}, w_{i2}, ..., w_{ij}, ..., w_{it}$, where t is the number of all unique terms and i is the document number. The fitness function described in this section is the mean absolute difference (MAD) based on the weighting scheme term frequency-inverse document frequency (TF-IDF) used as an evaluation measure to assess every candidate solution provided by meta-heuristic algorithms or different search strategies [8]. The selected subset with the highest MAD value is the best solution to the feature selection problem. MAD is calculated (see Eq. 5) as the absolute difference between the value and the mean value of the weight of the selected features.

$$\begin{array}{l} MAD_i = \frac{1}{a_i} \sum_{j=1}^{t} s_{i,j} |w_j - \bar{x_i}| \\ where \\ \bar{x_i} = \frac{1}{a_i} \sum_{j=1}^{t} s_{i,j} w_j \end{array} \quad (5)$$

MAD_i is the fitness function of the solution i, w_j is the weight of the jth term in the currently processed document, calculated using the TF-IDF scheme, where t is the total number of unique terms. $\bar{x_i}$ is the mean value of the selected weights in solution i for the current document.

4 Meta-heuristics Approaches

The feature selection is a challenging optimization problem, solved by selecting a new subset of informational text features to improve the effectiveness of the text clustering process. Recently, many studies have attempted to incorporate intelligent optimization algorithms, such as PSO, Genetic, and ant colony optimization approaches, into feature selection methods, due to their widespread use in the resolution of complex optimization problems [37].

The particle swarm optimization (PSO) approach is a swarm intelligence-based evolutionary computation algorithm introduced by Kennedy and Eberhart in 1995 [14]. It simulates the social behavior of birds flocking and fish schooling, inspired by earlier swarm intelligence metaheuristics. Because of its information sharing, quick convergence, and coding simplicity, the method has received attention in different fields, including bioinformatics [12], data clustering [15], and tuning learning algorithms [16]. Each candidate solution is referred to as a particle by the particle swarm optimization (PSO) algorithm. The particle moves through the search space based on the concept of the global best solution, the local best solution, and the velocity [1], which are updated in each iteration [2].

Another study in 1998 introduced the inertia weight with the primary objective of retaining some of the particle's former velocity [18].

Different tuning of particle swarm optimization (PSO) parameters was proposed in the literature to control the particle search process, such as particle swarm optimization in text categorization [30], four PSO variants, and two PSO models [17], an improved inertia weight PSO algorithm with the mean absolute difference (MAD) as a fitness function [4]. Other methods employed the PSO algorithm with an opposition-based mechanism [1], chaotic maps [15], and the catfish effect for feature selection [31]. Others suggest including PSO in SVM for parameter determination to solve the feature selection problem [16,32].

Two of the most commonly utilized metaheuristics for tackling the feature selection problem are the genetic algorithm (GA) and PSO. The genetic technique was introduced in 1970 by John Holland [19]. It simulated natural evolution theory using multiple operations such as crossover and mutation. Term weight (TF-IDF) is a typical objective function in evolutionary algorithms used to reduce the number of uninformative features and increase text clustering efficacy [20]. Several studies utilized the genetic algorithm to handle the feature selection problem [33,34], such as a genetic algorithm wrapper Approach with neural network pattern classifiers DistAl [35], and a genetic algorithm (GA) with adjusted term variance [36].

5 Experimental Results

The subsection below shows four typical evaluation criteria and several text standard datasets extensively used in the features selection problem. Multiple comparative analyses in the literature are reviewed, such as between particle swarm optimization (PSO) and various state-of-the-art feature selection optimization algorithms (i.e. Genetic algorithm GA, Harmony Search algorithm HS, and K-mean clustering without feature selection) [4]. Another comparison demonstrates the difference between different implementations of inertia weight and constriction factor in solving the feature selection problem [17]. Finally, a comparison between several integrations in the PSO algorithm, such as chaotic maps, adaptive inertia weight, and opposition-based initialization, illustrates how these improve the particle search process [1].

5.1 Datasets and Evaluation Criteria

Four text benchmark datasets are commonly used to test the performance of the features selection algorithm: Reuters-21578, 20Newsgroups, Classic4, and WebKB. These datasets have been pre-classified into several categories, but this information is not used during the text clustering process; instead, used to assess the correctness of the underlying algorithms. Accuracy (Ac), Precision (P), Recall (R), and F-measure (F) are the most commonly utilized evaluation criteria. These measurements evaluate the benefits and drawbacks of various feature selection approaches by comparing their clustering accuracy [13]. Precision and recall are typical measurements in text mining, and the F-measure is a harmonic combination of these measures, which computes the percentage of matched clusters. Accuracy is a commonly used metric for calculating the ratio of correct documents assigned to each group [4].

5.2 Results

The first experiment demonstrated that the Particle Swarm Optimization algorithm (PSO) outperforms other competitive optimization algorithms, such as the Genetic algorithm (GA) and the Harmony Search algorithm (HS), with the Genetic algorithm ranking second and the Harmony Search approach ranking third. The three outperform k-means without any feature selection, demonstrating the value of the feature selection preprocesses in improving clustering accuracy. However, PSO will achieve better outcomes due to particle information sharing and fast convergence. According to the convergence comparison, PSO has the quickest and the best final results. The HS algorithm was the slowest of the competing algorithms. Also, PSO has the best reduction ratio of feature space, which improves its computation speed and memory usage.

The second experiment compares different PSO variants to highlight the importance of inertia weight and constriction factor in increasing the particle swarm optimization algorithm's performance. In terms of effectiveness and stability, the asynchronous inertia weight and constriction factor outperformed the other systems. The improved inertia weight and constriction factor produced better results than the fixed versions, indicating the significant impact of tuning these parameters on the performance of the particle swarm optimization strategy. Furthermore, the inertia weight and constriction factor comparison are complicated because they update particle velocity and guide the search in the features space from different perspectives.

Finally, multiple particle swarm optimization algorithms with distinct exploration and initialization characteristics, such as opposition-based learning, chaotic map, and fitness-based dynamic inertia weight, are compared in the third experiment. The particle swarm optimization approach that merged them all performed the best in terms of text clustering accuracy. Thus, it is clear that opposition-based initialization is effective in assisting metaheuristic algorithms to start with a good position in the search space. Additionally, chaotic map and fitness-based dynamic inertia weight integration in the particle swarm optimization algorithm helps the technique escape from local optimal, avoid swarm stagnation, and search efficiently in the feature dimension space.

6 Conclusion

This work explored different literature comparisons, such as (1) a comparison of particle swarm optimization (PSO) and several state-of-the-art feature selection optimization approaches, (2) a comparison between various implementations of inertia weight and constriction factor to select informative subsets of features, as well as (3) a comparison of several PSO algorithm integrations, such as chaotic maps, adaptive inertia weight, and opposition-based initialization, to demonstrate how these improve the particle search process.

For the evaluation process, Experiments were carried out on several text standard datasets extensively used in the features selection problem. The results show that the Particle Swarm Optimization algorithm (PSO) outperforms other competitive optimization algorithms, such as the Genetic algorithm (GA) and the Harmony Search algorithm (HS). Also, the results highlight the significance of inertia weight and constriction factor tuning in improving the performance of the particle swarm optimization algorithm. Furthermore, additional exploration and initialization characteristics are significant to the effectiveness of metaheuristics search in the feature dimension space.

For future work, a hybrid particle swarm optimization algorithm that combines modified inertia weight or constriction factor and additional exploration components can be used to improve the exploitation search abilities of the original PSO, thus improving text clustering performance.

References

1. Bharti, K.K., Singh, P.K.: Opposition chaotic fitness mutation based adaptive inertia weight BPSO for feature selection in text clustering. Appl. Soft Comput. **43**, 20–34 (2016)
2. Abualigah, L.M., Khader, A.T., AlBetar, M.A., Hanandeh, E.S.: Unsupervised text feature selection technique based on particle swarm optimization algorithm for improving the text clustering. In: First EAI International Conference on Computer Science and Engineering, pp. 169–178. EAI (2017)
3. Guyon, I., Elisseeff, A.: An introduction to variable and feature selection. J. Mach. Learn. Res. **3**, 1157–1182 (2003)
4. Abualigah, L.M., Khader, A.T., Hanandeh, E.S.: A new feature selection method to improve the document clustering using particle swarm optimization algorithm. J. Comput. Sci. **25**, 456–466 (2018)
5. Li, J., Cheng, K., Wang, S., Morstatter, F., Trevino, R.P., Tang, J., Liu, H.: Feature selection: a data perspective. ACM Comput. Surv. (CSUR) **50**(6), 1–45 (2017)
6. Salton, G., Buckley, C.: Term-weighting approaches in automatic text retrieval. Inf. Process. Manage. **24**(5), 513–523 (1988)
7. Salton, G., Wong, A., Yang, C.S.: A vector space model for automatic indexing. Commun. ACM **18**(11), 613–620 (1975)
8. Bharti, K.K., Singh, P.K.: A three-stage unsupervised dimension reduction method for text clustering. J. Comput. Sci. **5**(2), 156–169 (2014)
9. Jafer, Y., Matwin, S., Sokolova, M.: Privacy-aware filter-based feature selection. In: 2014 IEEE International Conference on Big Data (Big Data), pp. 1-5. IEEE, Washington, DC (2014)
10. Bharti, K.K., Singh, P.K.: Hybrid dimension reduction by integrating feature selection with feature extraction method for text clustering. Expert Syst. Appl. **42**(6), 3105–3114 (2015)
11. Bai, X., Gao, X., Xue, B.: Particle swarm optimization based two-stage feature selection in text mining. In: 2018 IEEE Congress on Evolutionary Computation (CEC), pp. 1–8. IEEE, Rio de Janeiro (2018)

12. Saeys, Y., Inza, I., Larranaga, P.: A review of feature selection techniques in bioinformatics. Bioinformatics **23**(19), 2507–2517 (2007)
13. Uğuz, H.: A two-stage feature selection method for text categorization by using information gain, principal component analysis and genetic algorithm. Knowl.-Based Syst. **24**(7), 1024–1032 (2011)
14. Kennedy, J. and Eberhart, R.: Particle swarm optimization. In: Proceedings of ICNN'95-International Conference on Neural Networks, pp. 1942–1948. IEEE, Perth (1995)
15. Chuang, L.Y., Yang, C.H., Li, J.C.: Chaotic maps based on binary particle swarm optimization for feature selection. Appl. Soft Comput. **11**(1), 239–248 (2011)
16. Lin, S.W., Ying, K.C., Chen, S.C., Lee, Z.J.: Particle swarm optimization for parameter determination and feature selection of support vector machines. Expert Syst. Appl. **35**(4), 1817–1824 (2008)
17. Lu, Y., Liang, M., Ye, Z., Cao, L.: Improved particle swarm optimization algorithm and its application in text feature selection. Appl. Soft Comput. **35**, 629–636 (2015)
18. Shi, Y., Eberhart, R.: A modified particle swarm optimizer. In: 1998 IEEE international conference on evolutionary computation proceedings. In: IEEE World Congress on Computational Intelligence (Cat. No. 98TH8360), pp. 69–73. IEEE, Anchorage (1998)
19. Holland, J. H.: Adaptation in natural and artificial systems: an introductory analysis with applications to biology, control, and artificial intelligence. MIT Press (1992)
20. Abualigah, L.M., Khader, A.T., Al-Betar, M.A.: Unsupervised feature selection technique based on genetic algorithm for improving the text clustering. In: 2016 7th International Conference on Computer Science and information technology (CSIT), pp. 1–6. IEEE, Amman (2016)
21. Narendra, P.M., Fukunaga, K.: A branch and bound algorithm for feature subset selection. IEEE Trans. Comput. **26**(09), 917–922 (1977)
22. Kohavi, R., John, G.H.: Wrappers for feature subset selection. Artif. Intell. **97**(1–2), 273–324 (1997)
23. Arai, H., Maung, C., Xu, K., Schweitzer, H.: Unsupervised feature selection by heuristic search with provable bounds on suboptimality. In: Proceedings of the AAAI Conference on Artificial Intelligence, vol. 30, No. 1 (2016)
24. Peng, H., Long, F., Ding, C.: Feature selection based on mutual information criteria of max-dependency, max-relevance, and min-redundancy. IEEE Trans. Pattern Anal. Mach. Intell. **27**(8), 1226–1238 (2005)
25. Quinlan, J.R.: Induction of decision trees. Mach. Learn. **1**(1), 81–106 (1986)
26. Yang, Y.: Noise reduction in a statistical approach to text categorization. In: Proceedings of the 18th Annual International ACM SIGIR Conference on Research and Development in Information Retrieval, pp. 256–263 (2010)
27. Ferreira, A.J., Figueiredo, M.A.: Efficient feature selection filters for high-dimensional data. Pattern Recogn. Lett. **33**(13), 1794–1804 (2012)
28. Hsu, H.H., Hsieh, C.W., Lu, M.D.: Hybrid feature selection by combining filters and wrappers. Expert Syst. Appl. **38**(7), 8144–8150 (2011)
29. Zafra, A., Pechenizkiy, M., Ventura, S.: HyDR-MI: A hybrid algorithm to reduce dimensionality in multiple instance learning. Inf. Sci. **222**, 282–301 (2013)
30. Aghdam, M.H., Heidari, S.: Feature selection using particle swarm optimization in text categorization. J. Artif. Intell. Soft Comput. Res. **5**(4), 231–238 (2015)
31. Chuang, L.Y., Tsai, S.W., Yang, C.H.: Improved binary particle swarm optimization using catfish effect for feature selection. Expert Syst. Appl. **38**(10), 12699–12707 (2011)

32. Liu, Y., Wang, G., Chen, H., Dong, H., Zhu, X., Wang, S.: An improved particle swarm optimization for feature selection. J. Bionic Eng. **8**(2), 191–200 (2011)
33. Hong, S.S., Lee, W., Han, M.M.: The feature selection method based on genetic algorithm for efficient of text clustering and text classification. Int. J. Adv. Soft. Comput. Appl. **7**(1), 2074–8523 (2015)
34. Tan, F., Fu, X., Zhang, Y., Bourgeois, A.G.: A genetic algorithm-based method for feature subset selection. Soft. Comput. **12**(2), 111–120 (2008)
35. Yang, J., Honavar, V.: Feature subset selection using a genetic algorithm. In: Feature Extraction. Construction and Selection, pp. 117–136. Springer, Boston (1998)
36. Shamsinejadbabki, P., Saraee, M.: A new unsupervised feature selection method for text clustering based on genetic algorithms. J. Intell. Inf. Syst. **38**(3), 669–684 (2012)
37. Sharma, M., Kaur, P.: A comprehensive analysis of nature-inspired meta-heuristic techniques for feature selection problem. Archives Comput. Methods Eng. **28**(3), 1103–1127 (2021)

MYC: A Moroccan Corpus for Sentiment Analysis

Mouad Jbel[(✉)], Imad Hafidi, and Abdelmoutalib Metrane

Departement of Computer Science and Technology, ENSA Khouribga, Khouribga, Morocco
mouad.jbel@gmail.com

Abstract. Due to a shortage of resources for studying opinions and feelings in Arabic dialects, the task to identify the polarity of sentiments in the Arabic web is a challenging task. In this paper we present MYC a Moroccan YouTube Corpus of manually annotated comments with the aim of facilitating the task of sentiments analysis of Moroccan dialect in the web. Comments are collected from the wildly used website YouTube and manually annotated by several annotators. Using the voting approach, we created the largest Moroccan dialect subjectivity corpus of 20 000 comments labeled into positive and negative comments and including some other information (topic, likes and dislikes). This dataset could be a useful tool for the creation of Moroccan dialect-specific NLP applications in the future. In the trials, Support Vector Machines (SVM) and Naive Bayes (NB), two well-known supervised learning classifiers, were used to categorize comments as positive or negative, using a distinct set of parameters for each. Each classifier's recall, precision, and F-measure are computed. Both SVM and NB do well in terms of precision. The acquired results encourage us to move further with additional Moroccan comments from different videos in order to generalize our model.

Keywords: Computation and Language · Sentiment analysis · Machine learning · Natural Language Processing · Corpus

1 Introduction

The explosive expansion of social networking sites like Facebook, Twitter, and YouTube over the past few years has been well documented. Through these social networks, people and organizations may express and share their opinions on a range of subjects (products, political events, economics, restaurants, books, hotels, video clips, etc.). Sentiment analysis, or opinion mining [1], is a new field that combines data mining and Natural Language Processing/Natural Language Understanding (NLP/NLU) [2]. Its goal is to extract and analyze opinionated documents and categorize them into positive and negative classes [3], or more classes, as in sentiment analysis [4]. Sentiment analysis is concerned with extracting structured knowledge from unstructured data, as opposed to data mining, which tracks useful information from structured data [5]. Sentiment analysis and

N. Aboutabit et al. (Eds.): ICMICSA 2022, LNNS 656, pp. 59–68, 2023.
https://doi.org/10.1007/978-3-031-29313-9_6

emotion identification in Arabic have received a lot of attention. The majority of related work focuses on Modern Standard Arabic (MSA) [6–8], though a few studies have been conducted on Arab dialects. North African dialects, like Moroccan dialects, are less standardized than MSA. They now have a dynamic linguistic environment as a result of the various languages that have added to them over time. For the bulk of these languages, we also found a serious dearth of resources, including lexicons, dictionaries, and annotated corpora. In this paper, we focus on the Moroccan dialect's lack of resources for opinion and sentiment analysis connected to North African dialects. YouTube is one of the good websites for gathering opinions and sentiments because it has 1,850,000,000 unique monthly users and is the second most popular social media platform worldwide [9]. We present MYC (Moroccan YouTube Corpus) a corpus of manually annotated comments. The corpus is the largest Moroccan dialect annotated for both sentiment (positive & negative) and also containing a balanced quantity of comments written in both Arabic letters and Arabizi (Arabic or Arabic dialect written in Latin letters). As fellows, we have structured this paper as follows: the second section of the paper presents similar works and existing datasets. The third section of the paper is a summary of sentiment analysis existing techniques. Section four discusses Moroccan dialect challenges. The fifth section is devoted to presenting MYC corpus. In the sixth segment, we discuss the machine learning experiment and its results. The last section is to wrap up our work and discuss potential future work.

2 Related Works

With a focus on study cases from dialectal Arabic, we will provide research in the subject of Arabic sentiment analysis in this part.

Since Arabic is the fourth most widely spoken language and the language with the fastest-growing user base on the Internet, where the number of users is increasing at a rate of 6.6% annually, sentiment analysis in Arabic is regarded as an important research activity in the field of sentiment analysis. However, because Arabic is a morphologically complex language, sentiment analysis for Arabic is thought to be challenging. Many studies of SA have conducted on Arabic written text in the recent years (7, 8, 10), although most of them focused on modern standard Arabic, but there is also a few who done research on Arabic dialects (14, 17, 18).

SANA, a massive multilingual lexicon for sentiment analysis built on both MSA and informal Arabic, was created by Abdul-Mageed and Diab [10] (Egyptian and Levantine). The MSA and several Arabic dialects are covered by the 224,564 entries in the SANA. The authors manually labeled two-word lists from the Yahoo Maktoob and Penn Arabic Treebank. We used Google's translation API to generate lists of three existing English lexica: SentiWordNet, YouTube Lexicon, and GeneralInquirer. To increase the lexicon's coverage, they employed a statistical technique based on PMI to extract additional polarized tokens from the Twitter and conversation datasets. Despite the large size of the resulting resource, many of the entries are not lemmatized or dia-critized, limiting their lexicon's usability.

An electronic lexicon was developed by Diab et al. [11] and can be used for a number of NLP applications, including sentiment analysis in our situation. Three categories make up their lexicon: MSA, dialectal Arabic, and English. Tharwa, which can be used primarily for sentiment analysis in Egyptian dialects, was made available to the public by the authors.

[12] Rahab et al. (2017) put forth a technique for classifying Arabic comments taken from websites belonging to Algerian newspapers as good or negative. For this study, they created an Arabic corpus known as SIAAC (Sentiment polarity Identification on Arabic Algerian newspaper Comments). They tested Support Vector Machines (SVM) and Naive Bayes, two popular supervised learning classifiers (NB). To compare and assess the outcomes of their trials, they used a variety of factors and metrics (recall, precision and F-measure). The best results in terms of precision were generated by SVM and NB. It has been demonstrated that using a bi-gramme improves the accuracy of the two models. Additionally, SIAAC displayed competitive results when compared to OCA (Opinion Corpus for Arabic; Rushdi-Saleh et al., 2011).

[7] Raed Marji and Rehab M. Duwair built a framework to analyze Arabic tweets and some Arabic dialects and classify them into positive, negative and neutral sentiments. They used famous Machine learning classifiers such as SVM and NB and trained them by creating a dataset of 25000+ labelled tweet. In terms of accuracy the best results were obtained using NB, also crowdsourcing was proven to be a good method to collect large amounts of data.

[13] presented a Lexicon-Based Sentiment Analysis Approach for Saudi dialect in twitter. A manually created lexicon with a list of words and with their assigned polarities and some common phrases were proposed and tested for negative and positive classification. Compared to the bigger automatically created dictionary AraSenTi, SauDiSenti's sentiment analysis for tweets in the Saudi dialect demonstrated impressive results despite its small size. The performance of SauDiSenti can be enhanced by including dictionaries and introducing new themes.

[14] g. and g. built a lexicon to have a tangible component to use for the purpose of determining semantic orientation. Towards this end, a seed sentiment lexicon comprised of 380 words was manually constructed and used to collect more sentiment terms. Each of the terms in the lexicon was tagged as being either a verb, adjective, noun, adverb, or idiom/compound. After applying this procedure, a number of times, a sentiment lexicon of 4,392 terms was constructed from both Arabic and Algerian dialect terms. The lexicon resulted an accuracy of 83.

Guellil and co. [15] A straightforward polarity computation method was put forth for corpus annotation. It is a lexicon-based method, and the English lexicon "SOCAL" is used to automatically build the lexicon (Taboada & al., 2011). The polarity of the words was translated into Arabic but not their meaning. The corpus is then annotated using the produced lexicon.

[16] created a framework for Moroccan sentiment analysis where they implemented many techniques for SA. The authors build a training-set that consisted of 2000 Moroccan text annotated for positive and negative. However, the study only focused on text written in Arabic letters and didn't include a test on 'arabizi' text written in Latin words. DL classifiers gave better results using pre-trained word embedding. It is important to note that the majority of Arabic OEA resources and research articles are devoted to MSA. The majority of study work and funding in terms of Arabic dialects went to the Egyptian and Middle Eastern varieties. However, there hasn't been a lot of research done on the sentiment analysis of the Maghrebian dialects. Furthermore, there aren't many studies on Moroccan dialect, hence there aren't many resources. The proposed Arabic OEA techniques concentrate mostly on MSA, where only a small number of Arabic dialects have been studied.

[17] Nawaf A. Abdulla and Nizar A. Ahmed have explored the lexicon method and created a lexicon of polarity of Arabic words and Jordanian dialect. The process of creating this lexicon was by translating words from the famous SentiStrenght dictionary of sentiment analysis, the created lexicon contains 2500 words from MSA (Modern Standard Arabic) and Jordanian dialect. The lexicon was tested on extracted Arabic tweets. The outcome of this experiment illustrates that the bigger the lexicon is, the better the results are.

[18] They introduce ArSAS in this paper, an Arabic tweet corpus that has been annotated for sentiment analysis tasks. 21k Arabic tweets covering a wide range of topics were compiled, prepared, and annotated in a sizable collection. Twenty diverse themes from around the world that are anticipated to elicit spirited discussion among Twitter users were used to extract and collect the tweets in the corpus. The tweet collection did not rely on emotions or sentiment keywords, especially for the task of sentiment analysis, to avoid data bias to a given lexicon.

[19] HARD (Hotel Arabic-Reviews Dataset), a dataset of Arabic book reviews for arbitrary sentiment analysis and machine learning applications, was produced by the authors of this work. 490587 hotel reviews were compiled by HARD from Booking.com. The Arabic review text, the reviewer's star rating out of ten, and other details about the hotel and the reviewer are all included in each record. They used six popular classifiers to examine the datasets. The classifiers were put through their paces in terms of polarity and rating classification. The best results were obtained using logistic regression and SVM classifiers. For polarity classification, reported accuracy ranges from 94% to 97%. Studying comparable works has shown that there are little publicly accessible resources for sentiment analysis in MD. Therefore, we offer the first and largest Moroccan corpus with sentiment level annotations. Table 1 presents a comprehensive summary of the aforementioned datasets, providing a more clear overview of the similar resources that are currently available.

Table 1. Examples of some existing datasets.

Dataset Name	Size	Type	source
SANA	224,564	MSA/ Egyptian dialect	Yahoo/Youtube/SentiWordNet)
Tharwa	73,000	Egyptian dialects	Paper and electronic dictionaries
SIAAC	150	Algerian dialect	newspaper websites
SauDiSenti	4431	Saudi dialect	Twitter
ASA	2000	MSA/Moroccan dialect	Hespress website
ArSAS	21,000	MSA	Twitter)
HARD	49,0587	MSA	Booking.com

3 Challenges of Sentiment Analysis in Arabic Dialects

3.1 Morphological Analysis

Decomposing words into morphemes and assigning each morpheme morphological information, such as stem, root, POS (Part of Speech), and affix, is the basic goal of morphological analysis. It is significantly more challenging to complete these tasks in Arabic because of the language's intricate morphology. This complexity necessitates the creation of appropriate systems capable of dealing with tokenization, spell checking, and stemming.

3.2 Arabic Dialect

Arabic speakers typically use colloquial Arabic rather than MSA for communication (Moroccan dialect in our case). There are no language academies or standard orthographies for Moroccan dialect. As a result, processing Arabic dialects with methods and resources made for MSA yields noticeably subpar outcomes. Recently, researchers started creating parsers for certain dialects. These assessments are only made for specific dialects and still have poor accuracy. The efficiency of information retrieval will increase once this gap in Arabic processing is closed, particularly for data from social media.

3.3 Arabizi

Arabizi is the name of a Latin-based Arabic writing system. In social media sites, it is widely used to write MSA and Arabic dialects, this can cost a lot when it comes to accuracy and precision of detecting the polarity since same words can be written in different ways. Table 2 show an exemple of this writing technique.

4 MYC (Moroccan YouTube Corpus)

If this area of research is to advance, tools and resources are required. To guarantee the legitimacy of services, we created a forum for manual annotation using Python and the YouTube API. The following are the key contributions of our work:

Table 2. Examples of different encodings.

Encoding class	Text
English	Well done brother, can you give me the name of the application?
MSA	أحسنت أخي ممكن تعطيني اسم التطبيق؟
Moroccan dialect	مزيان خويا واش تقدر تغطيني اسم تطبيق؟
Arabizi (MD)	Mzyan khouya, t9dr t3tini smiya dyal tatbi9 ?

- Script to collect comments of videos using YouTube API.
- Comments taken from videos chosen manually (famous videos, sports, politics, trending videos and others).
- Voting technique to determine the comments polarity.

A schema of the work discussed in this article's description is shown in Fig. 1. As we can see, the annotation is provided at the comment level by the annotators and is applied to the comments gathered through the YouTube API. This contributes to the formation of a corpus. Then, these tools are used to perform polarity classification.

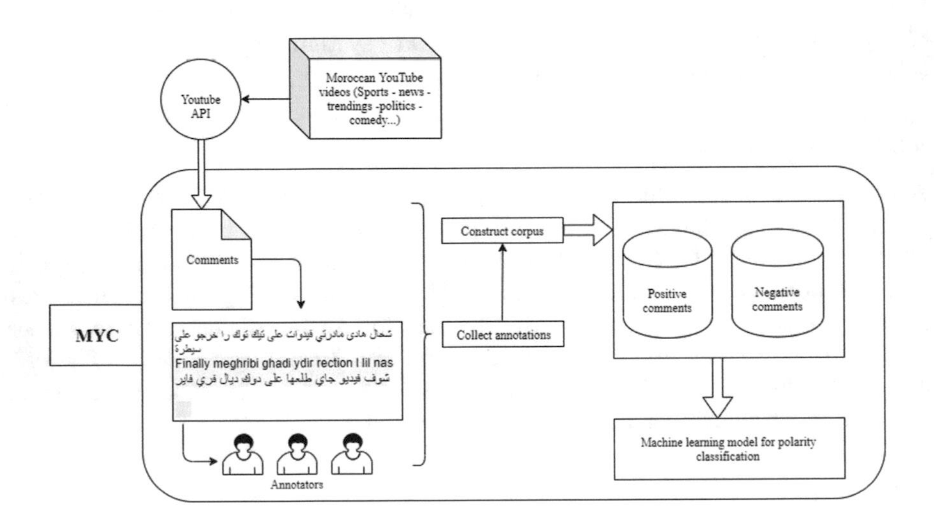

Fig. 1. MYC creation steps.

In the end, we compiled comments made between 2018 and 2021 on various Moroccan issues and locations. It was possible to create a corpus with the

Table 3. Examples of Moroccan dialect texts in Arabic letters and Latin letters.

Text in English	MD using Arabic letters	MD using Latin letters (Arabizi)
this is ugly	هادشي خايب	Hadchi khayb
I love summer	كيعجبني الصيف	kay3jbni sif

aid of five annotators; the labels of the corpus were assigned by majority vote; comments must use a label more than once in order for them to be included in the corpus. As previously indicated, the corpus was constructed using 20,000 annotated and sentiment-validated comments. In fact, there are 10,000 good and 10,000 negative comments in the corpus. There were an equal number of comments written in Latin letters and Arabic letters in each of the ten thousand comments (arabizi).

5 Experiments and Results

5.1 Comments Processing

A series of processing steps were carried out prior to feature extraction, which each comment goes through.

- Tokenization: Each token represents a word, and the spaces between words are simply used in this method.
- Stop Words removal: We used available MSA stop words list for this task, a further improvement for sentiment classification in Moroccan language can be to create stop words lists dedicated just to this dialect, one list for each way of writing (arabic and arabizi).
- Normalization: This process aids in increasing efficiency by minimizing the variety of information the computer must process. Lemmatization and stemming are examples of normalization procedures that aim to reduce a word's inflectional forms and occasionally derivationally related forms to a basic form that is shared by all words.
- Filtering Short tokens: Tokens with fewer than two letters were excluded from the mission since they were deemed to be of little significance.

5.2 Feature Selection

Here are various Arabic language devices that are employed as negation devices to reverse the sentiment of the comment. Nevertheless, depending on the context and where they are in the phrase, they may occasionally simply function as conjunctions. In our studies, we used unigrams and bigrams to lessen the impact of these two issues and test which was most applicable to our situation. While Bigrams break sentences into pairs of words by taking each word twice, once with the preceding word and once with the following word, Unigrams split each word on its own. By using the term frequency-inverse document frequency (tf-idf) feature vector, we divided each comment into unigrams and bigrams, then transformed them into numeric vectors.

5.3 Classifiers

Support vector machines (SVM) are a widely used classifier in a variety of disciplines due to their versatility in processing multi-dimensional input and high achieved accuracy. The Naive Bayes classifier, a well-known technique for classifying texts, is utilized. Given the context of the class, all attributes of the instances are assumed to be independent under the "Naive Bayes assumption." If a document is classified into multiple classes, each with a distinct probability, it is placed in the class with the highest posterior probability.

5.4 Results and Discussion

Accuracy, recall, and F-measure are the metrics used in this analysis to judge and evaluate the performance of the classifiers. Using several datasets, a comparison of the performances of various feature sets on various ML algorithms is conducted. The tables display the results of a comparative evaluation and analysis using n-gram features based on the collected metrics.

Table 4. Results using raw data

Feature	SVM(unigram)	NB (unigram)	SVM (bigram)	NB (bigram)
Accuracy (%)	71.5	69.57	70.64	70.2
Recall (%)	70.94	70.28	69.46	75.26
F-measure (%)	72.56	70.25	72.56	71.23

Our model, which includes several tools to help with text preparation for mining jobs, was created and implemented using Python and its text mining module (tokenization, stop word removal, and stemming, among others). Two different classifiers are implemented and their performance is estimated using Python (NB and SVM). Table 3 lists the results for each classifier in terms of recall, accuracy, and F-measure. We may evaluate the performance of the tested ML classifiers by comparing the experimental data displayed in Table 4. The outcomes show that SVM performed better than NB classifier in practically every evaluation metric. It reached 71% of accuracy.

Table 5. Results using pre-processed and cleaned data

Feature	SVM (unigram)	NB (unigram)	SVM (bigram)	NB (bigram)
Accuracy (%)	81.3	79.2	78.28	76.07
Recall (%)	79.49	73.98	74.96	75.89
F-measure (%)	80.01	76.1	76.92	72.14

Table 5 displays the results of the three classifiers, NB, SVM, and LSTM, on the comments using normalization and stop word removal. Accuracy, recall, and F-measure are used to assess performance. Best results are obtained in accuracy with TF-IDF using SVM classifier, NB also showed some promising results. We notice that the use of bigrams didn't increase the accuracy of our classification and it might be due to negation in Moroccan dialect is done using a prefix or suffix. The results show that using the pre-processing stage yields the best performance for all classifiers when compared to the results obtained without pre-processing. In fact, Table 5 demonstrates that the three classifiers perform best when stop words and emojis are eliminated. Given that pre-processing is an effective approach to reduce text noise, these results demonstrate that pre-processing is required in order to increase the accuracy and performance of the Moroccan dialect.

6 Conclusion

The Arabic language is distinguished by a large number of dialects. With the advent of the social web, people can now express themselves using these dialects. Moroccan Dialect varies from MSA at all linguistic stages such as phonology and morphology to lexicon and syntax. Because of the language's rich morphology, analyzing opinions and emotions in Moroccan dialect is difficult. Extraction of the massive number of comments and feedback presented on social media necessitates consideration of the Moroccan Dialect's peculiarities and characteristics (Arabizi, code-switching, etc.). In this paper we presented a quite large annotated corpus. That consist of twenty thousand Moroccan dialect comments. Several contentious themes were used to extract and compile the comments from the corpus (famous videos, sports, politics, trending videos and others). To avoid data bias to a certain lexicon, especially for the task of sentiment analysis, the comments collecting did not rely on emotions or sentiment keywords. At a reasonable cost, this technology can be used for sentiment analysis. We also remarked that it is balanced and deals with various web writing styles. As a last step, we used different machine learning models including SVM and NB to categorize the MD comments as either positive or negative, then we measured their performance to further investigate the validity and efficacy of MYC. For polarity categorization, reported accuracies vary from 79 to 81%. The good results of our models prove that our corpus can be useful for further sentiment analysis researches in sentiment analysis. Future plans include addressing the remaining issues, generating stop word lists specifically for this dialect, and experimenting with deep learning models to improve the findings. These resources and tools will be used for opinion and emotion analysis.

References

1. Liu, B.: Sentiment analysis and opinion mining. Synth. Lect. Hum. Lang. Technol. **5**(1), 1–167 (2012)
2. Jackson, P., Moulinier, I.: Natural Language Processing for Online Applications: Text Retrieval, Extraction and Categorization, vol. 5. John Benjamins Publishing Company, Amsterdam (2002)

3. Atia, S., Shaalan, K.: Increasing the accuracy of opinion mining in Arabic. In: Proceedings—1st International Conference on Arabic Computational Linguistics: Advances in Arabic Computational Linguistics ACLing 2015, pp. 106–113 (2015)
4. Cherif, W., Madani, A., Kissi, M.: Towards an efficient opinion measurement in Arabic comments. Procedia Comput. Sci. **73**(Awict), 122–129 (2015)
5. Salloum, S.A., Al-emran, M., Monem, A.A., Shaalan, K.: A survey of text mining in social media: facebook and twitter perspectives. Adv. Sci. Technol. Eng. Syst. J. **2**(1), 127–133 (2017)
6. Heikal, M., Torki, M., El-Makky, N.: Sentiment analysis of Arabic tweets using deep learning. Procedia Comput. Sci. **142**, 114–122 (2018)
7. Duwairi, R.M., Marji, R., Sha'ban, N., Rushaidat, S.: Sentiment analysis in Arabic tweets. In: 5th International Conference on Information and Communication Systems (ICICS) (2014)
8. Al-Tamimi, A.-K., Shatnawi, A., Bani-Issa, E.: Arabic sentiment analysis of YouTube comments. In: IEEE Jordan Conference on Applied Electrical Engineering and Computing Technologies (2017)
9. Ebimba.com. Top 15 Most Popular social Networking Sites (2021). http://www.ebizmba.com/articles/social-networkingwebsites. Accessed: 06 Mar 2021. (Don't DELETE)
10. Abdul-Mageed, M., Diab, M.T.: SANA: a large scale multi-genre, multi-dialect lexicon for Arabic subjectivity and sentiment analysis. In: LREC (2014)
11. Diab, M., et al.: Tharwa: a large scale dialectal Arabic-standard Arabic-English lexicon. In: Proceedings of the Language Resources and Evaluation Conference (LREC) (2014)
12. Rahab, H., Zitouni, A., Djoudi, M.: SIAAC: sentiment polarity identification on Arabic Algerian newspaper comments. In: Silhavy, R., Silhavy, P., Prokopova, Z. (eds.) CoMeSySo 2017. AISC, vol. 662, pp. 139–149. Springer, Cham (2018). https://doi.org/10.1007/978-3-319-67621-0_12
13. Al-Thubaity, A., Alqahtani, Q., Aljandal, A.: Sentiment lexicon for sentiment analysis of Saudi dialect tweets. Procedia Comput. Sci. **142**, 301–307 (2018)
14. El-Beltagy, S.R., Ali, A.: Open issues in the sentiment analysis of Arabic social media: a case study. In: 2013 9th International Conference on Innovations in Information Technology (IIT) (2013)
15. Guellil, I., Adeel, A., Azouaou, F., Hussain, A.: SentiALG: automated corpus annotation for Algerian sentiment analysis. In: Ren, J., et al. (eds.) BICS 2018. LNCS (LNAI), vol. 10989, pp. 557–567. Springer, Cham (2018). https://doi.org/10.1007/978-3-030-00563-4_54
16. Oussous, A., Benjelloun, F.-Z., Lahcen, A.A., Belfkih, S.: ASA: a framework for Arabic sentiment analysis. J. Inf. Sci. (2019)
17. Abdulla, N.A., Ahmed, N.A., Shehab, M.A., Al-Ayyoub, M., Al-Kabi, M.N., Al-rifai, S.: Towards improving the lexicon-based approach for Arabic sentiment analysis. Int. J. Inf. Technol. Web Eng. **9**(3), 55–71 (2014)
18. Elmadany, A.A., Hamdy Mubarak, W.M.: ArSAS: an Arabic speech-act and sentiment corpus of tweets. In: OSACT 3: the 3rd Workshop on Open-Source Arabic Corpora and Processing Tools, p. 20 (2018)
19. Elnagar, A., Khalifa, Y.S., Einea, A.: Hotel Arabic-reviews dataset construction for sentiment analysis applications. Studies Comput. Intell. 35–52 (2017)

Feature Selection for Text Classification Using Genetic Algorithm

Salma Belkarkor[1](✉), Imad Hafidi[1], and Mourad Nachaoui[2]

[1] Laboratory of Process Engineering, Computer Science, and Mathematics at the National School of Applied Sciences, Sultan Moulay Slimane University, Khouribga, Morocco
salma.belkarkour@usms.ac.ma, i.hafidi@usms.ma

[2] Laboratory of Mathematics and Applications, Team of Mathematics and Interactions, at the Faculty of Science and Technology, Sultan Moulay Slimane University, Beni-Mellal, Morocco
m.nachaoui@usms.ma

Abstract. Today's large amount of text data makes feature classification a difficult text processing challenge. High dimensionality is the primary challenge in text processing, and feature selection is a common method for reducing dimensions. The most crucial factors in text classification are a strong text representation and a very accurate classifier. As a result, choosing the right features is essential for using machine learning algorithms effectively. Different optimization techniques, including the Genetic Algorithm (GA), have been effectively used for dimensionality reduction in the field of text classification. To evaluate the performance of GA for Feature Selection (FS), we compared the GA for FS with other filtering methods to prove the efficiency of the GA for FS, for that, we used the NB classifier and three benchmark document collections: SMS, BBC, and 20Newsgroups.

Keywords: Feature selection · Text classification · Genetic Algorithm

1 Introduction

Data mining is the process of looking at large amounts of data, from many perspectives to find patterns and turn them into useful information [13].

One of the most important and common issues in data mining and machine learning is classification. Numerous real-world problems from various fields can be rephrased as classification issues. This covers a variety of tasks, like software quality assurance, text classification, medical diagnosis, and microarray data diagnosis [20].

Text classification is one of the fields of study that contributes to the resolution of many problems, such as sentiment analysis, searching by filters, and classification of spammer and no-spammer content.

Supported by LIPIM.

N. Aboutabit et al. (Eds.): ICMICSA 2022, LNNS 656, pp. 69–80, 2023.
https://doi.org/10.1007/978-3-031-29313-9_7

To automatically categorize text documents, text classification often relies on building a model through learning from training samples. Text documents are typically transformed into a vector of individual words in text classification methods. The corpus of documents can produce an enormous feature space due to the existence of every distinct word. As a result, a significant problem with text classification is the feature space's high dimension, which frequently degrades the effectiveness of learning techniques. One approach to overcoming these constraints is feature selection, which aims to minimize the quantity of the data by removing noisy data from a given training dataset [18,30].

High dimensionality can lead to issues like model overfitting and the dimensionality curse. Reduce dimensionality, eliminate unimportant data and improve learning performance by using FS. The feature in text classification issues is typically some sort of representation of a group of words. It's possible that a large fraction of characteristics taken from the text corpus won't be effective for classifying texts. These irrelevant features have the potential to reduce the classification models' effectiveness and accuracy [17,23].

In order to determine the optimum subset for feature selection, several methodologies have been offered to resolve the issue, including exhaustive search, greedy search, and random search. Premature convergence, tremendous complexity, and high computing costs are problems that plague the majority of approaches. As a result, metaheuristic algorithms receive a lot of attention to handle this kind of situation. [1].

Among the best-known metaheuristic algorithms in the field of feature selection, we can cite the GA. This algorithm takes inspiration from natural evolution and begins with a population of solutions that are generated at random. The best solutions are combined in this algorithm to produce new individuals. Mutation, crossover, and the best option are used to create new individuals [1].

The rest of the paper is structured as follows: Sect. 2 summarizes pertinent literature, incorporating the notion of feature selection approaches and GA for feature selection, in Sect. 3 experimental results of comparison performance metrics are reported and Sect. 4 concludes the work.

2 Background and Literature Review

2.1 Text Classification

The task of text classification is to determine which class(es) a given text belongs to [28]. Three basic subtypes of the classification issue may be distinguished: binary, multiclass, and multilabel. The issue is known as a binary classification problem if only two classes are predefined. A multiclass classification issue is one in which there are three or more declared classes, but each document can only belong to one of them. Finally, a multilabel classification issue is one in which each document might simultaneously belong to two or more classes (or labels).

For the goal of text classification, several algorithms have been created. The two primary categories of these algorithms are deep learning and traditional machine learning. Some traditional algorithms, such as k-Nearest Neighbors

(KNN), Support Vector Machines (SVM), and Naive Bayes (NB) are still frequently researched and employed by the scientific community. Deep Belief Network (DBN), Hierarchical Attention Network (HAN), and Convolutional Neural Network (CNN) are examples of deep learning-based architectures that are progressively being studied for text classification [16,23].

2.2 Feature Selection and Approaches

By reducing the dataset through feature selection, machine learning models become less complicated in terms of the time and computational resources needed to learn the data. Filter, wrapper, and embedding are frequent categories used to group feature selection algorithms [3,12,27].

Filter Approaches. In filter techniques, each feature is given a score based on how well it correlates with other characteristics as well as the target variable using a statistical measure, and the characteristics with the best ratings are then chosen [25]. Filter techniques, which are classifier-independent, provide excellent generalization control over a large variety of classifiers, allowing the chosen characteristics to be trainable across various classifiers, as a result, when the classifier is changed, the feature selection operation does not have to be repeated. Additionally, filter techniques are simple and straightforward to use with high-dimensional datasets. A duplicate subset of characteristics may be chosen in heavily corrected datasets, which may produce a subset of characteristics that is inappropriate for a classifier's training [12]. This is a significant disadvantage of filter techniques.

Wrapper Approaches. In wrapper techniques, the machine learning algorithm's classification capability is utilized to assess a set of characteristics, which are compared next to one another to choose the highest subset. Wrapper techniques frequently have a greater ability in terms of classification accuracy than other feature selection methods, but when used on high-dimensional datasets, they demand a lot of computing work [22]. Additionally, since the chosen set of characteristics is influenced in favor of the classifier employed in the wrapper technique, wrapper-based techniques have a lower generalization power than other classifiers [5].

Embedded Approaches. Filter and wrapper techniques are both utilized by embedded approaches, which attempt to choose highly discriminating features [4]. Through analysis and measurement of the impact of the chosen features on the creation of a classifier, embedded techniques eliminate noisy features. While building a model, embedded techniques identify the elements that best improve classification accuracy. Although embedded techniques are less difficult than wrapper approaches, they nonetheless are not appropriate for large-scale datasets since they carry over the limitations of wrapper techniques.

2.3 Genetic Algorithm

A population-based evolutionary algorithm called the GA was first suggested by Holland in 1975. A GA mimics the process of evolution that occurs in nature and is designed to look for the best answer to a particular issue by simulating natural selection. Many studies have shown the benefits of the GA in dealing with increased dimensions and feature selection difficulties [10].

2.4 GA for Feature Selection

The feature dimension space serves as the search space when using the GA as an FS technique. The most significant components to consider are the initialization of the initial population, the fitness function's definition, the selection strategy's architecture, and the definition and implementation of the mutation and crossover operators. These components are described below [18,26].

Steps of Genetic Algorithm

Initialization: This step's purpose is to provide an initial population of solutions. To accomplish this, each solution must be represented as a vector of values that illustrates the chosen characteristics.

Fitness assignment: the solutions are evaluated using a single objective or a multi-objective function.

Selection: The chromosome can be maintained via a variety of selection approaches. These include rank selection, proportional selection, steady-state selection, roulette wheel selection, and tournament selection.

Crossover: The primary goal of the crossover procedure is to choose the adjacent suitable chromosome (child). Different crossover approaches are applied in GA. The uniform, arithmetic, single-point, and two-point crossover are a few of the crossover methods. An arbitrary number is sufficient for two distinct chromosomes in a single-point crossover. To create a new child, the two chromosomes are switched based on the random value. The uniform crossover involves using a pair of random numbers or more to bring out additional characteristics from the chromosomes of the two parents to produce a new child. Arithmetic crossover, as opposed to a uniform crossover and single point, is employed to compare and contrast two distinct chromosomes that are produced in future stages by employing the OR, AND, and different operators.

Mutation: By flipping bit values or swapping multiple bits within a chromosome, the mutation procedure in GA entails enhancing the shade of a child's chromosomes.

2.5 Related Work on GA for Feature Selection

Uguz et al. [31] proposed combining a GA, an Information Gain (IG) filtering approach, and a Principal Component Analysis (PCA) to create a hybrid methodology. Each characteristic in the collections of textual data is given a rank in the first phase using the IG, and a specific proportion of the characteristics are then selected from all of the characteristics based on their rank. The GA and PCA each submit a separate application in the second stage, concentrating on the selected sub-ensembles of characteristics. C4.5 and KNN's average F-measure serves as the foundation for the fitness function utilized by GA to assess the chromosomes (feature subsets).

In this study [10], hybrid feature selection methods based on GA are proposed. In order to manage the high dimensionality of the feature space and better classification performance at the same time, this strategy employs a hybrid search methodology that merges the benefits of filter feature selection techniques with an Enhanced GA (EGA) in a wrapper approach. By enhancing the crossover and mutation operators, firstly the EGA was proposed. The crossover operation is carried out using the word and document frequencies of features, whereas the mutation is carried out using the original parent's performance classifiers and the significance of the feature. As a result, rather than employing probability and a random selection, mutation and crossover processes are carried out according to meaningful information. Secondly, the EGA was combined with six popular filter feature selection techniques (Class Discriminating Measure (CDM), Odd Ratio (OR), GSS [9], F-measure (FM), IG, and Inverse Term Frequency (TF-IDF)) to develop hybrid characteristics selection methods. In the hybrid technique, dimension reduction is done after the EGA is used for several characteristic subsets of various dimensions that are ordered in descending order according to their relevance. The most significant characteristics that have higher rankings were subject to EGA procedures.

In [18], a multi-objective text feature selection strategy based on filters was developed. The relative discriminative criterion is employed to identify the relevance of a text characteristic, whereas the correlation measure is used to determine redundancy.

Through the hybridization of a GA and straight multi-search for key quality feature selection, Li et al. [19] created a multi-objective feature selection approach called GADMS. It is a hybrid approach made up of a GA and a local search method called direct multi-search (DMS). Key quality characteristics selection is defined as a bi-objective issue of maximizing the importance of the efficiency characteristics subset and minimizing the size of the characteristic subset.

Nag et al. [21] proposed the characteristics extraction and selection strategy for parsimonious classifiers by using multi-objective genetic programming. The purpose of this research is to examine if genetic programming can choose linearly severable characteristics when the evolutionary process is directed to accomplish this and to make a proposal for an integrated system to achieve this goal. For the fitness functions, they proposed a fitness function according to Golub's index [11] to choose important characteristics and unfitness functions for characteristic nodes to exclude less commonly utilized characteristics.

A bi-objective genetic algorithm-based characteristics selection approach is suggested as a solution to the problem of feature selection in data mining. In the proposed study [7], two objective functions-Rough set theory's border area analysis and information theory's multivariate mutual information-are employed to filter out only accurate and instructive data from the data set. After sampling the data set with a replacement technique, the approach is used to discover non-dominated feature subsets from each sampled data set. In order to create a highly generalized feature subset, a collection of these bi-objective evolutionary algorithm-based feature selectors is constructed with the aid of concurrent implementations. In reality, individual feature selector outputs are combined to form the final feature subset using a new dominance-based strategy.

This study [27] presents a hybrid framework named filter/filter, a unique evolutionary-based filter feature selection method that is successively hybridized with the Fisher score filter algorithm. The suggested approach is based on an Asexual GA in conjunction with a long-term memory Tabu Search (TAGA). A unique mutation operator, a new tabu list encoding technique, a novel integer-encoded representation of the solution, and a fitness function based on an information theory-based criteria are all advantages of TAGA.

Tsai et al. [30] proposed a Biological Genetic Algorithm (BGA) for instance selection in text classification. The BGA changes the GA using an elite reserve zone, nonlinear conversion of fitness values, and migration. The proposed method has been compared with traditional instance selection techniques, including SVM, edited nearest neighbor, IB3 [2], and DROP3 [32] algorithms on k-NN.

Ewees et al. [8] proposed an enhanced metaheuristic optimization method that merges the traditional Arithmetic Optimization Algorithm (AOA) with GA operators. This method presents enhanced crossover and mutation operators for exploitation and exploration based on search.

In order to categorize microarray cancer data, this research [24] suggests a model that makes use of a hybrid or compound feature selection technique (Filter + Wrapper). To remove duplicate characteristics that will not significantly contribute to the categorization process, an initial filter approach known as IG is applied. After that, a micro GA is employed in evolutionary computing to discover the optimal minimal subset of necessary traits.

This article's [26] main objective is to identify spam accounts and content. It is suggested to use a hybrid method that uses the skills of different machine learning algorithms to distinguish between spammer and non-spammer contents and accounts. The numerous characteristics are first analyzed using the GA, which then chooses the best attributes that have the most effect on user profile behavior. These characteristics are then utilized to train classifiers.

3 Comparison Performance Metrics

In this section, we are going to present comparative studies between the GA for FS and the traditional methods: CHI2 and Mutual information (MI) filters. The algorithms have been implemented in Python and the experiments were done under windows 10 using a laptop equipped with an Intel I5 processor and 8 GB memory.

3.1 Text Representation

In many text mining applications, including information retrieval, text clustering, and document classification, text representation is a basic problem. Text representation in text classification is the conversion of the text into a vector of terms [14].

In this study, we use the TF-IDF (Term Frequency - Inverse Document Frequency) method to estimate the importance of each term. Details about this method can be found in the literature [16]. The TF-IDF formula is represented in Eq. 1. For each feature f in category C_i:

$$TF - IDF = TF(f, c_i) * log(N/DF(f)) \tag{1}$$

where N is the total number of training documents in the collection.

3.2 Classifier Used

In this study, we employed Naive Bayes (NB) classifier. The word features that the multinomial NB classifier [6,15,29] uses are discrete features. According to the subsequent process, the unknown document class label is assigned. Let the document be d = $\{t_1, t_2, t_3, ..., t_n\}$. The class label prediction of document d is:

$$class\_label(d) = \max_{i=1,2,...c} P(C_i) \prod_{j=1}^{n} P(t_j \mid C_i) \tag{2}$$

where P(C_i) is the probability of class C_i; P($t_j|C_i$) is the probability of the term t_j for a given class C_i.

3.3 Description of Datasets

In these experiments, we have used three distinct datasets (SMS, BBC, 20Newsgroups). These datasets in the preprocessing stage must go through three processes: tokenization, stop word removal, and stemming. In this section, we present the different datasets.

SMS Dataset. The SMS data set is a collection of SMS messages, labeled as spam or ham. It contains 5572 tagged SMS messages. The description of the SMS dataset is presented in Table 1.

Table 1. SMS dataset

SI. no	Category	Number of documents
1	SPAM	747
2	HAM	4825
	Total	5572

BBC Dataset. The BBC dataset contains 2225 textual documents from the BBC news website between 2004 and 2005, classified into five categories. The description of the BBC dataset is presented in Table 2.

Table 2. BBC dataset

SI. no	Category	Number of documents
1	Sport	511
2	Business	510
3	Politics	417
4	Technology	401
5	Entertainment	386
	Total	2225

20Newsgroups Dataset. The 20 newsgroups dataset contains approximately 18,846 topical documents, which are manually classified into 20 different newsgroups. Table 3 describes the 20 Newsgroup dataset.

Table 3. 20Newsgroups dataset

SI. no	Category	Number of documents
1	rec.sport.hockey	999
2	soc.religion.christian	997
3	rec.motorcycles	996
4	rec.sport.baseball	994
5	sci.crypt	991
6	rec.autos	990
7	sci.med	990
8	comp.windows.x	988
9	sci.space	987
10	comp.os.ms.windows.misc	985
11	sci.electronics	984
12	comp.sys.ibm.pc.hardware	982
13	misc.forsale	975
14	comp.graphics	973
15	comp.sys.mac.hardware	963
16	talk.politics.mideast	940
17	talk.politics.guns	910
18	alt.atheism	799
19	talk.politics.misc	755
20	talk.religion.misc	628
	Total	18846

3.4 The Experiments

For GA, we encode the solution with 65% of the features and create a population of 32 chromosomes at random. We choose the chromosomes for the crossover and mutation using the roulette wheel selection method. With a probability parameter of 0.9, one crossover point operator is chosen for crossover. The uniform mutation operator is used for mutation, and the probability parameter is set at 0.001. We utilize the NB classifier to calculate each chromosome's fitness. The maximum generation number, which is 200, serves as the termination criterion. Results of Accuracy, Precision, Recall, and F1-score values on SMS, BBC, and 20Newsgroups datasets are shown in the tables below. As we can see from the tables, the GA achieved the best results for all datasets using only 65% of the total features (Table 4, 5 and 6).

Table 4. Performance measurements of the NB classifier with GA for FS and CHI2 and MI filters on SMS dataset

		65% of features		
Evaluation measures	NB with 100% of features	CHI2 Filter	MI filter	GA
Accuracy	95.69%	95.60%	94.88%	**96.23%**
Precision	97.59%	97.54%	97.16%	**97.88%**
Recall	85.54%	85.24%	82.83%	**87.34%**
F1-score	90.31%	90.08%	88.17%	**91.67%**

Table 5. Performance measurements of the NB classifier with GA for FS and CHI2 and MI filters on BBC dataset

		65% of features		
Evaluation measures	NB with 100% of features	CHI2 Filter	MI filter	GA
Accuracy	97.07%	97.52%	96.85%	**99.77%**
Precision	97.02%	97.35%	96.75%	**99.79%**
Recall	96.84%	97.34%	96.57%	**99.75%**
F1-score	96.82%	97.28%	96.59%	**99.77%**

Table 6. Performance measurements of the NB classifier with GA for FS and CHI2 and MI filters on 20 newsgroups dataset

		65% of features		
Evaluation measures	NB with 100% of features	CHI2 Filter	MI filter	GA
Accuracy	86.31%	86.47%	86.41%	**87.66%**
Precision	87.82%	87.93%	87.83%	**89.09%**
Recall	85.40%	85.57%	85.51%	**86.72%**
F1-score	84.89%	85.08%	85.02%	**86.14%**

4 Conclusion

The quantity and variety of research that investigates feature selection demonstrate the challenge and significance of the FS approach. In this paper, we focus on the selection of text features for classification using the genetic algorithm. Many studies have shown the benefits of applying GA to high-dimensional problems. The main differences between these studies may be in the modification of GA operators, the hybridization of GA with other FS approaches, and the utilization of several objective functions.

The intention was to demonstrate the usefulness of genetic algorithms for FS for text classification. In our comparison between some traditional FS methods and the GA, we can notice that with the GA all the performance measures of the NB algorithm have higher values either before the feature reduction (the use of 100% features) or after the use of CHI2 and MI filters. For our future work, we're interested in utilizing hybrid solutions to enhance textual document classification.

References

1. Agrawal, P., Abutarboush, H.F., Ganesh, T., Mohamed, A.W.: Metaheuristic algorithms on feature selection: a survey of one decade of research (2009–2019). IEEE Access **9**, 26766–26791 (2021)
2. Aha, D.W., Kibler, D., Albert, M.K.: Instance-based learning algorithms. Mach. Learn. **6**(1), 37–66 (1991)
3. Bolón-Canedo, V., Sánchez-Marono, N., Alonso-Betanzos, A., Benítez, J.M., Herrera, F.: A review of microarray datasets and applied feature selection methods. Inf. Sci. **282**, 111–135 (2014)
4. Canul-Reich, J., Hall, L.O., Goldgof, D.B., Korecki, J.N., Eschrich, S.: Iterative feature perturbation as a gene selector for microarray data. Int. J. Pattern Recognit Artif Intell. **26**(05), 1260003 (2012)
5. Chandrashekar, G., Sahin, F.: A survey on feature selection methods. Comput. Electr. Eng. **40**(1), 16–28 (2014)
6. Chen, J., Huang, H., Tian, S., Qu, Y.: Feature selection for text classification with naïve bayes. Expert Syst. Appl. **36**(3), 5432–5435 (2009)
7. Das, A.K., Das, S., Ghosh, A.: Ensemble feature selection using bi-objective genetic algorithm. Knowl.-Based Syst. **123**, 116–127 (2017)
8. Ewees, A.A., et al.: Boosting arithmetic optimization algorithm with genetic algorithm operators for feature selection: case study on cox proportional hazards model. Mathematics **9**(18), 2321 (2021)
9. Galavotti, L., Sebastiani, F., Simi, M.: Experiments on the use of feature selection and negative evidence in automated text categorization. In: Borbinha, J., Baker, T. (eds.) ECDL 2000. LNCS, vol. 1923, pp. 59–68. Springer, Heidelberg (2000). https://doi.org/10.1007/3-540-45268-0_6
10. Ghareb, A.S., Bakar, A.A., Hamdan, A.R.: Hybrid feature selection based on enhanced genetic algorithm for text categorization. Expert Syst. Appl. **49**, 31–47 (2016)
11. Golub, T.R., et al.: Molecular classification of cancer: class discovery and class prediction by gene expression monitoring. Science **286**(5439), 531–537 (1999)

12. Guyon, I., Elisseeff, A.: An introduction to variable and feature selection. J. Mach. Learn. Res. **3**, 1157–1182 (2003)
13. Han, J., Kamber, M., Pei, J.: Outlier detection. Data mining: concepts and techniques, pp. 543–584 (2012)
14. Hong, S.S., Lee, W., Han, M.M.: The feature selection method based on genetic algorithm for efficient of text clustering and text classification. Int. J. Advance Soft Comput. Appl. **7**(1), 2074–8523 (2015)
15. Kim, S.B., Han, K.S., Rim, H.C., Myaeng, S.H.: Some effective techniques for Naive Bayes text classification. IEEE Trans. Knowl. Data Eng. **18**(11), 1457–1466 (2006)
16. Kowsari, K., Jafari Meimandi, K., Heidarysafa, M., Mendu, S., Barnes, L., Brown, D.: Text classification algorithms: a survey. Information **10**(4), 150 (2019)
17. Kumbhar, P., Mali, M.: A survey on feature selection techniques and classification algorithms for efficient text classification. Int. J. Sci. Res. **5**(5), 9 (2016)
18. Labani, M., Moradi, P., Jalili, M.: A multi-objective genetic algorithm for text feature selection using the relative discriminative criterion. Expert Syst. Appl. **149**, 113276 (2020)
19. Li, A.D., Xue, B., Zhang, M.: Multi-objective feature selection using hybridization of a genetic algorithm and direct multisearch for key quality characteristic selection. Inf. Sci. **523**, 245–265 (2020)
20. Nag, K., Pal, N.R.: A multiobjective genetic programming-based ensemble for simultaneous feature selection and classification. IEEE Trans. Cybern. **46**(2), 499–510 (2015)
21. Nag, K., Pal, N.R.: Feature extraction and selection for parsimonious classifiers with multiobjective genetic programming. IEEE Trans. Evol. Comput. **24**(3), 454–466 (2019)
22. Naghibi, T., Hoffmann, S., Pfister, B.: A semidefinite programming based search strategy for feature selection with mutual information measure. IEEE Trans. Pattern Anal. Mach. Intell. **37**(8), 1529–1541 (2014)
23. Pintas, J.T., Fernandes, L.A., Garcia, A.C.B.: Feature selection methods for text classification: a systematic literature review. Artif. Intell. Rev. **54**(8), 6149–6200 (2021)
24. Pragadeesh, C., Jeyaraj, R., Siranjeevi, K., Abishek, R., Jeyakumar, G.: Hybrid feature selection using micro genetic algorithm on microarray gene expression data. J. Intell. Fuzzy Syst. **36**(3), 2241–2246 (2019)
25. Ruiz, R., Riquelme, J.C., Aguilar-Ruiz, J.S., García-Torres, M.: Fast feature selection aimed at high-dimensional data via hybrid-sequential-ranked searches. Expert Syst. Appl. **39**(12), 11094–11102 (2012)
26. Sahoo, S.R., Gupta, B.B.: Classification of spammer and nonspammer content in online social network using genetic algorithm-based feature selection. Enterp. Inf. Syst. **14**(5), 710–736 (2020)
27. Salesi, S., Cosma, G., Mavrovouniotis, M.: TAGA: TABU asexual genetic algorithm embedded in a filter/filter feature selection approach for high-dimensional data. Inf. Sci. **565**, 105–127 (2021)
28. Schütze, H., Manning, C.D., Raghavan, P.: Introduction to Information Retrieval, vol. 39. Cambridge University Press, Cambridge (2008)
29. Thirumoorthy, K., Muneeswaran, K.: Optimal feature subset selection using hybrid binary Jaya optimization algorithm for text classification. Sādhanā **45**(1), 1–13 (2020)
30. Tsai, C.F., Chen, Z.Y., Ke, S.W.: Evolutionary instance selection for text classification. J. Syst. Softw. **90**, 104–113 (2014)

31. Uğuz, H.: A two-stage feature selection method for text categorization by using information gain, principal component analysis and genetic algorithm. Knowl.-Based Syst. **24**(7), 1024–1032 (2011)
32. Wilson, D.R., Martinez, T.R.: Reduction techniques for instance-based learning algorithms. Mach. Learn. **38**(3), 257–286 (2000)

Machine Learning Methods in the Detection of Type 2 Diabetes Mellitus Risk Factors

Boutayeb Wiam[1](✉), Badaoui Mohammed[2], Al-Ali Hannah[3], Boutayeb Abdesslam[4], and Doukali Mouhssine[4]

[1] Superior School of Education and Training, LaMSD, University Mohammed first, Oujda, Morocco
wiam.boutayeb@gmail.com

[2] Superior School of Technologies, LaMSD, University Mohammed first, Oujda, Morocco
med.badaoui@ump.ac.ma

[3] Emirates Aviation University, Dubai, UAE
hannah.alali@emirates.com

[4] Faculty of Sciences, LaMSD, University Mohammed first, Oujda, Morocco
x.boutayeb@gmail.com

Abstract. Big Data Analysis techniques are some of the most used tools in healthcare industry. They allow the prediction of outcomes and interpretation of results from the study of huge data sets. In order to detect the main risk factors of Type 2 Diabetes Mellitus, we compare the performance of three supervised data mining algorithms (Decision trees, Bayesian networks, Neural networks) according to the accuracy rate and the model sensitivity.

Our comparative study is applied on "Gulf Coast" cleansed Database.

Keywords: Data mining tools · Type 2 Diabetes · Gulf Coast Database

1 Introduction

According to the last International Diabetes Federation (IDF) report, most countries will not fulfill the World Health Organisation (WHO) 2025 target of stopping the rise of Type 2 Diabetes Mellitus (TD2M). The large majority (around 90%) of diabetes worldwide is represented by T2DM. Unfortunately, this type of diabetes can lead to short or long-term complications including premature death [1].

Out of all IDF Regions, MENA region had the highest age-adjusted diabetes prevalence (12.2%) in 2019. Therefore, the number of diabetics is estimated to increase by 91% from 2019 to 2045 in this region.

In particular, the prevalence of diabetes in Middle East region reaches (14.6%), with highest prevalence in Kuwait, Saudi Arabia, and UAE (> 21%) [2].

N. Aboutabit et al. (Eds.): ICMICSA 2022, LNNS 656, pp. 81–91, 2023.
https://doi.org/10.1007/978-3-031-29313-9_8

Consequently, important efforts are needed in order to prevent or avoid the evolution towards this type of diabetes by acting on its risk factors [1].

Artificial Intelligence(AI) is defined as an imitation of intelligent human behaviour using a computer program (machine) [3]. Artificial Intelligence has been widely used in a large variety of areas including medical sector, in which its application allows the machine to process data. Accordingly, the machine has to understand, learn, predict then proceed [4]. Consequently, Doctors can act more quickly to avoid complications and premature death.

In artificial intelligence, data mining is used in order to extract implicit patterns between the items in the data set to predict the results [5,6]. The most common techniques of data mining are: Detection of anomalies, Classification, Clustering and Association [7]. In the last decades, the use of data mining tools has known a successful improvement in health care [8–11]. Several studies have focused on finding risk factors leading to T2DM using data mining tools. Boutayeb et al. used Decision Tree classifier to determine the main risk factors of T2DM in Gulf countries [12]. Saman Hina et al. applied different algorithms of classification on "Pima Indians Diabetes Database of National Institute of Diabetes and Digestive and Kidney Diseases datasets" that include records of 768 patients to find the most suitable algorithm that allows to detect T2DM risk factors [13]. Esmaeily et al. analysed the performance of three data mining techniques (Artificial Neural Network, Support Vector Machines, Multiple Logistic Regression) to seek potential risk factors of T2DM in a large population composed by 9582 subjects [14].

In order to figure out the causes of increasing diabetes in adult patients, Akkarapol Sa-ngasoongsong et al. applied and compared Artificial Neural Network, Logistic Regression, and Decision three models in a categorized analysis into three different focuses. The groups are based on the patients' healthcare costs [15].

In this paper we compare the performance of three supervised data mining algorithms (DDecision trees, Bayesian networks, Neural networks) according to the accuracy rate and the model sensitivity, in order to detect the main risk factors of T2DM in four Gulf countries (Bahrain, Kuwait, Saudi Arabia, and UAE).

Our comparative study is applied on the cleaned Gulf Coast Database that contains 3372 rows.

2 Methods

2.1 Dataset

In this work,"Gulf Coast" dataset is used in order to determine T2DM risk factors in Gulf countries.

As described in our latest paper [12], clinical informations were gathered from the Gulf Coast registry obtained by surveys of patients suffering from Acute Coronary Syndrome (ACS), admitted between the years 2012 and 2013. "Gulf Coast" dataset contains 317 variables collected from 4061 patients. However, after data cleaning, only 33 variables were kept to be studied on 3327 remained individuals.

2.2 Variables Description

Retained variables are given in the following table (see Table 1)

Table 1. Variables description

Variable	Type	Signification	Modality
Country	Categorical	Country of residence	Bahrain, Kuwait, Saudi Arabia, or UAE
Gender	Categorical	Gender	Male or Female
Age	Numerical	Patient's age	from 18 to 112
Marital_Status	Categorical	Marital status	Married, single, Divorced or Widowed
Work	Categorical	Work status	No, Yes Full time, Yes Part time
Cardiac_Arrest_Admission	Categorical	Admitted because of cardiac arrest	Yes or No
Hypertension	Categorical	Hypertension	Yes or No
Dyslipidemia	Categorical	Dyslipidemia	Yes or No
DM	Categorical	Type 2 Diabetes Mellitus	Yes or No
Year_DM_Diagnosed	Numerical	Year when DM was discovered	from 1970 to 2012
DM_Duration	Numerical	Duration of DM	from 0 to 42
Family_History_DM	Categorical	Family history of DM	Yes or No
Smoking_History	Categorical	Smoking history	Current smoker or Never smoked or Past smoker or Recent smoker
Waist	Numerical	Waist size	from 25 to 152 (cm)
BMI	Numerical	Body Mass Index	from 15 to 61 $(\frac{kg}{m^2})$
Admission_Blood_Glucose_Value_SI_Units	Numerical	Admission blood glucose	from 0.5 to 6 $(\frac{g}{l})$
Fasting_Blood_Glucose_Value_SI_Units	Numerical	Fasting blood glucose	from 0.5 to 5 $(\frac{g}{l})$
HbA1C_Admission_Value	Numerical	Percentage of glycated haemoglobinse	from 3% to 20%
Lipid_24_Collected	Boolean	Rate of Lipid during 24 storage	True or False
Cholesterol_Value_SI_Units	Numerical	Cholesterol level	from 1 to $10(\frac{g}{l})$
Triglycerides_Value_SI_Units	Numerical	Triglycerides level	from 0.5 to $17.6(\frac{g}{l})$
LDL_Value_SI_Units	Numerical	Rate of Low Density Lipoproteins	from 0.4 to $9(\frac{g}{l})$
HDL_Value_SI_Units	Numerical	Rate of High Density Lipoproteins	from 0.3 to $4.29(\frac{g}{l})$
Creatinine_Clearance	Numerical	Creatinine clearance	from 30 to $300(\mu mol)$
Education	Categorical	Level of education	No school or High school or College
Sleep_Apnea	Categorical	Sleep apnea	Yes or No
Heart_Rate	Numerical	Heart rate	from 40 to 180(beats per minute)
Systolic_BP	Numerical	Systolic blood pressure	from 42 to $253(mmHg)$
Diastolic_BP	Numerical	Diastolic blood pressure	from 29 to $168(mmHg)$
DM_treatment	Categorical	DM treatment	Non diabetic or Oral drugs or Diet or Insulin
DM_type	Categorical	Type of diabetes	Type2 or Non diabetic
Non_Cardiac_Condition	Categorical	Non Cardiac Condition	Yes or No
Stress	Categorical	Stress	Yes or No

3 Results

3.1 Bivariate Analysis

Chi-square test of independence is carried out in order to study the relationship between T2DM and Dyslipedemia. It shows that the association between the two variables is significant, χ^2 $(1, N = 3372) = 357.4347$ and $p < .00001$. Consequently, people with dyslipidemia are more likely to be diabetic (see Fig. 1).

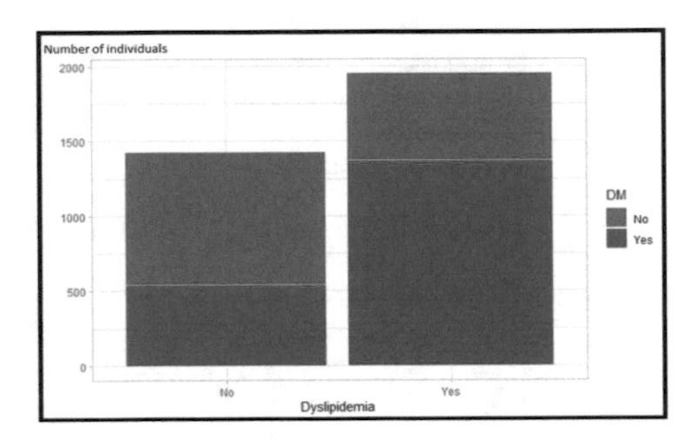

Fig. 1. Bar chart of Dyslipidemia and Diabets Mellitus

A cross sectional study was performed by Jeong Han Yeom et al. in a general hospital in Ulsan South-Korea to study the impact of shift work on hypertension. Their results showed that shift work affects blood pressure [16]. However, chi-square test of independence shows that hypertension is significantly associated to non work as: χ^2 $(2, N = 3372) = 233.2469$ and $p < 0.00001$. Consequently, hypertension is more likely to be developed in non-workers (see Fig. 2).

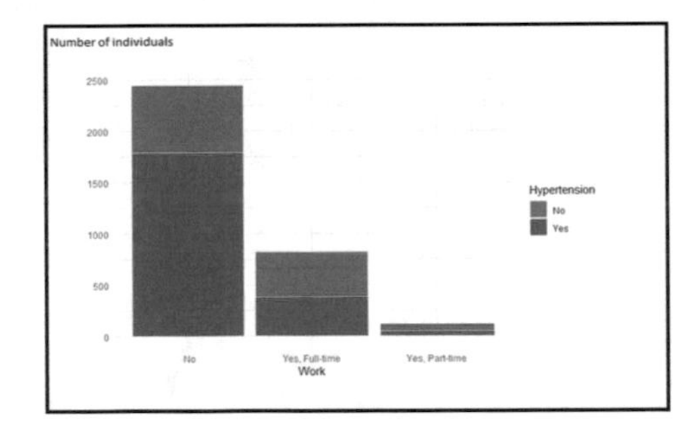

Fig. 2. Bar chart of Work and Hypertension

Young Gan et al. achieved a meta-analysis of observational studies to confirm that shift work increases risk of diabetes mellitus [17], but our bivariate analysis shows different results:

Non-workers are more expected to developp T2DM than workers (result is significant at $p < .05$).

3.2 Data Mining

In order to extract knowledge from data, machine learning algorithms have been integrated into data mining processing steps [18].

B. He et al. gathered different works related to the application of machine learning and data mining in the diagnosis and treatment of diabetes mellitus, they found that supervised learning methods are the most common algorithms for diabetes prediction [19].

Sossy S et al. achieved a comparative study of four classification algorithms (Naive Bayes, Neural network, Support vector machines (SVM) and Decision tree) applied on a women dataset to detect diabetes. Results showed that SVM is the best technique in terms of accuracy, sensitivity and precision [20].

In this work, Three different classification techniques are used, namely, Decision trees, Bayesian networks, Neural networks.

Decision Tree. In order to detect the risk factors of diabetes, Cart algorithm is applied on our cleaned dataset.

Variable "DM" is the output variable representing a decision in figure (see Fig. 3).

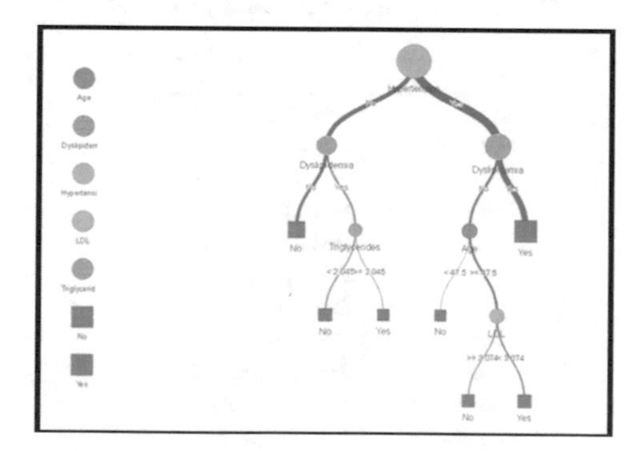

Fig. 3. Decision tree: Cart algorithm

As shown in figure (see Fig. 3), the main risk factors of T2DM are hypertension and dyslipidemia. Moreover, hypertensive middle aged adults can also develop diabetes if LDL is less than 3.07, even in absence of dyslipidemia. However, T2DM can be avoided if the patient doesn't have hypertension, dyslipidemia or triglycerides issues.

Bayesian Network. Conditional dependencies between our variables are portrayed by Bayes network as shown in figure (see Fig. 4).

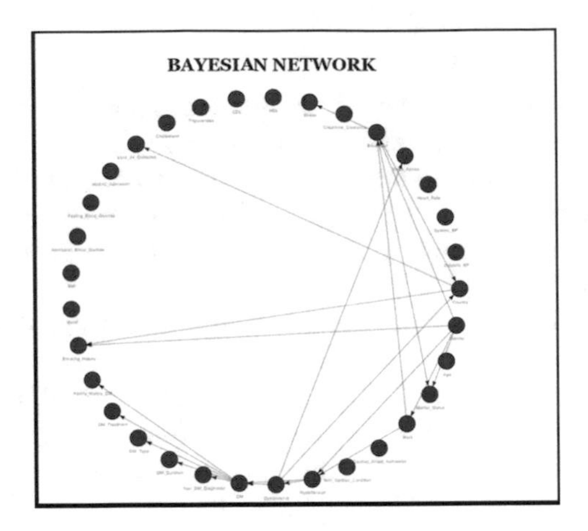

Fig. 4. Bayesian network

Probabilistic relationships between our different variables are summarized in the following table (see Table 2):

Table 2. Variables dependencies

From	To
DM	DM_Type
DM	DM_Treatement
DM	Family_History_DM
DM	Year_DM_Diagnosed
DM	DM_Duration
Work	Education
Work	Gender
Work	Hypertension
Gender	Smoking_History
Gender	Marital_Status
Gender	Hypertension
Hypertension	Dyslipidemia
Hypertension	DM
Hypertension	Cardiac_Condition
Dyslipidemia	DM
Dyslipidemia	Country
Education	Stress
Education	Country
Education	Gender
Country	Smoking_History
Country	Lipid_24_Collected

The table shows that:

- Diabetes is related to hypertension and dyslipedimia.
- Hypertension is associated to work activity. In fact, non workers have more risk of being hypertensive.

– Diabetes is indirectly influenced by low level of education, absence of work activity and gender.

Neural Network. The use of neural networks in diabetes prediction is very recurrent [21–23]. Consequently, in our comparative study we use which is the Multilayer Perceptron (MLP).

In order to ensure a good interpretation, variables that have no effect on the results, are eliminated using Olden and Decision tree methods.

Olden Method. Based on Garson algorithm [24], Olden et al. proposed to calculate variable importance as the product of the raw input-hidden and hidden-output connection weights between each input and output neuron and sums the product across all hidden neurons [25,26].

The obtained results are presented in the following table (see Table 3):

Table 3. Variables dependencies

Variable importance	Variable
6452	DM_Type
-196	DM_Treatment
-5089	Family_History_DM
5598	Year_DM_Diagnosed
5803	DM_Duration
-240	Education
-303	Gender
221	Hypertension
-225	Smoking_History
-520	Marital_Status
244	Dyslipidemia
86	non_Cardiac_Condition
-373	Country
309	Stress
401	BMI
470	Age
603	Admission_Blood_Glucose
909	Fasting_Blood_Glucose
1898	Hb1ac_Admission
-179	Heart_Rate
-171	Cholesterol
-364	Lipid_24_Collected
-615	Creatinine_Clearance
-584	Work
-518	Sleep_Apnea
-513	LDL
-465	Cardiac_Arrest_Admission
-387	Dyastolic
-330	Systolic
-269	HDL
-228	Waist
-223	Triglycerides
-196	DM_Treatment

From the results in table (see Table 3), variables with positive sign connection weights are maintained, which gives the following neural network (see Fig. 5):

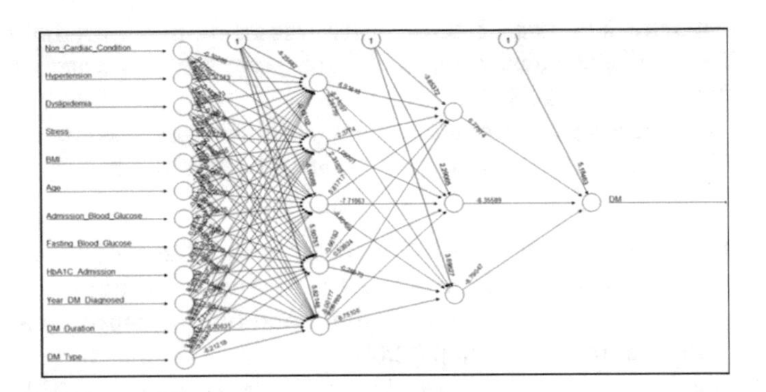

Fig. 5. Multilayer Perceptron

Variables that have the most influence on T2DM are:
Non Cardiac Condition, Hypertension, Dyslipidemia, Stress, BMI, Age, Admission Blood Glucose, Fasting Blood Glucose, HbA1C Admission, Year DM Diagnosed, DM Duration, DM Type.

Decision Tree Method. Another elimination of variables with less influence on our output variable is carried out by cart algorithm presented in Sect. 3.2. The result are shown in figure (see Fig. 6):

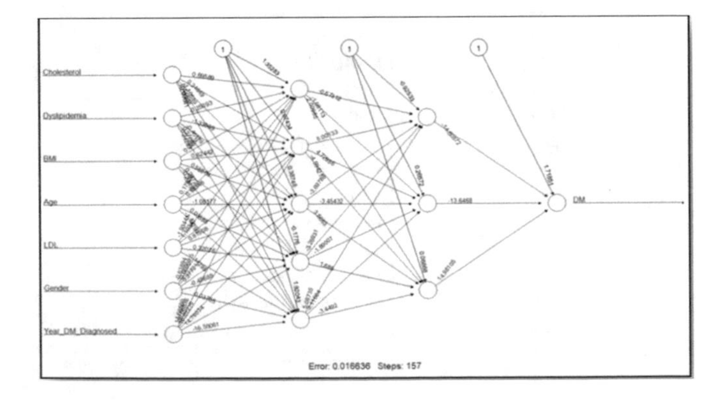

Fig. 6. Multilayer Perceptron

Consequently, variables that are maintained by both methods are: 'Dyslipidemia', 'Hypertension', 'Age', 'BMI', 'Cholesterol' and 'LDL'.

3.3 Methods Comparison

In order to compare the three applied methods (Decision trees, Neural networks, Bayesian networks) in terms of accuracy and sensitivity, the 80/20 training/test split is used on our dataset.

The accuracy rate for decision tree model is represented by figure (see Fig. 7).

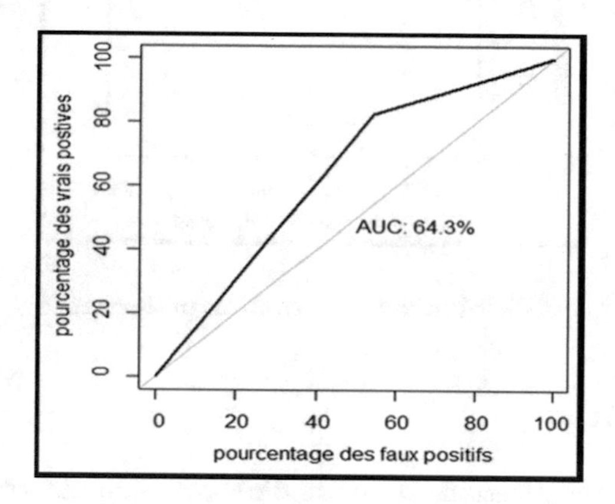

Fig. 7. Decision tree performance

As illustrated in figure (see Fig. 7) the accuracy rate is equal to 64.3%. Accuracy rate of bayesian network model is shown in figure (see Fig. 8).

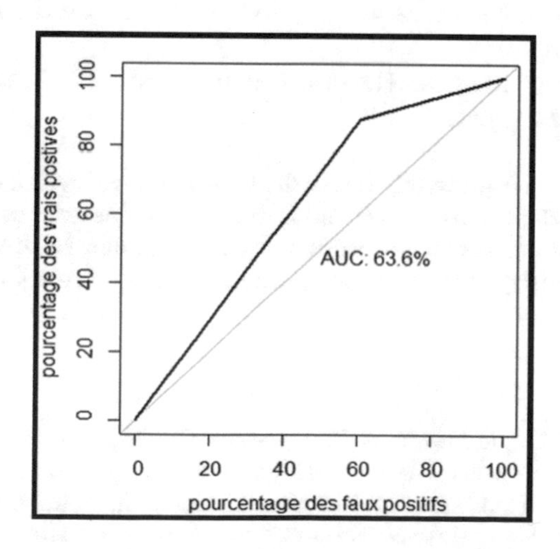

Fig. 8. Bayesian network's performance

According to the Bayesian algorithm, the accuracy rate is equal to 63.6%. Whereas, the accuracy rate of neural network model is equal to 100% as shown in figure (see Fig. 9).

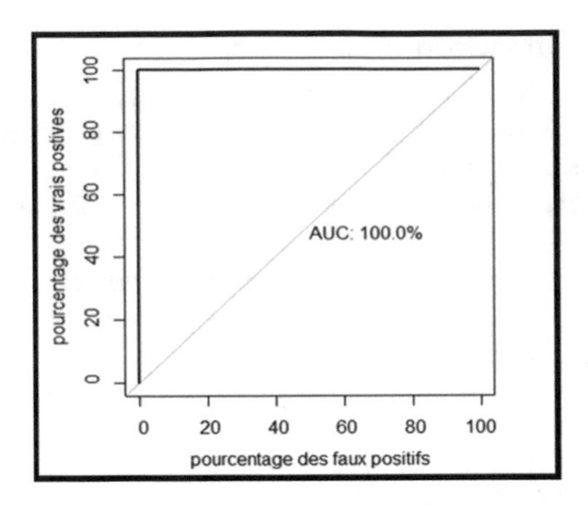

Fig. 9. Multilayer perceptron performance

4 Conclusion

Several diabetes complications occur if diabetes is unidentified or untreated, like neurophaty, nephropathy, cardiovascular disease or even stroke. Accordingly, efforts are needed to prevent diabetes and favour its early detection. In order to avoid the burden of health care costs, data mining approaches are used to build a model that can predict the risk factors of diabetes. In this work, a comparison between three data mining algorithms namely: the Decision Trees, Bayesian Networks and Neural Networks was carried out to choose the most appropriate algorithm for our dataset.

Multilayer Perceptron gives the best prediction of T2DM risk factors in terms of sensitivity and accuracy.

Acknowledgment. We gratefully thank the Principal Investigator of Gulf Coast **Professor Mohammad Zubaid** for providing the **Gulf Coast dataset**. In addition, we would like to express our sincere gratitude to "The Mohammed Bin Rashed University" and "Gulf Coast registry" for giving the permission to use **Gulf Coast dataset**.

References

1. de Lapertosa, S.G., de Moura, A.F., Decroux, C., Duke, L.: International diabetes federation diabetes atlas, Elsevier, 9th ed., vol. 157, pp.1–164 (2019)
2. Farmanfarma, K.K., Ansari-Moghaddam, A., Zareban, I., Adineh, H.A.: Prevalence of type 2 diabetes in middle-east: systematic review and meta-analysis. Prim. Care Diabetes **14**(4), 297–304 (2020)
3. Kok, J.N., Boers, E.J.W.: Artificial intelligence: definition, trends, techniques, and cases. Artif. Intell. **1**, 270–299 (2009)
4. Built In Beta. https://builtin.com/artificial-intelligence/artificial-intelligence-healthcare. Accessed 29 sept 2021
5. Brighterion AI. https://brighterion.com/data-mining/. Accessed 28 sept 2021

6. JavaTpoint. https://www.javatpoint.com/data-mining-vs-machine-learning. Accessed 28 sept 2021
7. Jothi, N., Husain, W.: Data mining in healthcare-a review. Procedia Comput. Sci. **72**, 306–313 (2015)
8. Koh, H.C., Tan, G.: Data mining applications in healthcare. J. Healthcare Inf. Manage. **19**(2) 65 (2011)
9. Kaur, H., Wasan, S.K.: Empirical study on applications of data mining techniques in healthcare. J. Comput. Sci. **2**(2), 194–200 (2006)
10. Ahmad, P., Qamar, S., Rizvi, S.Q.A.: Techniques of data mining in healthcare: a review. Int. J. Comput. Appl. **15**, 120 (2015)
11. Durairaj, M., Ranjani, V.: Data mining applications in healthcare sector: a study. Int. J. Sci. Technol. Res. **2**(10), 29–35 (2013)
12. Boutayeb, W., Badaoui, M., Al Ali, H., Boutayeb, A., Lamlili, M.N.M.: Use of data mining in the prediction of risk factors of type 2 diabetes mellitus in Gulf countries. Math. Model. Comput. **4**(8), 638–645 (2021)
13. Hina, S., Shaikh, A., Sattar, S.A.: Analyzing diabetes datasets using data mining. J. Basic Appl. Sci. **13**, 466–471 (2017)
14. Esmaeily, H., Tayefi, M., Ghayour-Mobarhan, M., Amirabadizadeh, A.: Comparing three data mining algorithms for identifying the associated risk factors of type 2 diabetes. Iran. Biomed. J. **22**(5), 303 (2018)
15. Sa-ngasoongsong, A., Chongwatpol, J.: An analysis of diabetes risk factors using data mining approach. Oklahoma state university, USA (2012)
16. Yeom, J.H., Sim, C.S., Lee, J.: Effect of shift work on hypertension: cross sectional study. Annal. Occupat. Environ. Med. **1**(29), 1–7 (2017)
17. Gan, Y., Yang, C., Tong, X.: Shift work and diabetes mellitus: a meta-analysis of observational studies. Occup. Environ. Med. **1**(72), 72–78 (2015)
18. Dagliati, A., Marini, S., Sacchi, L.: Machine learning methods to predict diabetes complications. J. Diabetes Sci. Technol. **2**(12), 295–302 (2018)
19. He, B., Shu, K.I., Zhang, H.: Machine learning and data mining in diabetes diagnosis and treatment. IOP Conf. Ser. Mater. Sci. Eng. **490**(4), 042–049 (2019)
20. Sossi Alaoui, S., Aksasse, B., Farhaoui, Y.: Data mining and machine learning approaches and technologies for diagnosing diabetes in women. In: Farhaoui, Y. (ed.) BDNT 2019. LNNS, vol. 81, pp. 59–72. Springer, Cham (2020). https://doi.org/10.1007/978-3-030-23672-4_6
21. El_Jerjawi, N.S., Abu-Naser, S.S.: Diabetes prediction using artificial neural network. Int. J. Adv. Sci. Technol. **121**, 55–64 (2018)
22. Zecchin, C., Facchinetti, A.: A new neural network approach for short-term glucose prediction using continuous glucose monitoring time-series and meal information. In: 2011 Annual International Conference of the IEEE Engineering in Medicine and Biology Society, pp. 5653–5656. IEEE (2011)
23. Pappada, S.M., Cameron, B.D., Rosman, P.M.: Neural network-based real-time prediction of glucose in patients with insulin-dependent diabetes. Diabetes Technol. Therap. **13**(2), 135–141 (2011)
24. Garson, D.G.: Interpreting neural network connection weights, pp. 47–51 (1991)
25. Olden, J.D., Jackson, D.A.: Illuminating the "black-box": a randomization approach for understanding variable contributions in artificial neural networks. Ecolog. Modell. **154**, 135–150 (2002)
26. Olden, J.D., Joy, M.K., Death, R.G.: An accurate comparison of methods for quantifying variable importance in artificial neural networks using simulated data. Ecol. Model. **178**, 389–397 (2004)

A Survey of Deep Learning Based Natural Language Processing in Smart Healthcare

Zineb El M'hamdi[1], Mohamed Lazaar[1](✉), and Oussama Mahboub[2]

[1] ENSIAS, Mohammed V University in Rabat, Rabat, Morocco
zineb_elmhamdi@um5.ac.ma, mohamed.lazaar@ensias.um5.ac.ma
[2] ENSA, Abdelmalek Essâadi University, Tetuan, Morocco
mahbouboussama@gmail.com

Abstract. Natural language processing (NLP) is the subfield of artificial intelligence that has the potential to make human language analyzable by computers. NLP is increasingly proving its importance in the medical field where a huge amount of data remains unstructured (free text) stored as electronic medical records (EMR); discharge summaries, lab reports, clinical notes, pa-thology reports, etc. Traditional Machine learning (ML) based approaches have been widely used for medical NLP tasks, but these methods require a set of manual work and still suffer in terms of accuracy. However, deep learning (DL) based methods have made significant improvement. The main goal of this study is to present the state-of-the-art DL based NLP tech-niques in healthcare. We started by presenting word embedding techniques and popular deep learning models used in this area, and then reviewed ap-plications of NLP tasks in medical domain such as classification, predic-tion, and information extraction. We concluded our study with analyzing cited architectures and showing the promising results of CNN and BiLSTM and BERT fine-tuning.

Keywords: Natural language processing · Deep Learning · Word Embedding · Neural networks · Medical text · CNN · BiLSTM · BERT

1 Introduction

Over the years, the health sector has been generating a larger volume of free-text data such as clinical notes, discharge summaries, pathology reports and radiology reports, carrying useful and valuable unstructured information. On the other hand, healthcare systems are always suffering from several issues and in need of improvements. Here comes the role of natural language processing [35]. NLP is a subdomain of artificial intelligence which deals with extracting knowledge from unstructured textual data using machine learning and deep learning models. Deep learning-based NLP techniques have recently proved their higher performance and showed promising results compared to traditional techniques for multiple NLP tasks like part of speech tagging, named entity recognition and sentiment analysis. Clinical domain is challenging for NLP researchers because

N. Aboutabit et al. (Eds.): ICMICSA 2022, LNNS 656, pp. 92–107, 2023.
https://doi.org/10.1007/978-3-031-29313-9_9

of many challenges like document organization and structure which are not easy to handle, non-unified medical jargon and requirement of protecting patient privacy by anonymizing data [34]. In the context of medicine, NLP applications are relying on these common steps: Preprocessing, Feature extraction, Modeling. Preprocessing is a combination of tasks such as tokenization, stemming, lemmatization, stop-word-removal, etc., which aims to prepare and normalize the input text for the next steps. Feature extraction aims to transform the input text to feature vectors to be exploitable by machine learning algorithms. Taking on consideration that the medical language is semantically distinct from general language, in this paper we are going to focus on feature extraction based on word embedding which gives an input representation with preserving semantical and contextual information. This step is the key for creating models for different NLP tasks such as classification, information extraction, etc. [33]

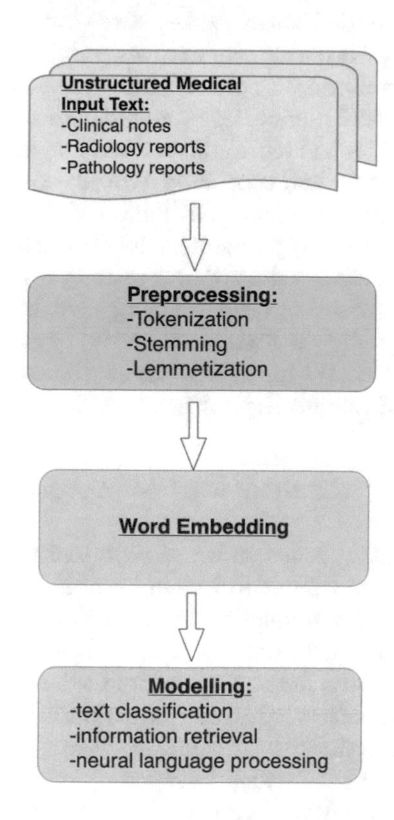

Fig. 1. Medical language processing steps.

Research strategy:

This article was guided by three research questions:

(a) How an input text is represented?
(b) What are most useful deep learning models for NLP?
(c) What are recent applications of deep NLP in medicine?

In order to find resources that will allow us to answer these questions we followed Two steps:

- Defining suitable keywords for selecting articles
- Filtering articles.

The first one is important to find articles that will allow us to answer research ques-tions. We assumed that this list of keywords will be relevant for the first and second question: Natural language processing, Input representation, word embedding, Deep natural language processing, neural natural language processing. For the third ques-tion we have added these keywords; medical NLP, electronic health records NLP deep clinical NLP, NLP for healthcare. This step gave us an idea about popular preliminar-ies, popular techniques of input representation (GLOVE, WORD2VEC, BERT) and deep natural language processing (CNN and RNN), and added them as keywords for our research. The filtering phase was done by prioritizing the articles published since 2019, select-ing articles explaining the preliminaries of NLP and choosing articles presenting applications of NLP in medical domain using popular techniques mentioned above. This benchmark allowed us to identify 30 relevant articles for our study; 10 explaining word representation and especially Word2Vec and glove methods, 5 for BERT model and its variants in medical field and 15 for applications of deep learning models (CNNs, RNNs) in medical NLP. In the second section of this paper, we present an overview of the popular techniques of word embedding. The next section presents a survey of deep learning models used in NLP problems. In the fourth one we provide a literature review of deep learning-based NLP applications in the medical. We finish this paper with an analysis and discussion of applications cited and perspectives (Figs. 1, 2, 3, 4 and 5).

2 Distributed Representation: Word Embedding

Natural language processing relies on input representation which aims to represent a word with a vector in order to be exploitable by machine. This step has proven its importance to determine how much useful information we can extract from data and its importance for improving the efficiency and robustness of NLP performance [1]. Local representation is the traditional method used to assign a unique representation to each word. It is easy to implement and understand due to its simplicity but not efficient for many reasons. First, local representation cannot detect the semantic relations between words. Second, the output is a high-dimension sparse matrix that can be time and memory consuming. Third, this representation is proportional to input size [2]. Distributed representation DR is then proposed to overcome these challenges by encoding a word's meaning in a compact, dense, and low-dimensional representation. With DR each concept is represented by many neurons, and each neuron participates in representation of many concepts [3], and the semantic relatedness between words is represented with the geometric distance between their representations, and most importantly, we can learn new concepts without adding new units due to generalization ability of the network mimicking human brain. In summary, DR is an unsupervised approach for input representation that suits most of NLP benchmarks due to its unreliability to annotated data and its capacity of pretraining data to automatically generate significant representations for new inputs. In this section we will dive into distributed representations namely word embedding introducing in detail important implementations used in deep learning for NLP.

2.1 Word2Vec

Tomas Mikolov in 2013 [4], proposed the Word2Vec embedding technique which is a two layers neural network that inputs text (words), iterates over it, learns associations and dependencies, calculates similarity between words using the cosine similarity metric, and generates a set of vectors namely feature vectors. Word2Vec is a predict based method which can learn the word vectors via two distinct neural network-based variants: Continuous Bag of Words (CBOW) and Skip-Gram. With CBOW, the neural network predicts the current word w(t) given the context C of text taken as input. For Skip-Gram, the neural network takes one word w(t) as input and predicts words that are close to its context. The comparison of model architectures between the two models was done by Mikolov and AI. has shown that CBOW is quick and more appropriate for larger corpus and frequent words, while Skip-gram has shown its efficiency for infrequent words. In medical research, many researchers have adopted Word2Vec method in the input representation phase either by combining it with other methods or relying just on it. To automatically annotate chest CT radiology reports, Banerjee I and Al. [5], developed an intelligent word embedding IWE method combining Word2Vec and domain specific semantic dictionary mapping technique. IWE methods is formed of three steps. First, mapping word of the corpus to terms derived from a domain specific dictionary and replace terms. Second, create word embedding by training Word2Vec based on Skip-Gram architecture

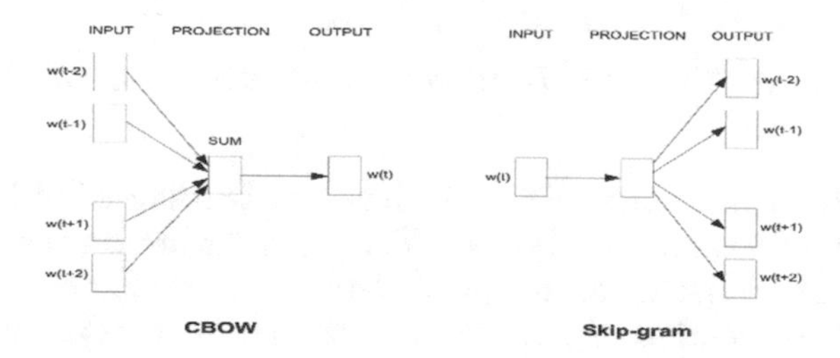

Fig. 2. CBOW and Skip-Gram architectures [4].

with a vector dimension of 300 and a window size of 30. First, mapping word of the corpus to terms derived from a domain specific dictionary and replace terms. Third, creation of document vector following this formula. The first step was to handle the limitation of Word2Vec with unknown words. Due to IWE based on Word2Vec, the annotation of the radiology reports has achieved highest accuracy with F1 score of 0.97, and outperformed similar studies based on other embedding techniques. In the same context, the semantic search tool Snomed2Vec was developed by Martinez Soriano et Al. [6]. Snomed2Vec relies on Word2Vec to create a vector space model representing the Systematized Nomenclature of Medicine-Clinical Terms (SNOMED-CT) terms by training it with general and local corpus, and a method that calculates cosine distance between the vector of a query text and vectors of space model and outputs the most similar vectors is then implemented. Snomed2Vec provides a set of most similar concepts to help specialists to unify reports codification.

2.2 Global Vectors Glove

Glove is a word embedding model developed by Pennington and AI. [7]. While Word2Vec is only considering the local context of the input (we were either predicting the target word from its context window or predicting context words from the target word), Glove is taking advantage of the statistics from the whole dataset by using co-occurrence matrix. This is a Count based method which relies on co-occurrence counts from the corpus, the co-occurrence matrix obtained is factorized using the stochastic gradient descent (SGD), the meaning of a word is then conceived from the words that co-occur with it in multiple contexts. According to Pennington and Al., Glove performed better than related models in semantic relatedness tasks, named entity recognition and word analogies. In contrast to Word2Vec, there is no publicly available word embeddings model based on Glove trained on a medical corpus [8] and many recent researches in the medical field proved that Word2Vec and other methods outperform Glove embeddings that failed in capturing semantics [9,10].

2.3 Bidirectional Encoder Representations from Transformers (BERT)

In this part, we will introduce BERT which is a contextualized technique of word embedding that relies on transformers. Transformers are conceived in 2017 by Google researchers [11], they are a type of neural network that employs positional encoding and self-attention. A positional encoder is a vector that has information on distances between words in the sentence. After passing the input text to the embedding layer, positional encoding is then applied to get vector of positional information (context). For self-attention block, we get an attention vector which captures contextual relationships between words in the input text. The main advantage of this architecture is that it can be parallelized.

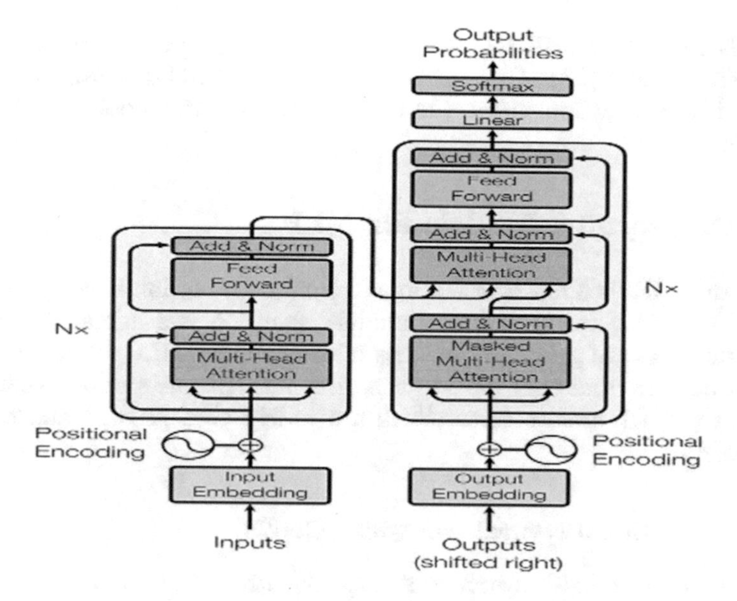

Fig. 3. Transformers architecture [11].

BERT was released by Google researchers in 2018. As its name indicates, BERT is a pretrained language model (on Wikipedia and Google's Books Corpus) that combines bidirectional training and transformer architecture. Unlike other language models which predict a word based on its precedents, BERT uses Masked LM technique to hide words randomly and trains to predict the hidden word using the whole context of the sentence. It is also using Next Sentence Prediction NSP technique by taking in two sentences and determining if the second one follows the first in kind of a binary classification problem, this helps BERT understand context across the sentences. MLM and NSP are the key for BERT to understand language. The training of the model generates features that can be used for training other models. Since its release, BERT proved its efficiency in the most common NLP tasks such as named entity recognition, sentiment analysis, and word sense disambiguation and has achieved great attention by NLP researchers, so many variants were developed by making some adjustments on the original model to get better results. RoBERTa [12] is one of these variants which was trained on a larger corpus of data and larger mini-batches, made an adjustment on the masking technique and did not use the NSP technique. In the medical context, many variants of BERT pretrained on a biomedical or health records corpus have been released. BioBERT [13] is a variant pretrained on biomedical research articles from PubMed and designed for medical text mining tasks. Leila Rasmy et Al. Conceived Med-BERT [14] model by adapting BERT to medical context by training on a structured electronic health record EHR of 28,490,650 patients, then they conduct experiments to evaluate the model by fine-tuning on two complex disease prediction tasks: Prediction of heart failure among diabetic patients and prediction of onset of pancreatic cancer, using two large EHR databases. The experiments have proven a higher accuracy compared to models without Med-BERT. Kexin Huang et Al. [15] conceived a pretrained model namely ClinicalBERT, a modified variant of BERT pretrained on clinical notes and electronic health records and fine-tuned the model for readmission prediction task.

3 Deep Learning Models for NLP

Deep learning DL is a machine learning technique using multi-layer neural networks which are powerful learning models requiring less human intervention. DL architectures are gaining a growing interest among NLP researchers due to the great advance that they gave to the discipline. In this section, we are going to present two DL architectures which are widely used in NLP and especially clinical NLP.

3.1 Convolutional Neural Networks CNNs

CNNs are a type of multi-layer neural network with convolutional layers, pooling layers and fully-connected layers. The convolution layer takes word embedding as an input and uses a convolution filter to create a feature map, the feature vector is then obtained by adding a bias term to the feature map, then an activation function (ReLU) is applied to ensure non-linearity as the data moves through each layer. The pooling layer creates fixed-length feature vectors using variable- length feature vector of the previous layer to remove the less-relevant information to make processing much fast-er. Fully-connected layer allows to learn non-linear combinations of the features. CNNs require less preprocessing compared to other classification methods, and improving the number of trainings can improve results.

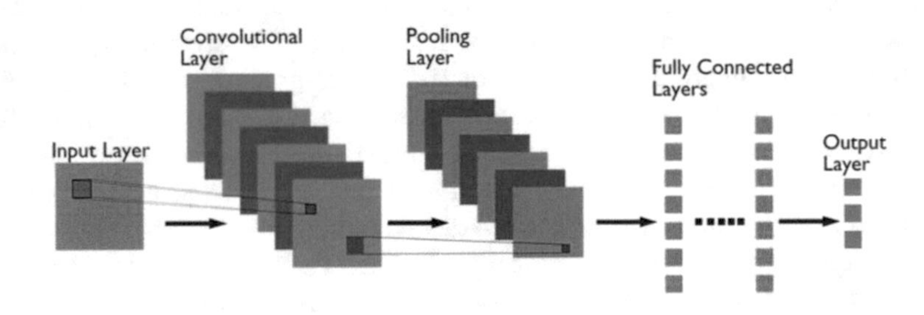

Fig. 4. CNN architecture [32].

3.2 Recurrent Neural Networks RNNs

RNNs are a type of neural networks with a built-in feedback loop and although this neural network may look like it only takes a single input value, the feedback loop makes it possible to use sequential input values (like stock market prices collected over time to make predictions), this is making it the only neural network with varying length of input. Each time step in RNN network has a cell that processes its input the output of the previous steps to generate an output to be passed to the next cell and another one that can be used for decision making at that time step.

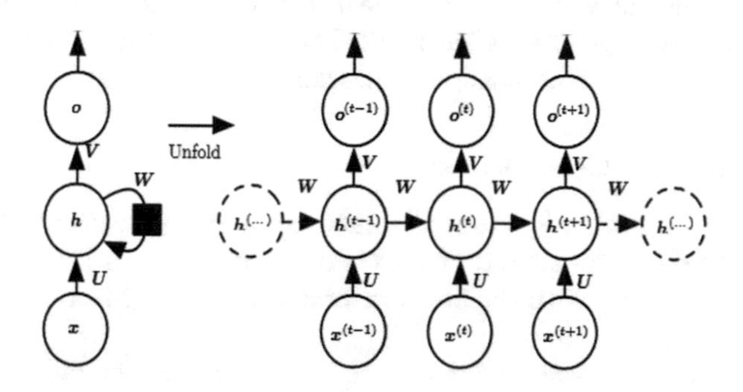

Fig. 5. RNN architecture [31].

The previous figure shows RNN architecture where xt, yt, and ht, respectively represent input, output, and hidden state at the time step t. W, U, and V denote respectively the input, hidden, and output layer weights shared across time steps.

The standard RNNs struggled to capture long-term dependencies because of the problem of vanishing gradients, and were not able to learn dependencies in two directions of language assuming that the meaning of a word cannot be affected by the next one. For these reasons, other RNNs architecture were developed such as LSTM, Gated recurrent unit GRU and bidirectional RNN.

Long-Short Term Memory LSTM. It is a special version on RNN which solves the problem of vanishing gradients and short-term memory by using internal mechanisms called gates. Each LSTM cell has an input gate, forget gate, and output gate, and a memory cell. Gates regulated the flow of information by learning which data in a sequence is important to keep or throw away and by doing that, it learns to use relevant information to make predictions. Memory cells help LSTM to retain information for a longer time which helps to solve the problem of long-term dependencies.

Gated recurrent unit GRU is also a gated-based architecture like LSTM with a smaller number of parameters and less cost of computation. GRU uses only two gates namely update and reset gate to modulate the flow of information, and it doesn't have a separated cell memory.

These two architectures are solving the problem of short-term dependencies, but the problem related to the two directions mentioned before is resolved by the bidirectional RNN (BiLSTM and BiGRU) architecture which aims to make the final predictions by combining the results of a forward and a backward recurrent neural network. This type of architecture is much needed for NLP problems where the words surrounding a word must be taken into account.

Variants of RNNs are widely used for natural language processing to solve several problems such as information retrieval and classification and have shown higher performances [18].

4 Applications of Deep Natural Language Processing in Healthcare

In this section, we present a literature survey of deep natural language processing applications in the healthcare sector. Wo focused on three major NLP tasks which have shown high prevalence in medical NLP applications: Medical text classification, clinical predictions, and information extraction.

4.1 Medical Text Classification

Medical text classification is one of the most popular applications of NLP in the medical field that inspired many researchers to conduct studies developing ML and DL models. Models with the best performance can save healthcare personnel from tedious manual work.

In recent years, DL based models gave important results and have attracted attention of many researchers. Bayrak et Al. [19] conducted a study that aims to classify unstructured epilepsy radiology reports using neural networks. After the preprocessing, feature vectors are generates to be an input for LSTM, CNN, and BiLSTM architectures. The comparative study of the architectures shows that CNN is time consuming, and the BiLSTM gives the best classification results.

In the actual year, Chaib et Ai. [20] have conceived a model called GL-LSTM for multi label classification of cardiovascular disease reports. GL-LSTM is a system based on Glove for creating the embedding matrix that captures the global and local context, and LSTM for classification task. GL-LSTM is compared to other models that use the same medical dataset and showed higher result.

BERT model can also be used not only for text representation but also for a text classification task by fine-tuning and it has shown promising results compared to traditional classifiers [29].

4.2 Clinical Predictions

Clinical predictions consist of going through health records and identifying the prob-ability of a disease, hospital readmission, or adverse events. Automation of clinical predictions is giving healthcare providers best assistance.

Mao et Al. [21] used AKI-BERT, a variant of BERT pretrained on clinical notes of patients of Acute kidney injury (AKI) and fine-tuned it for disease prediction task by adding new layer to the architecture and updating parameters, and the fine-tuning was followed by a linear classifier. Authors compared AKI-BERT with other variants of BERT for prediction of AKI and concluded that the proposed model have considerable performance.

Liu et AI. [22] conducted a study for predicting heart failure general readmission and 30-days readmission using unstructured clinical notes. Text data were embedded using a pretrained Word2Vec model and generate input vectors to train a CNN model. The CNN model shows exceptionnal performance compared to other baseline models using machine learning algorithms.

Borjali et Al. [17] studied adverse events prediction in the case of a total hip re-placement from a huge dataset containing radiology notes and follow-up telephone notes using two methods LSTM and CNN. The CNN model implemented has two 1-D convolution layers with filter sizes 3 and 5 respectively and has achieved the best performances outperforming the LSTM model and other models cited in the research paper.

A CNN model was also used by Jia et Al. [16] to predict weaning from mechani-cal ventilation for intubated patients to prevent the risk of over prolonged ventilation or premature extubating.

4.3 Information Extraction

Information Extraction (IE) is an NLP task for getting targeted structured information from unstructured textual data. IE is heavily needed in the medicine sector because of the nature of most of generated data. Extracting information from medical records is a necessary task for several NLP applications. Named Entity Recognition is a subtask of IE that aims to determine if a token corresponds to the target named entities.

Schneider et Al. [23] conceived a Portuguese language model for clinical NER namely BioBERTpt. Authors used a voluminous dataset containing Brazilian clinical notes to develop BioBERTpt by passing clinical notes and biomedical abstracts (from Pubmed and Scielo) to a pertained BERT multilingual model and generated BioBERTpt(clin) (trained on clinical notes), BioBERTpt(bio) (trained on biomedical text) and BioBERTpt(all) (trained on all datasets), these three models are fine-tuned to extract information from the Portuguese clinical notes. To evaluate these models, authors used two annotated corpuses and the results were in favor of BioBERTpt models.

Khalafi et Al. [24] used two hybrid deep learning approaches to extract phenotypic informations from clinical notes. The proposed models are a combination of a bidirectional sequence model (BiLSTM or BiGRU) and CNN. After the preprocessing step, data are passed to the model which first layer is the word embedding layer, its output is then passed to the sequence model layer to capture semantic dependencies and the output is fed to convolutional layer that extracts semantic features considering long-term dependencies then the output is generated by passing through a fully connected layer with a sigmoid activation function. To extract more features related to phenotypes, a CNN layer is tuning in parallel of the model and improved its performance (a higher F1-score). In the embedding layer, authors used FastText and Word2Vec to evaluate them separately. This approach has proven the effectiveness of CNN and sequence model (BiLSTM and BiGRU) in extracting comprehensive information from text and outperformed non hybrid architectures.

To extract daily dosage from medical prescriptions, Mohajan et Al. [25] were the first to automate this task and used Clinical-BERT and BiLSTM for automatic entity extraction and a pre-built lexicon for normalization and dosage calculation (Table 1).

4.4 Summary Table

Table 1. Summary table of all applications mentioned

NLP task	Paper	Data	Goal	Methods	Model with best results
Medical text Classification	[19]	Small dataset of radiology reports (122)	Classification of unstructured epilepsy radiology reports	• CNN • LSTM • BiLSTM	BiLSTM with an accuracy= 88.89%
	[20]	Public dataset (oshumed) of 13929 medical summaries	Multi-label classification of cardiovascular disease reports	-Glove + LSTM	GL-LSTM has best Accuracy= 0.927
	[29]	33268 Deutsch radiology reports	Classification of radiology reports of orthopedic trauma	-BERT fine-tuning -traditional machine learning	BERT fine-tuning with an accuracy of 0.96 for simple reports and 0.93 for complex ones
	[30]	1921 Chinese clinical notes	Classification of reports in two classes: lumbar disc herniation (LDH) and lumbar spinal stenosis (LSS)	LSTM compared to XGBoost	LSTM has better capacity outperforming XGBoost with an AUC = 0.7565
	[27]	TREC dataset	Medical short-text classification	Word-cluster embedding and classification using CNN and LSTM.	Both models outperformed state-of-the-art models applied for medical and general short text classification. LSTM outperformed CNN with minim difference
Clinical predictions	[21]	16560 clinical notes of patients with risks of AKI	Predicting risk of Acute kidney Injury	Pretraining BERT base and medical variants on AKI notes (AKI-BERT) + Fine-tuning AKI-BERT for prediction	AKI-BERT improved performance of BERT its medical variants for AKI prediction
	[22]	3411 clinical notes of intensive care unit patients with heart failure	Predicting heart failure general readmission and 30-days readmission	Word2Vec+CNN	Accuracy is 0.76 for general readmission and 0.72 for 30-days readmission

(*continued*)

Table 1. (*continued*)

Clinical predictions	[16]	2229 historical ICU data	Prediction of weaning from mechanical ventilation	CNN	Accuracy = 0.86
	[17]	3014 radiology notes+ 783 telephone notes	Prediction of adverse events in the case of a total hip replacement	• Word2Vec+ CNN • Word2Vec+ BiLSTM	CNN model with kappa=0.97 % for radiology notes and kappa=100% for follow-up phone notes.
Information extraction	[23]	2,100,546 brazilian clinical notes and biomedical abstracts (from Pubmed and Scielo)	-Extract information from the Portuguese clinical notes using named entity recognition	-BioBERTpt (Based on BERT)	BioBERTpt improved F1 score of BERT model
	[24]	The discharge reports in the MIMIC III	-Extracting phenotypic informations	• Word2Vec+ CNN+BiLSTM • Word2Vec+ CNN+BiGRU • FastText+CNN+BiLSTM • FastText+CNN+BiGRU	-Hybrid models with Word2Vec embedding and BiGRU has the higher F1 score on evaluation corpus
	[25]	505 clinical notes for train,development,and test. 1000 medication orders of 427 patients for validation.	Extracting daily dosage from medical prescriptions	Clinical-BERT+BiLSTM+FC layer for automatic entity extraction and pre-built lexicon for normalization and dosage calculation	The system shows high performance with an accuracy of 96.2%
	[26]	320 french clinical notes for training and evaluation	Extracting drug mentions and their information	Contextual word embedding using the language model ELMo trained on 100K clinical notes + BiLSTM	BiLSTM model outperformed the compared models with overall F-measure 89.9%.

5 Analysis and Discussion

Following the analysis of articles presented in this paper, we mainly find:

- Deep learning based natural language processing has powerful performances and gained more attention than traditional machine learning approaches.
- Since its release BERT and its variants offer best contextual embedding helping to encode a word taking in consideration its semantic and contextual meaning whereas Word2Vec and GloVe are still used for embedding layer in some recent researches due to their cheap computation cost or the lack of suitable data to train BERT.
- Natural language processing has helpful applications for better diagnoses and decision making.
- Convolutional and bidirectional recurrent neural networks are recently the most popular deep learning approaches for clinical natural language processing tasks.
- CNN proved its efficiency for clinical predictions and compete RNNs in classification and information retrieval.
- BiLSTM is widely used for medical information extraction and text classification, it outperformed compared models due to its bidirectional architecture.
- BERT is not only used for embedding layer, but fine-tuned on medical NLP tasks and gave a state-of-the-art result for medical text classification.
- Fine-tuned BERT models require a small set of data and a GPU is necessary for training.
- Designed architectures cited in this paper require high expertise for choice of parameters and model validation.
- Every approach is limited to only one language.
- Every architecture is designed for the language of healthcare institution where data came from. Medical jargon and language styles differ from healthcare system to other.
- BERT variants can be pretrained on a specific disease domain.
- Medical domain is very critical and need a very high level of precision.

6 Conclusion and Perspectives

This paper studies deep natural language processing and its application in medicine. We presented relevant techniques of input presentation, reviewed deep learning archi-tectures CNN and RNNs and their variants having more attention in clinical NLP research, cited recent applications of these techniques for medical text classification, clinical predictions and information extraction. This study showed that deep learning-based NLP techniques are very helping for clinical decision making and avoiding laborious manual work. In the perspective, to benefit from the progress of natural language processing in our country, it is preferable to guide an original comparative study of deep learning approaches using Moroccan data and design specific architecture for local medical language style and apply it for medical NLP tasks.

References

1. Liu, Z., Lin, Y., Sun, M.: Representation learning and NLP. In: Representation Learning for Natural Language Processing, pp. 1–11. Springer, Singapore (2020). https://doi.org/10.1007/978-981-15-5573-2_1
2. Liu, Z., Lin, Y., Sun, M.: Word representation. In: Representation Learning for Natural Language Processing, pp. 13–41. Springer, Singapore (2020). https://doi.org/10.1007/978-981-15-5573-2_2
3. Plate, T.: Distributed Representations. Cognitive Science, pp. 1–15 (2003)
4. Mikolov, T., Chen, K., Corrado, G., Dean, J.: Efficient estimation of word representations in vector space. arXiv, arXiv:1301.3781 (2013)
5. Banerjee, I., Chen, M.C., Lungren, M.P., Rubin, D.L.: Radiology report annotation using intelligent word embeddings: applied to multi-institutional chest CT cohort. J. Biomed. Inform. **77**, 11–20 (2018). https://doi.org/10.1016/j.jbi.2017.11.012
6. Soriano, I.M., Castro, J.L., Fernandez-Breis, J.T., Román, I.S., Barriuso, A.A., Baraza, D.G.: Snomed2Vec: Representation of SNOMED CT Terms with Word2Vec, pp. 678–83. IEEE Computer Society (2019). https://doi.org/10.1109/CBMS.2019.00138
7. Pennington, J., Socher, R., Manning, C.D.: Glove: Global Vectors for Word Representation, vol. 14, 1532–1543 (2014). https://doi.org/10.3115/v1/D14-1162
8. Kalyan, K.S., Sangeetha, S.: SECNLP: A Survey of Embeddings in Clinical Natural Language Processing. J. Biomed. Inform. **101**, 103323 (2020). https://doi.org/10.1016/j.jbi.2019.103323
9. Khattak, F.K., Jeblee, S., Pou-Prom, C., Abdalla, M., Meaney, C., Rudzicz, F.: A survey of word embeddings for clinical text. J. Biomed. Inf. 100(1-4) 100057 (2019). https://doi.org/10.1016/j.yjbinx.2019.100057
10. Habib, M., Faris, M., Alomari, A., Faris, H.: AltibbiVec: a word embedding model for medical and health applications in the Arabic language. IEEE Access **9**, 133875–88 (2021). https://doi.org/10.1109/ACCESS.2021.3115617
11. Vaswani, A., et al.: Attention Is All You Need. arXiv (2017). https://doi.org/10.48550/arXiv.1706.03762
12. Liu, Y., et al.: RoBERTa: a robustly optimized bert pretraining approach. arXiv (2019). https://doi.org/10.48550/arXiv.1907.11692
13. Lee, J., et al.: BioBERT: a pre-trained biomedical language representation model for biomedical text mining. Bioinform. 36, btz682 (2019). https://doi.org/10.1093/bioinformatics/btz682
14. Rasmy, L., Xiang, Y., Xie, Z., Tao, C., Zhi, D.: Med-BERT: PRetrained contextualized embeddings on large-scale structured electronic health records for disease prediction. npj Digit. Med. **4**(1), 1–13 (2021). https://doi.org/10.1038/s41746-021-00455-y
15. Huang, K., Altosaar, J., Ranganath, R.: ClinicalBERT: modeling clinical notes and predicting hospital readmission. ArXiv (2019)
16. Jia, Y., Kaul, C., Lawton, T., Murray-Smith, R., Habli, I.: Prediction of weaning from mechanical ventilation using convolutional neural networks. Artif. Intell. Med. **117**, 102087 (2021). https://doi.org/10.1016/j.artmed.2021.102087
17. Borjali, A., Magnéli, M., Shin, D., Malchau, H., Mu-ratoglu, O.K., Varadarajan, K.M.: Natural language processing with deep learning for medical adverse event detection from free-text medical narratives: a case study of detecting total hip replacement dislocation. Comput. Biol. Med. **129**, 104140 (2021). https://doi.org/10.1016/j.compbiomed.2020.104140

18. Guan, M., Cho, S., Petro, R., Zhang, W., Pasche, B., Topaloglu, U.: Natural language processing and recurrent network models for identifying genomic mutation-associated cancer treatment change from patient progress notes. JAMIA Open **2**(1), 139–149 (2019). https://doi.org/10.1093/jamiaopen/ooy061
19. Bayrak, S., Yucel, E., Takci, H.: Epilepsy radiology reports classification using deep learning networks (2022). https://doi.org/10.32604/cmc.2022.018742
20. Chaib, R., Azizi, N., Schwab, D., Gasmi, I., Chaib, A.: GL-LSTM Model for multi label text classification of cardiovascular disease reports (2022). https://easychair.org/publications/preprint/BMRx
21. Mao, C., Yao, L., Luo, Y.: AKI-BERT: a pre-trained clinical language model for early prediction of acute kidney injury. arXiv (2022). https://doi.org/10.48550/arXiv.2205.03695
22. Liu, X., Chen, Y., Bae, J., Li, H., Johnston, J., Sanger, T.: Predicting heart failure readmission from clinical notes using deep learning. In: 2019 IEEE International Conference on Bioinformatics and Biomedicine (BIBM), pp. 2642–48 (2019). https://doi.org/10.1109/BIBM47256.2019.8983095
23. Schneider, E.T.R., et al.: BioBERTpt - a portuguese neural language model for clinical named entity recognition. In: Proceedings of the 3rd Clinical Natural Language Processing Workshop. Association for Computational Linguistics (2020). https://doi.org/10.18653/v1/2020.clinicalnlp-1.7
24. Khalafi, S., Ghadiri, N., Moradi, M.: Hybrid deep learning methods for phenotype prediction from clinical notes. arXiv (2022). https://doi.org/10.48550/arXiv.2108.10682
25. Mahajan, D., Liang, J.J., Tsou, C.-H.: Extracting daily dosage from medication instructions in EHRs: an automated approach and lessons learned. arXiv (2021). https://doi.org/10.48550/arXiv.2005.10899
26. Jouffroy, J., Feldman, S., Lerner, I., Rance, B., Burgun, A., Neuraz, A.: MedExt: combining expert knowledge and deep learning for medication extraction from french clinical texts (Preprint) (2020). https://doi.org/10.2196/preprints.17934
27. Shen, Y., Zhang, Q., Zhang, J., Huang, J., Lu, Y., Lei, K.: Improving medical short text classification with semantic expansion using word-cluster embedding (2018). https://doi.org/10.48550/arXiv.1812.01885
28. Hsu, E., Malagaris, I., Kuo, Y.-F., Sultana, R., Roberts, K.: Deep learning-based NLP data pipeline for EHR scanned document information extraction. arXiv (2021). https://doi.org/10.48550/arXiv.2110.11864
29. Olthof, A.W., et al.: Machine learning based natural language processing of radiology reports in Orthopaedic trauma. Comput. Methods Programs Biomed. **208**, 106304 (2021). https://doi.org/10.1016/j.cmpb.2021.106304
30. Ren, G.R., et al.: Differentiation of lumbar disc herniation and lumbar spinal stenosis using natural language processing-based machine learning based on positive symptoms. Neurosurg. Focus **52**(4), E7 (2022). https://doi.org/10.3171/2022.1.FOCUS21561
31. https://www.deeplearningbook.org/contents/rnn.html
32. Kumar, A.: Different types of CNN architectures explained: examples. Data Analytics (blog) (2022). https://vitalflux.com/different-types-of-cnn-architectures-explained-examples
33. Kumar, E.S., Jayadev, P.S.: Deep learning for clinical decision support systems: a review from the panorama of smart healthcare. In: Deep Learning Techniques for Biomedical and Health Informatics (2020) https://doi.org/10.1007/978-3-030-33966-1_5

34. Sandeep Kumar, E., Satya Jayadev, P.: Deep learning for clinical decision support systems: a review from the panorama of smart healthcare. In: Dash, S., Acharya, B.R., Mittal, M., Abraham, A., Kelemen, A. (eds.) Deep Learning Techniques for Biomedical and Health Informatics. SBD, vol. 68, pp. 79–99. Springer, Cham (2020). https://doi.org/10.1007/978-3-030-33966-1_5
35. Adnan, K., Akbar, R., Khor, S.W., Ali, A.B.A.: Role and challenges of unstructured big data in healthcare. In: Sharma, N., Chakrabarti, A., Balas, V.E. (eds.) Data Management, Analytics and Innovation. AISC, vol. 1042, pp. 301–323. Springer, Singapore (2020). https://doi.org/10.1007/978-981-32-9949-8_22

Machine Learning-Based Classification of Leukemia Comparative Study

Zineb Skalli Houssaini(✉), Omar El beqqali, and Jamal El Riffi

Faculty of Sciences Dhar-Mahraz Sidi Mohamed Ben Abdellah University, Fez, Morocco
{zineb.skallihoussaini,omar.elbeqqali,jamal.riffi}@usmba.ac.ma

Abstract. Leukemia disease designates a cancer of the bone marrow and lymphatic system. It occurs when certain blood cells acquire changes i.e. or mutations in their genetic material. Leukemias are classified according to their rate of progression and the type of cells involved. Acute Lymphocytic Leukemia (ALL), Acute Myelogenous Leukemia (AML), Chronic Lymphocytic Leukemia (CLL), and Chronic Myelogenous

Leukemia are the four main kinds of leukemia (CML). The classification of the type of Leukemia is very important to diagnose the disease and determine its progression. In this context, we have used the classifiers of machine learning to identify different forms of leukemia., which facilitates the task of doctors and patients. The main objective of this paper is to determine the most effective methods for the detection of leukemia. According to this context, we have established a comparative study between five classifiers (Support Vector Machine, Random Forest, Logistic Regression, K-Nearest Neighbors, and Naïve Bayes). We have evaluated our system with four metrics: Precision, Accuracy, Recall, and F1-score. The experimental results on Gene Expression Dataset demonstrate that the Support Vector Machine classifier obtains the highest accuracy; however, this accuracy varies depending on the algorithm used to classify the types of leukemia and also on the shape and size of the sample.

Keywords: Leukemia · Gene Expression · Machine Learning · KNN · SVM · Random Forest · Naïve Bayes · Logistic Regression

1 Introduction

Cancer is one of the top causes of death worldwide, according to the World Health Organization (WHO). It has been responsible for millions of deaths in recent years. According to the WHO, the term "cancer" refers to a broad class of diseases that can affect any area of the body. Neoplasms and malignant tumors are other words that are used. The rapid development of aberrant cells that outgrow their normal borders and can infect nearby body parts and spread to other organs is one of the traits of cancer. Leukemia is severe cancer, mostly fatal hematopoietic malignancy. Is a blood cell cancer caused by abnormal production of white

N. Aboutabit et al. (Eds.): ICMICSA 2022, LNNS 656, pp. 108–115, 2023.
https://doi.org/10.1007/978-3-031-29313-9_10

blood cells in the bone marrow [1]. There are 4 common types of leukemia, Acute lymphoblastic leukemia (ALL), Acute myeloblastic leukemia (AML), Chronic lymphocytic leukemia (CLL), and Chronic myeloid leukemia (CML). Leukemia is a type of cancer that can be life-threatening, and its treatment requires an accurate and timely diagnosis. The traditional techniques for analyzing, diagnosing, and predicting symptoms have evolved into automated computerized tools. The practice of medicine grows every year and evolves toward the most efficient and automated systems that improve diagnostic results [2].

Artificial intelligence plays a very important role in the detection of diseases [3]. AI also has a positive impact on the medical care system, assisting doctors and providing second opinions to improve diagnosis accuracy. Machine learning (ML) is a branch of artificial intelligence that is broadly defined as a machine's ability to mimic intelligent human behavior [1]. Artificial intelligence systems are used to solve complex problems in a manner similar to how humans solve problems. ML consists of programming to improve a performance criterion using example data or previous experience. The predictive model's accuracy and the value of a fitness or evaluation function could both be the optimized criterion [4]. In cancer research, several learning methods, such as genomic sequencing, are used to detect and identify patterns in input values and effectively diagnose cancer types [1].

In this study, five different machine learning methods for detecting the two types of leukemia (AML and ALL) were compared using the Gene Expression Dataset [5]. The remainder of the paper will be structured as follows: the second section mentions Related Work, the third section is about Leukemia Dataset, Sect. 4 describe the classifiers used in detecting Leukemia, and Sect. 5 shows the findings of the experiments and their discussion. The conclusion comes in the final section of the research.

2 Related Works

Golub et al. [6] represented a generic approach to identifying new cancer classes (class discovery) or assigning tumors to known cancer classes (class prediction).In his article, he represented the classification of cancers centered on the monitoring of gene expression by Deoxyribonucleic acid (DNA) microarrays. A method of class discovery allowed us to concretely reveal without prior knowledge of its classes, distinguish between acute myeloid leukemia (AML) and acute lymphoblastic leukemia (ALL), which demonstrated the possibility of the classification of cancers based only on the monitoring of gene expression. Ratley et al. [7] investigated various image processing and machine learning techniques used for leukemia detection classification. Alrefai1 et al. [8] applied an ensemble learning method i.e. (Support Vector Machine (SVM), k-Nearest Neighbor (kNN), Naive Bayes (NB), and Decision Tree (C4.5) classifiers were combined.) with Particle Swarm Optimization (PSO) on leukemia microarray gene expression, to ensure the best number of meaningful genes that lead to improved leukemia cancer diagnosis. Ghaderzadeh et al. [9] presented that several studies demonstrate the strength of the use of ML methods for the processing of images of leukemia smears. These methods which are now adopted in applications and tools of laboratories, can improve accuracy and provide a faster diagnosis. HUANG et al. [10] illustrated that support vector machine (SVM) learning, is robust and the algorithm is a potent classification approach that has been applied to the categorization or subtyping of cancer genomics. The effectiveness of the SVM technique can be attributed in part to the strength of the SVM algorithm as well as the adaptability of the kernel approach to data representation. P. M. Gumble proposed the identification of leukemic blood cells through morphological analysis of microscopic images.

3 Gene Expression Dataset

The data originates from proof-of-concept research that Golub et al. published in 1999 [6]. It demonstrated how gene expression monitoring (using a DNA microarray) might be used to categorize new cancer cases and established a general method for determining new cancer classes and classifying tumors. Acute lymphoblastic leukemia (ALL) and acute myeloid leukemia (AML) patients were classified using these statistics (ALL). The database consists of 7130 features (7129 features (Genes) and 1 class attribute) for 72 instances (47 for ALL class, 25 for AML class), all are numerical values except the Last column is the class (two classes: ALL, AML) [5].

4 Adopted Classifiers

4.1 Support Vector Machine (SVM)

Among the machine learning algorithms there is the SVM; a support vector machine; which belongs to the family of supervised learning algorithms and methods used for classification and regression. It works with large dimensions and guarantees good results achieved in practice. Also, SVMs are characterized by their simplicity of use. This method of classification aims to separate the positive examples from the negative ones in all examples. The latter then looks for the hyperplane that divides the positive examples from the negative ones, making sure that the margin between the positive and the nearest negative is as large as possible [11].

4.2 Random Forest (RF)

Random Forests, also known as Random Decision Forests, are a classification and regression ensemble learning method. This classifier is made up of a huge number of randomly generated decision trees, with each node in the decision tree calculating the output using a random subset of features [12].

4.3 K-Nearest Neighbor (KNN)

The K-Nest Neighbor algorithm is used to differentiate between leukemia cells and normal blood cells, and it was discovered that k-NN is a widely used classification tool with good scalability. It ranks new objects using similarity measures, assuming that similar things exist nearby [13].

4.4 Naïve Bayes (NB)

The goal of the Naïve Bayes algorithm is to detect blast cells. Bayesian classifiers are statistical classifiers. The algorithm is based on Bayes' theorem. To simplify the calculation, they assume that the values of the other attributes have no bearing on how an attribute's value affects a particular class. It is considered "naive".

4.5 Logistic Regression (LR)

A statistical model called logistic regression is used to examine the associations between a group of qualitative factors called Xi and a qualitative variable called Y. A logistic function is used as the link function in a generalized linear model. Based on the optimization of the regression coefficients, a logistic regression model may also forecast the likelihood of an event occurring (value of 1) or not (value of 0) [14].

5 Experimentations and Results

Our system was run on an Intel(R) Core (TM) i7 computer with a 64-bit operating system. The code was developed on the anaconda application with the Python programming language. This section summarizes the results of the classification in which we applied the classifiers to our dataset. To evaluate our approach, first, we should define some well-known performance measures. A typical confusion matrix is shown in Table 1.

5.1 Performance Metrics

5.1.1 Confusion Matrix

A classification model's prediction outcome is summarized by a confusion matrix or error matrix.

5.1.2 Accuracy

The metric uses precision to show the efficiency of a classifier, which is generated using Eq. 1 [15] and represents accurately categorized values in a set.

$$Accuracy = \frac{TP + TN}{TP + TN + FP + FN} \tag{1}$$

Table 1. A Typical 2*2 Confusion matrix

Actual Class (Observation)	Predicate Class (Prediction) Positive	Predicate Class (Prediction) Negative
Positive	TP (True Positive)	FN (False Negative)
Negative	FP (False Positive)	TN (True Positive)

5.1.3 Recall

When a classifier divides the entire number of true positives by the sum of the total number of true positives and false negatives, it performs a recall., which is presented in Eq. 2 [16].

$$Recall = \frac{TP}{TP + FN} \tag{2}$$

5.1.4 Precision

It figures out how many classifiers made accurate positive predictions. The number of true positives is divided by the total number of true positives and false positives, as shown by Eq. 3 [17].

$$Precision = \frac{TP}{TP + FP} \tag{3}$$

5.1.5 F1-score

The harmonic mean of recall and precision is known as the F1-score. The characteristics of both measures are combined into one in the F1. A value that is close to the precision or recall values is calculated by the score using Eq. 4 [17].

$$F1 - score = \frac{2 * Recall * Precision}{Recall + Precision} \tag{4}$$

5.1.6 Experimental Results

Table 2 shows an analysis of the results of using the KNN algorithm.

Table 3 shows an analysis of the results of using the SVM algorithm.

Table 4 depicts an analysis of the results of using the Random Forest algorithm.

Table 5 displays an analysis of the results of using the Naïve Bayes algorithm.

Table 6 shows an analysis of the results of using the Logistic Regression algorithm.

Table 7 shows the comparison between the five classifiers depending on the accuracy of the classifier.

Table 2. System assessment using KNN classifier

Class	Precision	Recall	F1 Score
ALL	0.53	1.00	0.70
AML	1.00	0.30	0.46
Weighted Avg.	0.79	0.61	0.57

Table 3. System assessment using SVM classifier

Class	Precision	Recall	F1 Score
ALL	0.80	1.00	0.89
AML	1.00	0.80	0.89
Weighted Avg.	0.91	0.89	0.89

Table 4. System assessment apply Random Forest classifier

Class	Precision	Recall	F1 Score
ALL	0.57	1.00	0.73
AML	1.00	0.06	0.12
Weighted Avg.	0.76	0.58	0.46

Table 5. System assessment using Naïve Bayes classifier

Class	Precision	Recall	F1 Score
ALL	0.50	1.00	0.67
AML	1.00	0.20	0.33
Weighted Avg.	0.78	0.56	0.48

Table 6. System assessment using Logistic Regression classifier

Class	Precision	Recall	F1 Score
ALL	0.62	1.00	0.76
AML	1.00	0.50	0.67
Weighted Avg.	0.83	0.72	0.71

Table 7. Comparison of results by the accuracy

Classifier	Accuracy
Logistic Regression	72.00%
SVM	89.00%
KNN	61.00%
Random Forest	58.00%
Naïve Bayes	56.00%

5.1.7 Discussion

This article was a comparison study that employed various algorithms to discover the best outcomes with the highest level of accuracy achievable when detecting leukemia malignancy from a gene expression dataset. A comparison of the outcomes for the accuracy of each classifier is summarized in Table 7. An accuracy evaluation is performed using a confusion matrix. When five classifiers are compared, the experimental result shows that Naive Bayes achieves an accuracy of 56.00%, Random Forest achieves 58.00%, the KNN algorithm achieves 61.00%, Logistic Regression achieves 72.00%, and the SVM reaches 89.00%.

5.2 Conclusion

Leukemia is a very serious disease that needs a lot of effort to diagnose and classify. We have used the classifiers of machine learning to detect the 2 types of leukemia, Acute Lymphocytic Leukemia (ALL), Acute Myelogenous Leukemia (AML), using the Gene expression Dataset. The results show that the Support Vector Machine (SVM) classifier has the highest accuracy 89.00% whereas the Naïve Bayes has the lowest accuracy 56.00%. The last three classifiers respectively are KNN, Logistic Regression, and Random Forest with accuracy ratios of 61.00%,72.00%,58.00%

References

1. Ali, N.O.: A Comparative study of cancer detection models using deep learning - leukemia. Deep Learn. 1–48 (2020)
2. Pham, T., Tran, T., Phung, D., Venkatesh, S.: Predicting healthcare trajectories from medical records: a deep learning approach. J. Biomed. Inform. **69**, 218–229 (2017). https://doi.org/10.1016/j.jbi.2017.04.001
3. Kumar, Y., Koul, A., Singla, R., Ijaz, M.F.: Artificial intelligence in disease diagnosis: a systematic literature review, synthesizing framework and future research agenda. J. Amb. Intell. Hum. Comput. 1–28 (2021). https://doi.org/10.1007/s12652-021-03612-z
4. Larrañaga, P., et al.: Machine learning in bioinformatics. Brief. Bioinform. **7**(1), 86–112 (2006). https://doi.org/10.1093/bib/bbk007
5. Golub, K., et al.: Gene Expression Dataset. https://www.kaggle.com/crawford/gene-expression
6. Golub, T.R., et al.: Molecular classification of cancer: class discovery and class prediction by gene expression monitoring," Science (80-.)., **286**(5439), 531–527 (1999). https://doi.org/10.1126/science.286.5439.531
7. Ratley, A., Minj, J., Patre, P.: Leukemia disease detection and classification using machine learning approaches: a review. In: 2020 1st Int. Conf. Power, Control Comput. Technol. ICPC2T 2020, pp. 161–165 (2020). https://doi.org/10.1109/ICPC2T48082.2020.9071471
8. Alrefai, N.: Ensemble machine learning for leukemia cancer diagnosis based on microarray datasets. Int. J. Appl. Eng. Res., **14**(21), 4077–4084 (2019). http://www.ripublication.com

9. Ghaderzadeh, M., Asadi, F., Hosseini, A., Bashash, D., Abolghasemi, H., Roshanpour, A.: Machine learning in detection and classification of leukemia using smear blood images: a systematic review. Sci. Program. **2021** (2021). https://doi.org/10.1155/2021/9933481
10. Huang, S., Nianguang, C.A.I., Penzuti Pacheco, P., Narandes, S., Wang, Y., Wayne, X.U.: Applications of support vector machine (SVM) learning in cancer genomics. Cancer Genomics Proteom. **15**(1), 41–51 (2018). https://doi.org/10.21873/cgp.20063
11. Gandhi, R.: Support Vector Machine - Introduction to Machine Learning Algorithms. https://towardsdatascience.com/support-vector-machine-introduction-to-machine-learning-algorithms-934a444fca47
12. Kwong, G.A., Ghosh, S., Gamboa, L., Patriotis, C., Srivastava, S., Bhatia, S.N.: Synthetic biomarkers: a twenty-first century path to early cancer detection. Nat. Rev. Cancer **21**(10), 655–668 (2021). https://doi.org/10.1038/s41568-021-00389-3
13. Italia Joseph Maria, D.R., Devi, T.: Machine Learning Algorithms for Diagnosis of Leukemia
14. "regression-logistique-quest-ce-que-cest." https://datascientest.com
15. "Confusion Matrix in Machine Learning, 23 February 2020." https://www.geeksforgeeks.org/confusion-matrix-machine-learning/
16. "(2019, August 5)." https://www.harrisgeospatial.com/docs/CalculatingConfusionMatrices.html. Accessed 05 August 2019
17. Brownlee, J.: How to calculate precision, recall, and f-measure for from, imbalanced classification. https://machinelearningmastery.com/precision-recall-and-f-measure-for-imbalanced-classification/

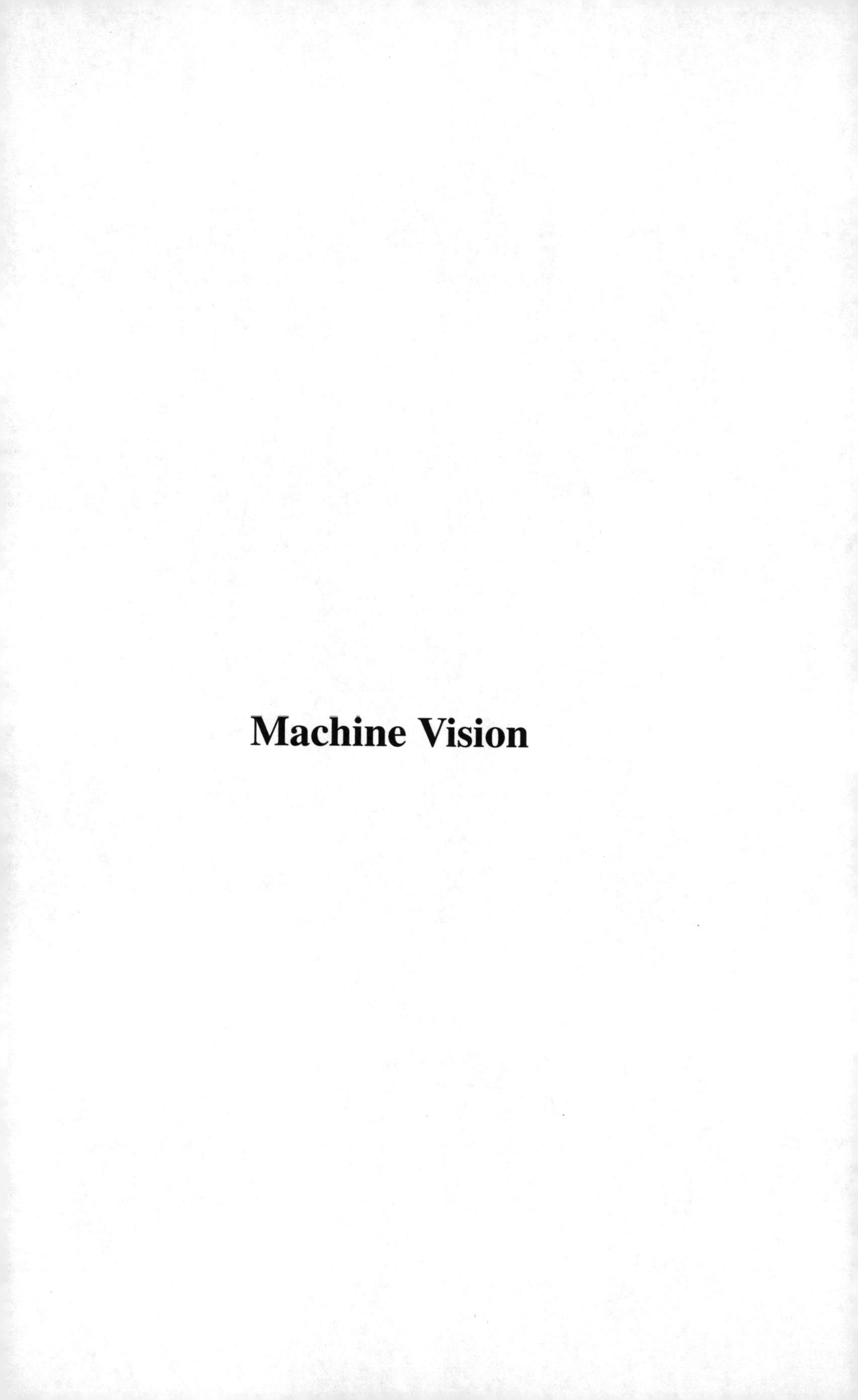

Machine Vision

A Deep Learning Approach for Hand Gestures Recognition

Fatima Zohra Ennaji[1(✉)] and Hamada El Kabtane[2]

[1] Laboratory of Process Engineering, Computer Science and Mathematics (LIPIM), National School of Applied Science - USMS, Khouribga, Morocco
f.ennaji@usms.ma

[2] SMARTE Systems and Applications (SSA) - National School of Applied Science - UCAM, Marrakesh, Morocco
h.elkabtane@uca.ma

Abstract. Hand gestures are part of communication tools that allows people to express their ideas and feelings. Those gestures can be used to insure a communication not only between people but also to replace traditional devices in human-machine interaction (HCI). This last leads us to use this technology in the E-learning domain. COVID'19 pandemic has attest the importance of E-learning. However, the Practical Activities (PA), as an important part of the learning process, are absent in the majority of E-learning plateforms. Therefore, this paper proposes a convolution neural network (CNN) method to ensure the detection of the hand gestures so the user can control and manipulate the virtual objects in the PA environment using a simple camera. To achieve this goal two datasets have been merged. Also the skin model and background subtraction were applied to obtain a performed training and testing datasets for the CNN. Experimental evaluation shows an accuracy rate of 97,2.%.

Keywords: Hand Gesture Recognition · CNN · deep learning · E-learning

1 Introduction

The communication between humans is not limited only to the use of oral speech, but also of the body gestures (Body Language), facial grimaces (Facial Expressions) and hands gestures. All those last can be employed to transmit and explain an idea and/or a feeling. Using those techniques to interact with machines might be beneficial, since the human being uses these communication means intuitively and intentionally. The voice command, for example, requires speech recognition. However, the noisy environments and the voice intonation in addition of the way that the same word is pronounced are all challenging tasks. In the other hand, the body language is considered as the most powerful human-computer communication method. The body language can involve different types of expressions, such as hand gestures, body poses, and facial emotions. The hands represent one of the most expressive parts comparing with the other parts of the body.

N. Aboutabit et al. (Eds.): ICMICSA 2022, LNNS 656, pp. 119–128, 2023.
https://doi.org/10.1007/978-3-031-29313-9_11

The hand gestures can be employed to insure a communication between human and machines. In Human-computer interaction (HCI) systems, hand gesture-based interfaces are applied in several practical applications, such as robots control [14], sign language recognition [10], exploration of medical image data [8], control of mouse and keybord [3], gaming technology [11] and interaction between human and vehicles [6].

The detection of hand gestures has attracted many researches. Therefore, many solutions have been proposed, generally categorized as "Sensors-based" and "Vision-based". The first category is based on the use of sensors to recognize, to detect and to track the hand (ex: the use of Leap Motion, Kinect and Myo sensor). The second category is based on the use of simple camera and it is based on making a machine that performs the process of the recognition, the detection and the tracking of the hand using traditional RGB cameras.

E-learning is one of the most powerful concepts, that stood out with the COVID-19 pandemic, where all the schools and the universities were locked. The apparition of the term E-learning was in 1960 by Bruner and called E-learning [1]. In the past, it was referring the use of electronic machines to learn or teach somethings, but now and with the widespread use of the Internet. E-learning offers a lot of opportunities to resolve issues related to media availability, presence of learner in the classrooms, etc. [16]. With the use of E-learning, the learner can collaborate with his/her classmate anytime and from anywhere.

Practical Activities (PA) aims to complete the theatrical knowledge built during the courses [5]. Unfortunately, the limitation of some E-learning platforms is the lack of PAs. To fix this deficiency some solutions have been proposed like the use of video where the teacher/instructor records her/his experimentation and uploads it or directly projects it to the learners. The major limitation of this solution is that the learner is always a passive actor s/he is just a spectator and cannot manipulate the objects used in the PA.

The Augmented Reality (AR) can be a promising technology that can be used in the learning approach to provide an interactivity for the users (instructors/learners) [7]. The manipulation of the virtual objects in an AR system can be a challenging task and this is the main goal of our system, wich aims to recognize a hand gesture from a RGB video streaming.

The remaining of this paper is structured as follows: Sect. 2 presents some previous works that focused on hand gesture recognition. Section 3 is dedicated to describe the used dataset in addition of the proposed process. The next section aims to expose the obtained results. Finally, the conclusion is presented in the fifth section.

2 Related Works

The communication between people is not only based on the use of words, but also on gestures and body language. Thus, proposing a system that allows people to interact with computing devices might be attractive and interesting rather than the classic means.

In order to make practical activities entertaining and comparable to the real ones, 3D objects have to be used and the manipulation using real body gestures. The major concern and challenge here is how to make these gestures understandable by machines. So, many solutions have been proposed that can be divided on two categories:
- Sensors-based
- Vision-based

The first category is based on the use of sensors; the most used are Kinect, Leap Motion and Myo Armband.

- The Kinect [13] is a motion device that was created first for Xbox360 in 2009 by Microsoft and soon a version for Windows was released followed by a second version (Kinect v2) in 2014 for Windows and Xbox one. It performs a RGB-D detection, since it captures using RGB colors camera and a depth sensor. It gives the possibility to interact with computer or Xbox games using their body and even their voice because it contains a microphone. The Kinect can detect the parts of the users' body and even can recognize the faces. In 2011, a Software Development Kit (SDK) was developed to give the scientists and the developers the possibility to experiment what the Kinect can offer.
- The Leap Motion [9] is a small device with a dimension of $80 \times 30 \times 11\,\mathrm{mm}$ that provides a high precision in the hands monitoring and gestures recognition unlike the Kinect it cannot monitor the body or the face of the user. It contains three infrared LEDs and two infrared cameras (as mentioned above). The user can connect the Leap Motion directly to the computer and use it to enhance the interaction in the Virtual Reality (VR) so the user can put it on the VR headsets like HTC Vive and Oculus Rift.
- The MYO Armband [17] is a Thalmic Labs device, created in 2012. It is a wireless device; the user can put it in her/his forearm. The Myo is composed of eight electromyography (EMG) sensors, a triple-axis accelerometer, a gyroscope and a magnetometer. Those sensors recognize the hand gestures from the muscles activity. The Armband, then, is wirelessly connected to the computer and other devices especially with Bluetooth 4.0.

The vision-based gesture recognition category is based, generally, on the use of a camera. This last records a video of the hands, then it is decomposed into a set of images (frames), so the process of recognition of the hand and its gestures can be applied. There are many common gestures recognition methods. Geometric features have been used to recognize the gesture structure, edge, contour and other features [2]. On the other hand, the Hidden markov model is known for describing the spatial and temporal changes of gestures, but unfortunately, it's recognition speed is not satisfactory [15].

A description of the techniques is included in the following section.

3 Proposed Solution

Deep learning is an extension of neural network architecture. It is used to automate the process of extraction of the features from data by passing through some hierarchical hidden layers. In this paper, we will use an architecture of deep learning using convolutional neural network (CNN) to recognize hand gestures, preceded by a preprocessing step that aims to enhance the input of our model.

3.1 Dataset

In this paper we combined two datasets:

- American ASL dataset [4]: is a dataset that contains 87000 RGB images. Those last have a dimension of 200 × 200 pixels and there resolution is 72 pixels/in. This dataset has 29 classes or folders that present 26 letters A to Z and three additional classes: Space, Delete and Nothing (empty). The size of the dataset is 1.11 GB. It is available in Kaggle [4].
- Our dataset: our dataset is composed of 7783 images. Those images were token from 26 participants, 9 females and 17 males, the volunteers were from different age (between 19 years and 51 years). The participants have used their own camera-phone. The different images were token with different angles, lighting, background and distance from the camera (depth). All the images are in RGB format but in different variation and different dimensions. This problem required a preprocessing step (Sect. 3.2) to facilitate and ensure the suitable recognition and classification (Fig. 1).

Fig. 1. Gesture of alphabets in the American ASL

In our case, we will keep from the 'American ASL' dataset only the folders that contain the needed gestures mentioned in the Table 1. To do so, we kept the 'A' folder for the "Fist", the 'B' folder for the "Palm", the 'C' Folder for the "C" gesture. The folder 'G' was used for three gestures: for the "index finger", the images in the folder were rotated 90°, for the "Previous" gesture the images have been used directly and for the "Next" gesture the images in the folder were reversed horizontally. The total number of the images is 18000 images.

Table 1. List of chosen gesture

Hand gesture	Label	Action
	Palm	Select and release the selected virtual objects
	Fist	Grab an object
	Previous	Decreasing values of the used objects in the PAs
	Next	Increasing values of the used objects in the PAs
	Index finger	Click on buttons if they are used in the PA
	C	Rotate of objects

3.2 Pre-processing

The pre-processing step is an essential step in the hand gesture recognition's domain, considering that it will directly affect the performance and final results. In this phase, we define the necessary techniques and methods to ensure the localization and the segmentation of the hand from the rest of the image. Also in this step, we must minimize the noises and clarify in the images of the resulting dataset.

In our paper, the pre-processing step was applied on the datasets. All the images in our database were token in RGB format, and since each participant took these own photos by their own device, the dimensions of the images were different and also a classification of the images were required to get the suitable recognition. A manual classification was applied on the images to put them in the right folders. Thus, we have six folders; one of each gesture.

Color space is a mathematical model to represent color information as three or four different color components [12]. The colors information is used to focus on an object in the images, in our case the center of interest is the hand. In our solution, the chosen method is based on the use of the hue and the saturation of the HSV color space to segment the hand from the entire image. The reason behind our choice is because that the RGB color space is not preferred for color based detection and color analysis and that due to the mixing of color (chrominance) and intensity (luminance) information and its non-uniform characteristics. Therefore, we need to convert the space color of the images from RGB color space into HSV color space, the equations below were used.

$$H = \arccos(\frac{1/2(2R - G - B)}{(\sqrt{(R-B)^2 - (R-B)*(G-B)})}) \quad (1)$$

$$S = \frac{\max(R,G,B) - \min(R,G,B)}{\max(R,G,B)} \quad (2)$$

$$V = \max(R,G,B) \quad (3)$$

The problem in this method is that we have to precis the ranges of the saturation and the hue. To do so, we have performed experiences on the images to determinate the adequate ranges of the skin color region. We applied a manual verification on the images, unfortunately, in some pictures the hand detection has gone bad either due to lighting conditions or other causes. Thus, these images were destroyed. The first step helps us to remove the background of the image; it is based on the skin's color detection, unfortunately this solution can detect also the face (if it exists), other parts of the body and even other object in the background that have the same color of the hand. Therefore, the step of the detection of the hand shape is necessary to focus on the hand.

The final result of the handshape detection step keeps only the hand in the image. To improve the result, the third step aims to eliminate noise from the image. The token images are in RGB format, which required a pre-processing to make it suitable for classification and recognition. The images in the dataset were resized to a 64×64 dimension and converted from RGB to gray scale images, with a range of pixel values from 0 to 255. The next step is the contrast adjustment to get a better quality of the image.

Fig. 2. Conversion from Gray-scale to Binary-scale

As results of all the previous steps, the hand remains as the only object in the images since we have eliminated the background of the latter. To accelerate to learning process, all the images are converted from Gray-scale (from 0 to 255 values) to Binary-scale (0 or 1) (Fig. 2). To ensure this conversion, we relied on the use of thresholding that is a method of image segmentation. We select a threshold value and all the values below the chosen threshold value is classified as 0 (for the background) and other values are classified as 1. Consequently, the gray-value function becomes:

$$g(x,y) = \begin{cases} 1 \text{ if f(x,y)} \geqslant T \\ 0 \text{ Otherwise} \end{cases}$$

3.3 Model Architecture

The CNN architecture for the gesture classes considered in our study is shown in Fig. 3. The model is constructed with an input layer, three convolution layers along with ReLu and maxpooling layers for feature extraction, one softmax output layer and a final fully connected output layer for classification of gestures.

At first, an input comes into the input layer so it can after that rendered and passes into the hidden layers. These later are several in number and size, and are interconnected convolutions. These convolutional layers determine a special value, which is the value of confidence. In our case, the value of the confidence threshold helps to determine whether the application has successfully determined what object it has encountered (greater than 0.5) or not. After each hidden layer, an exclusion layer was inserted, the exclusion value was set to 0.25, 0.25 and 0.5 respectively to avoid the overfitting.

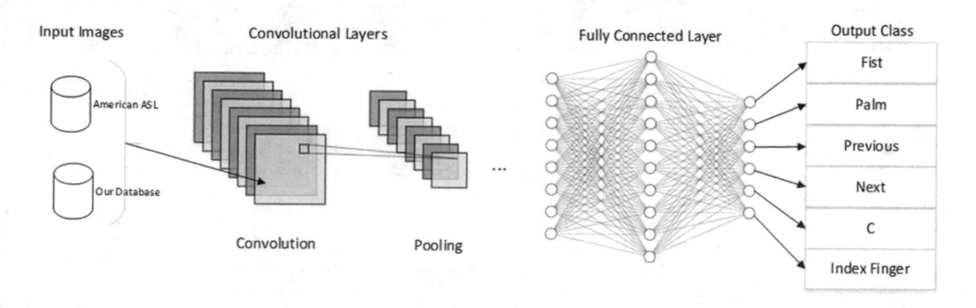

Fig. 3. Architecture of the proposed CNN classifier for hand gesture recognition

4 Results and Discussion

The proposed hand gesture recognition methodology is simulated using Python as programming language version 3.7.3. Anaconda was used as distribution of Python and the Python software package includes keras, panda and theano modules. These packages and modules are license free.

The proposed methodology is applied on two datasets. The first one is openly available and we created a second one to create a bigger dataset that contains a large number of images with hands. The resulting dataset contains six different gestures with 31738 images. The Table 2 shows the number of the images used in the training set for each gesture, to avoid the over-fitting problems the number of images in each gesture is quit the same. The training set contains 22738 gesture images and 9000 images are used as the testing set.

To evaluate the performance of the proposed methodology Accuracy, Specificity, Precision and Recall measures have been considered.

Table 2. Training set and the Testing set

	American ASL	Our Dataset	Training set	Testing set
Palm	3000	2210	3710	1500
Fist	3000	2427	3927	1500
Next	3000	2251	3751	1500
Previous	3000	2251	3751	1500
Index finger	3000	2309	3809	1500
C	3000	2290	3790	1500
Total	18000	13738	22738	9000

Table 3. Results of matrix of confusion for all the hand gestures

		Predicted class	
		Palm	Other
Original class	Palm	1440	60
	Other	75	7425

		Predicted class	
		Fist	Other
Original class	Fist	1395	105
	Other	126	7374

		Predicted class	
		Next	Other
Original class	Next	1389	111
	Other	131	7369

		Predicted class	
		Previous	Other
Original class	Previous	1372	128
	Other	111	7389

		Predicted class	
		Index finger	Other
Original class	Index finger	1357	143
	Other	130	7370

		Predicted class	
		C	Other
Original class	C	1345	155
	Other	129	7371

Based on the Table 3, which presents the information of the confusion's matrix of the different hand gestures, we were able to determine the performance of the method of predicting hand gestures. Our proposition achieves 97.4% of Accuracy and 92.2% of Precision (see Table 4).

Table 4. Performance of our hand gesture recognition method

Performance analysis parameters	Experimental results (%)
Accuracy	97,4%
Precision	92.2%
Recall	92.2%
Specificity	98.44%

5 Conclusion and Future Works

E-learning platforms provide the possibility to make a self-evaluation during the studies to get information about their progressing simply and precisely. However, These platforms suffer of the lake of Practical Activities (PAs). To overcome this issue, using Augmented reality and 3D objects might be an interesting solution.

In order to manipulate the 3D objects (used in the PA), hand gestures are used to simulate the real PAs. This approach have been achieved with a precision score of 92,2% and an accuracy score of 97,4%.

As a future work, we are attempting to enhance the manipulation of these virtual objects so as to make the virtual PAs easier and more manageable to carry out.

References

1. Learning styles and e-learning. Ph.D. thesis (2008)
2. Hand gesture recognition based on HU moments in interaction of virtual reality (2012). https://doi.org/10.1109/IHMSC.2012.42
3. Tran, D.-S., Ho, N.-H., Yang, H.-J., Kim, S.-H., Lee, G.S.: Real-time virtual mouse system using RGB-D images and fingertip detection. Multimedia Tools Appl. **80**(7), 10473–10490 (2020). https://doi.org/10.1007/s11042-020-10156-5
4. Akash: ASL Alphabet
5. Auer, M., Pester, A.: Toolkit for Distributes Online-Lab Kits. Adv. Remote Lab. e-learn. Exp. **6**, 285–296 (2007)
6. Dong, G., Yan, Y., Xie, M.: Vision-based hand gesture recognition for human-vehicle interaction. In: Proc of the International conference on Control Automation and Computer Vision (2000)
7. El Kabtane, H., El Adnani, M., Sadgal, M., Mourdi, Y.: Virtual reality and augmented reality at the service of increasing interactivity in MOOCs. Educ. Inf. Technol. **25**(4), 2871–2897 (2020). https://doi.org/10.1007/s10639-019-10054-w
8. Gallo, L., Placitelli, A.P., Ciampi, M.: Controller-free exploration of medical image data: experiencing the Kinect. In: Proceedings - IEEE Symposium on Computer-Based Medical Systems. pp. 1–6 (2011). https://doi.org/10.1109/CBMS.2011.5999138
9. Inc, L.M.: Leap Motion
10. Padmalatha, E., Sailekya, S., Ravinder Reddy, R., Anil Krishna, C., Divyarsha, K.: Machine learning methods for sign language recognition: a critical review and analysis. Intell. Syst. Appl. 12 (2021). https://doi.org/10.35940/ijrte.C4565.098319
11. Rautaray, S.S., Agrawal, A.: Interaction with virtual game through hand gesture recognition. In: 2011 International Conference on Multimedia, Signal Processing and Communication Technologies, IMPACT 2011 (2011). https://doi.org/10.1109/MSPCT.2011.6150485
12. Shaik, K.B., Ganesan, P., Kalist, V., Sathish, B.S., Jenitha, J.M.M.: Comparative study of skin color detection and segmentation in HSV and YCbCr color space. Procedia Comput. Sci. (2015). https://doi.org/10.1016/j.procs.2015.07.362
13. Soares Beleboni, M.G.: A brief overview of Microsoft Kinect and its applications. In: Interactive Multimedia Conference 2014. p. 6 (2014)

14. Stančić, I., Musić, J., Grujić, T.: Gesture recognition system for real-time mobile robot control based on inertial sensors and motion strings. Eng. Appl. Artif. Intell. **66**, 33–48 (2017). https://doi.org/10.1016/j.engappai.2017.08.013
15. Sugandi, B., Octaviani, S.E., Pebrianto, N.F.: Visual tracking-based hand gesture recognition using backpropagation neural network. Int. J. Innov. Comput. Inf. Control **16**(1), 301–313 (2020). https://doi.org/10.24507/ijicic.16.01.301
16. Sun, P.C., Tsai, R.J., Finger, G., Chen, Y.Y., Yeh, D.: What drives a successful e-Learning? An empirical investigation of the critical factors influencing learner satisfaction. Comput. Educ. **50**(4), 1183–1202 (2008)
17. Thalmic Labs: Myo

A Review on Video-Based Heart Rate, Respiratory Rate and Blood Pressure Estimation

Hoda El Boussaki(✉), Rachid Latif, and Amine Saddik

Laboratory of Systems Engineering and Information Technology LISTI,
National School of Applied Sciences, Ibn Zohr University,
80000 Agadir, Morocco
hodaelboussaki@gmail.com

Abstract. Contactless vital signs measurement has a lot of benefits and can be applied in different environments. The purpose is to achieve an estimation as precise as a regular monitor. First of all, blood pressure (BP), respiratory rate (RR) and heart rate (HR) are three of the most commonly measured clinical parameters and their values are major determinants of clinical decisions. The volume of blood that is ejected by the heart into the arteries is a key factor in calculating blood pressure. In addition to that, The elasticity of the arterial walls and the flow rate of blood through the arteries are also considered. An adult's resting heart rate ranges from 60 to 100 BPM and his respiratory rate ranges from 12 to 25 breaths per minute. These parameters can be retrieved from facial video-based photoplethysmographic signals. Photoplethysmography (PPG) is an optical technique used to measure changes in blood volume. It is used in many contact-based applications such as smartwatches and fitness trackers. Remote photoplethysmography (rPPG) enables the detection of the pulse-induced subtle color variations on the human skin surface with a camera. The purpose of this review is to summarize the studies made on contactless heart rate, respiratory rate and blood pressure monitoring.

Keywords: Heart rate · blood pressure · respiratory rate · photoplethysmography · non-contact

1 Introduction

Estimating vital signs is necessary to determine a person's condition. Various methods exists to estimate these vital signs without any contact between the monitoring system and the person. In this paper, we focus on three major vital signs: heart rate, blood pressure and respiratory rate. Heart rate is the number of heart beats per minute. The heart allows the blood to circulate through the body. According to the American Heart Association, a normal resting heart rate is between 60 and 100 bpm [1]. When the heart rate is too fast it is called tachycardia or is too slow it is called bradycardia. Blood pressure (BP) is the ratio of systolic BP that is the pressure exerted on the arterial walls when the

N. Aboutabit et al. (Eds.): ICMICSA 2022, LNNS 656, pp. 129–140, 2023.
https://doi.org/10.1007/978-3-031-29313-9_12

heart contracts and diastolic BP that is the pressure exerted on the arterial walls when the heart relaxes [2]. Blood pressure's values depend on different factors such as age, genetics, weight and arterial stiffness etc. According to the central of disease control and prevention, a normal BP is less than 120 mmHg systolic BP and 80 mmHg diastolic BP [3]. As for the third vital sign, a normal respiratory rate ranges from 12 to 20 breaths per minute [4]. The respiratory system allows the body to use oxygen. An increase or decrease in the number of breaths is an indicator of stress to the body.

Monitoring vital signs is an important determinant to a person's health. A. Bella et al. 2021 discussed the importance of contactless monitoring during the Covid-19 pandemic. Regular vital signs monitoring requires a close distance being kept between the patient and the medical staff which can be dangerous with a wild-spread virus. The purpose of these kind of systems in hospitals can reduce the contact between them [7]. Various studies were found on non-contact vital signs surveillance. Y. lee et al. 2018 used Impulse-Radio Ultra-Wideband (IR-UWB) Radar Technology to monitor vital signs [8]. The system was tested on 6 healthy subjects and 16 with atrial fibrillation (AF). The values obtained agreed with those obtained with an ECG with a percentage mean error of 2.3 %. W. Lv et al. 2021 used frequency-modulated continuous-wave (FMCW) Millimeter Wave Radar in the 120 GHz Band on 8 subjects with less than 3% of error [9]. The radar-based method consists of reducing the interference caused by the reflection of other objects and improving the signal-to-clutter ratio (SCR). Other studies like [11] used an RGB camera for the same purpose where the absolute error for two subjects was 2.11 and less than 4.06 for five subjects. Radar technique were also implemented in blood pressure monitoring. T. Ohato et al. 2019 proposed a method relying on a Doppler radar [10] with correlation coefficients of 0.79 for systolic BP and 0.88 for diastolic BP. The method was tested on 4 subjects. The Doppler radar transmits a microwave signal and receives a signal containing information on blood pressure. Different techniques exist to measure the respiratory rate. P. Marchionni et al. 2013 used a laser Doppler vibrometer to measure the chest movements with less than 3% of error for BP and less than 6% for HR [12]. Optical methods were also used. K.B. Gan et al. 2016 used a contactless displacement sensor on 32 healthy subjects in four breathing conditions: rest, coughing, talking and in movement [13]. The Root Mean Square Error (RMSE) was 0.33 breaths per minute which means less than 5%. Another method is to use an RGB camera as in [14]. It was tested on 16 subjects and the RMSE was at 1.5 breaths per minute.

In this work, we will focus on some of the different methods that exist on contactless respiratory rate, blood pressure and heart rate monitoring. A general description will be given on the several algorithms that were implemented, while concentrating on techniques that use an RGB camera. Generally to estimate vital signs through a camera, we focus on the changes of a the pixels' values. Every pixel contains 3 parameters: red, green and blue. The variations in these parameters give us different colors. By measuring these variations, it is possible to construct a signal containing that information. Signals extracted from images can be used to estimate different vital signs.

This paper is structured as follows: the first part gives an overview on photoplethysmography and on how both blood pressure, respiratory rate and heart rate are measured with traditional methods. The second part introduces the different studies made on non-contact vital signs monitoring and explains the advances made in this field. Finally, we will end with a discussion followed by a conclusion.

2 Contact-Based Heart Rate, Blood Pressure and Respiratory Rate Monitoring

2.1 Photoplethysmography

Photoplytesmography (PPG) is a non-invasive method for monitoring vital signs. PPG signals are extracted by measuring the change of light in the capillary vessels. The capillary vessels are blood vessels that ensure the nutritional supply to the tissues of the organism. They are thin-walled, small in caliber and their number is extremely large. Figure 1 represents how PPG sensors work. The intensity of the reflected light contains information about the blood volume of that area.

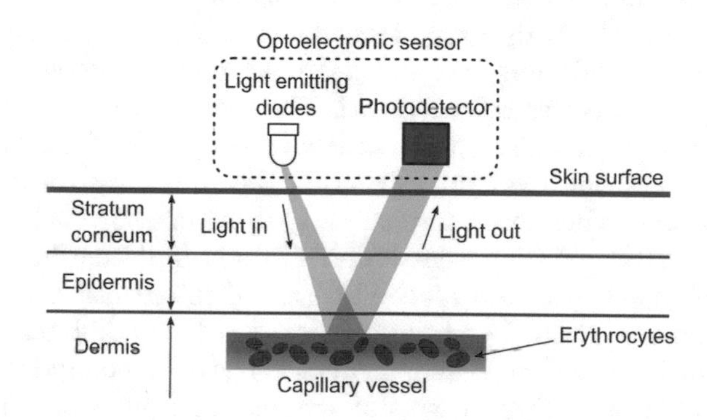

Fig. 1. PPG sensors [15].

PPG is an alternative to solve the limitations of ECG that requires several electrodes placed on the skin at certain locations to measure heart rate [16]. Figure 2 shows the differences between the ECG and the PPG waveforms. The ECG signal consists of the P wave, the QRS complex, and the T wave. They represent the depolarization and the re-polarization of the components of the heart. On the other hand the PPG signal consists of the systolic peak which indicated the contraction of the heart, the diastolic peak which indicates ventricular relaxation and the dicrotic notch that separates between the systolic phase and the diastolic phase [17].

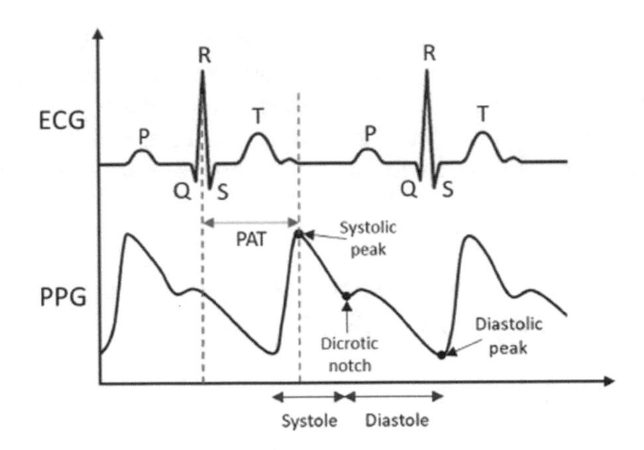

Fig. 2. PPG and ECG waveforms [17].

2.2 Traditional Vital Signs Monitoring

An irregular heart beat, respiratory rate and blood pressure can indicate many health conditions. Monitoring them can be life saving or used for prevention. Their fluctuations is the body's response to stress. The body adapts to environmental and health conditions and the heart activity responds to this adaptation. Therefore, the heart rate is modified. The heart beats autonomously but is required to adjust the frequency of its beats to the needs of the body [18].

Blood pressure should be kept at a stable level. It rises during systole and falls during diastole. Therefore there is the systolic pressure, which is the maximum pressure reached during ventricular ejection and the diastolic pressure, which is the minimum pressure reached during diastole just before ventricular ejection. Blood pressure can be measured directly with an inflatable cuff called a sphygmomanometer and a stethoscope which is the most commonly used, with a digital blood pressure monitor or with an oximeter. In the oscillometric method, the cuff is placed around the arm, slightly above the elbow crease, and then inflated with air until the pressure is so high that it crushes the arteries below the cuff. The blood can then no longer circulate. The cuff is then slowly deflated and the pressure is measured. A digital blood pressure monitor uses the same concept. However, the cuff is placed on the wrist or the arm and does not need a stethoscope or another person to do the measurement.

Many types of heart rate monitoring exist. One of the most common is the electrocardiogram (ECG). The ECG records the electrical phenomena that occurs inside the heart. The electrical signals give a representative trace of the cardiac activity recorded with the electrodes [18]. Pulse oximeters are also used. They are smaller and compact. They use a light source and a photodetector. The change in blood volume in every cardiac cycle affects the amount of light reflected on the photodetector. Therefore, heart rate is measured by measuring the intensity of the reflected light. The other type of heart rate monitors is smartwatches and they use essentially photoplytesmography (PPG). In order to estimate the respiratory rate, several contact based techniques exist. The basic method is to count the number of inhalations for one minute. Another method is to place a nasal cannula below the nostrils to record the pressure [19].

3 Methods for Non-contact Vital Signs Estimation

3.1 General Approach

Remote photoplethysmography (rPPG) allows the detection of the small color variations caused by the blood flow's variations on the human skin with a RGB camera. The face is captured then it is detected and analyzed to determine facial landmarks. Generally, the forehead, where most of the blood vessels are concentrated, is selected as the region of interest. T. Ysehal Abay et al. 2019 discussed the accuracy of PPG signals extracted from the forehead and showed that the signals from the forehead are less affected by vasoconstriction [20]. The average of each of the three channels (red, green, blue) of the region is measured over time. Then, the raw signal extracted from the image is filtered. By detecting its peaks, inter-beat intervals are measured and then the heart rate, respiratory rate and blood pressure are estimated.

3.2 Region of Interest Detection and Tracking

The first step in video-based monitoring is the detection and tracking of the region of interest. D. Lu et al. 2021 discussed the Haar classifier [21]. It is also known as the Viola-Jones algorithm or the OpenCV method. It extracts haar features from the image. Haar features are black and white rectangles that defines the image's pixels' values. The values are compared with the cascade to determine if the face is detected. [22] used the Yolov3 algorithm for detection and the Mousse algorithm for tracking. The Yolov3 was introduced by Redmon and Farhadi [23]. It uses features learned by a deep neural network to detect the face. Yolov3 is known to be fast and efficient in real time detection and uses DarkNet-53 to detect features by using convolutional layers [24]. The Multiscale Online Union of Sub-Spaces Estimation (MOUSSE) on the other hand is a probabilistic multi-scale learning algorithm for tracking [25]. There is also the Kanade Lucas Tomasi (KLT) algorithm that tracks the ROI using trained features by detecting facial features. Another method is the Bounded Kalman filter (BKF) algorithm. [26] introduced this algorithm that has a better accuracy with motion.

3.3 Signal Extraction

W. Verkruysse et al. 2008 explained that the PPG signal can be extracted from the red, green and blue channels of a frame using a source of light. It was demonstrated that the green channel G contains the strongest plethysmographic signal because hemoglobin absorbs green light better than red and blue [27]. Lam et al. 2015 also used the green spectrum to extract PPG signals [28]. After extracting the average of each channel, different methods can be used to obtain the final signal. The first is the Blind source Separation-based (BSS) methods. The BSS methods are the Independent Component Analysis (ICA) and a principal component analysis (PCA). An analyzer of ICA can be applied on the mean of the pixels' channels, to extract three independent components, then a band pass filter is used to eliminate undesired frequencies [29]. A similar method is the PCA that transforms a large data to smaller set. The ICA and PCA methods are relatively similar but have estimation latencies. The second method is the model-based methods like the CHROM technique and the Plane-Orthogonal-To-Skin (POS). [30] introduced a new method that is chrominance-based. It gave a superior performance compared to the BSS methods on different points. [31] improved the CHROM method and introduced the POS technique that puts the RGB signals on an orthogonal plane in a normalized skin. The third method is the Spatial Subspace Rotation (2SR or SSR). It was introduced by [32] and measures the rotation of the spatial subspace of the pixels.

3.4 Blood Pressure Estimation

Contactless blood pressure estimation methods can be characterized into two distinct categories. The first one relies on extracting blood pressure from two PPG signals, whereas in the second one, one PPG signal is required. In recent years, many studies were made on cuff-less blood pressure measuring methods. The cuff-less methods are based on pulse transit time (PTT). The PTT method is an alternative for blood pressure estimation. The PTT is the time it takes for a blood pressure pulse to travel from one place in a vessel to another. Generally the chest, where the ECG's electrodes are, is used as the first location and the finger is used as the second location. The arrival of the pulse at this location is usually detected by the use of a PPG sensor. Figure 3 represents different ways used to extract a PPT signal from a PPG and an ECG signal. The signal in blue represents an ECG signal and the red signal represents a PPG signal. The PTT can either defined as the difference between the ECG peak and the PPG peak or the ECG peak and the middle of the PPG peak or the ECG peak and the foot of the PPG signal. By monitoring the change of the PTT it is possible to track the changes in blood pressure. The BP estimation depends on the mathematical model used. In the case of linear model, a logarithmic model or an inverse model for example, the relation between the PTT and the BP is respectively as follows:

$$BP = a + b \cdot PTT \tag{1}$$

$$BP = a + b \cdot ln(PTT) \tag{2}$$

$$BP = a + \frac{b}{PTT} \tag{3}$$

where a and b are constants. [33] used the PTT-based method. They extracted two PPG signals, one from the forehead and the other from the right cheek.

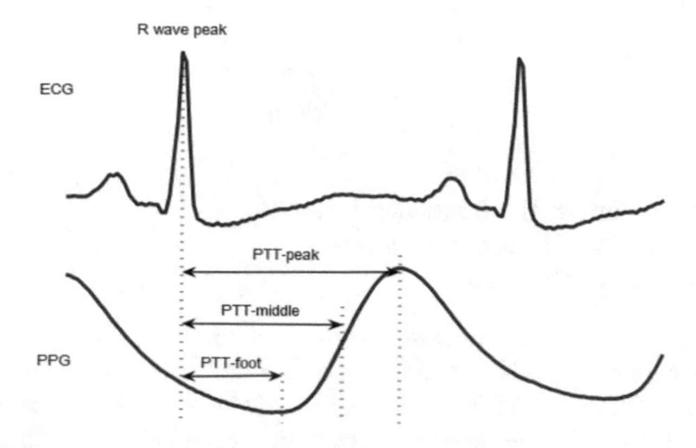

Fig. 3. PPG signal [15].

They studied the correlation between blood pressure and 1/PTT. X. Fan et al. 2018 used the Kernelized correlations filters (KLT) for tracking the palm and OpenFace toolkit for the face. They chose the chrominance-based method to extract the PPG signal. And, an empirically determined nonlinear function estimated the BP using the extracted PTT [34]. In the second category, one PPG signal is required. [35] was able to estimate blood pressure using partial least-squares (PLS) regression. Features were extracted from the second derivative of the PPG signal, and a regression model was used to estimate blood pressure. M. Charmi et al. 2020 used the Viola-Jones algorithm for tracking and ICA method for signal extraction. They used a whole-based method to estimate blood pressure from the extracted signal. First, the peaks were detected. Then, the amplitude values between two consecutive peaks of the PPG signal were put in a zero vector. A Random Forest Regressor (RFR) algorithm was applied on the vector to calculate blood pressure [36]. Q. -V. Tran et al. 2020 used the Yolov3 and Mousse methods for detection and tracking of the ROI. They introduced a method to remove color distortion, after extracting color changes in the image, called Adaptative Pulsatile Plane (APP). The APP method's main advantage is that it adapts to the lighting's changes. They also proposed a method called MLP based on machine learning and a pre-trained model to estimate blood pressure.

3.5 Heart Rate Estimation

Heart rate estimation requires one PPG signal. The peaks of the signal are used to calculate heart rate. P. Rouast et al. 2016 has proposed several techniques to estimate heart rate such as the green channel method and the principal component analysis (PCA) for signal extraction. the Viola-Jones algorithm and a deep neural network (DNN) are used for face detection. The heart rate is measured using frequency analysis either a discrete Fourier transform (DFT) or a fast Fourier transform (FFT). The heart rate is calculated with the following equation [37]:

$$BPM = \frac{F \cdot fps \cdot 60}{Total} \tag{4}$$

where:

F is the maximum of the frequency index
fps is the number of frames per second
Total is the size of the signal

K. Bensalah et al. 2021 proposed a method that applies a trained conventional neural network (CNN) for skin detection and used the Green method (G) for signal extraction. The Discrete wavelet transform (DWT) was employed for frequency analysis. Afterwards, the heart rate was calculated with the following formula [38]:

$$BPM = \frac{F \cdot 60}{int} \tag{5}$$

where:

F is the sampling rate
int is the distance between the peaks

3.6 Respiratory Rate Estimation

C. Massaroni et al. 2018 measure the respiratory rate by extracting the breathing pattern from the variations in light at the collar bones with an average mean absolute error of 1.53 breaths per minute [39]. The signal is extracted from the images using the following equation:

$$f = \frac{1}{x_{ROI}} \sum_{x=1}^{x_{ROI}} \left(\sum_{R,G,B} I \right) \tag{6}$$

where:

f is the signal
I is e intensity components of each channel

The signal f is then detrended then normalized using the following equation:

$$nf = \frac{f - m}{sd} \tag{7}$$

where:

nf is the normalised signal
f is the signal
m is the mean of the signal f
sd is the standard deviation if the signal f

The breathing rate is extracted from nf either in frequency or time domain. M.A. Hassan et al. 2017 also based their estimation on the respiratory motion. The signal obtained was converted to the frequency domain and the frequency that corresponds to the highest value was considered the respiratory frequency [40]. To estimate the number of breaths per minute, the respiratory frequency was multiplied by 60.

$$RPM = F_s \cdot 60 \tag{8}$$

where:

F_s is the respiratory frequency

4 Discussion

We selected 16 papers in total that focused on some or all of the steps mentioned above. The three first papers are about detection and tracking algorithms: Haar classier, Yolov3, Mousse and BFK. They are not about vital signs monitoring but were used in many studies. The four papers that follow are about the algorithms used in extracting a PPG signal from an image that are the G method, the Chrominance-based method, the POS and the SSR methods. They are the most used to our knowledge. On the other hand, the APP is a new method that still hasn't been implemented or studied in different articles. Table 1 summarizes the existing studies.

A. Lam et al. 2015 used the MAHNOB-HCI Database that consists of 61 FPS RGB videos. They have a mean absolute error (MAE) of 4.7 and a root mean squared error (RMSE) of 8.9 and an absolute error of less than 5% bpm [28]. P. Rouast et al. 2016 used videos of 8 females and 12 males and achieved a less average root mean square error of 7.32 bpm [37]. K. Bensalah et al. 2020 used the UBFC-rPPG signal datasets that is composed of 43 videos of 60 s. They attained a better MAE of 3.24 and a better RMSE of 5.64 [38].

To evaluate the accuracy of non-contact blood pressure's values. The error coefficients of both systolic and diastolic blood pressure were calculated. X. Fan et al. 2020 used the Vicar dataset and have a mean absolute difference (MAD) and of 8.42 for SBP and 12.34 for DBP and a standard difference (SD) of 8.81 for SBP and 7.1 for DBP [34]. O. R. Patil et al. 2019 used recorded videos of 4 subjects and achieved a correlation of d in the range of 0.64 to 0.94 between PTT and SBP [33]. Q. Tran et al.2020 tested the algorithm on 82 recorded videos and reached a MAE of 3.083 and a SD of 7.320 for SBP and a MAE of 2.626 and a SD of 7.463 for DBP [22]. M. Charmi et al. 2020 used a recorded database of 40 people and achieved a better mean error of 0.45 and a SD of 12.39 mmHg for SBP and a better mean error of 0.2 and a SD of 6.41 mmHg for DBP [36].

Two papers estimate respiratory rate. [39] used a dataset of 6 females and 6 males and reached a MAE of 0.55 breaths per min. [40] reached a close MAE of 0.58 and used the MAHNOB-HCI database of 27 subjects.

Table 1. Summary of the existing work on face detection and signal extraction.

Work	Face detection and tracking	Signal extraction	Feature extracted	Reference
D. Lu et al. 2021	Haar classifier	–	–	[21]
J. Redmon et al. 2018	Yolov3 and mousse tracking	–	–	[25]
S. K. A. Prakash et al. 2018	BFK	–	–	[26]
W. Verkruysse et al. 2008	–	G	–	[27]
G. De Haan et al. 2013	–	Chrom	–	[30]
W. Wang et al. 2017	–	POS	–	[31]
W. Wang et al. 2016	–	SSR	–	[32]
A. Lam et al. 2015	Classifier	ICA	HR	[28]
X. Fan et al. 2020	OpenFace toolkit and KLT	Chrom	BP	[34]
O. R. Patil et al. 2019	–	ICA	HR and BP	[33]
M. Charmi et al. 2020	Haar classifier	ICA	BP	[36]
Q. Tran et al. 2020	Yolov3 and mousse tracking	APP	BP	[22]
P. Rouast et al. 2016	Haar classifier and DNN	G and PCA	HR	[37]
K. Bensalah et al. 2021	CNN	G	HR	[38]
C. Massaroni et al. 2018	Manually	Average of the pixel's channels	RR	[39]
M.A. Hassan et al. 2017	Viola-Jones	G	RR	[40]

5 Conclusion

This work presents different methods used in video-based heart rate monitoring, respiratory rate, and blood pressure estimation. It lays out the different algorithms used for face detection, signal tracking, and feature extraction. There are several methods for face detection, such as the Viola-Jones algorithm, also known as the Haar classifier, the Yolov3, the mousse tracking, and the bounded Kalman filter. After face detection, a signal containing the needed information is extracted using either the G method, the chrom method, the blind source separation method, the plane-orthogonal-to-skin method, the spatial substance rotation, the independent component analysis, or the principal component method. The choice of which algorithm to use comes back to the application and what advantages each algorithm has, and the challenges it presents. Some algorithms show a better mean error when estimating vital signs than others. In future work, we aim to apply a Hardware/Software Co-Design approach to propose an optimal implementation of different algorithms presented in the literature. This approach allows us to exploit a high-performance embedded architecture for real-time implementation, especially for vital signs such as HR, RR, and BP.

References

1. Avram, R., et al.: Real-world heart rate norms in the Health eHeart study. NPJ Digit. Med. **2**(1) (2019)
2. Oparil, S., et al.: Hypertension. Nat. Rev. Disease Primers **4**(1) (2018)
3. Centers for Disease Control and Prevention: High Blood Pressure Symptoms, Causes, and Problems. https://www.cdc.gov/bloodpressure/about.htm. Accessed 23 Aug 2022
4. National Library of Medicine: Physiology, Respiratory Rate. https://www.ncbi.nlm.nih.gov/books/NBK537306/. Accessed 23 Aug 2022

5. World health organisation Homepage. https://www.who.int. Accessed 15 June 2022
6. Tervo, T., Räty, E., Sulander, P., Holopainen, J.M., Jaakkola, T., Parkkari, K.: Sudden death at the wheel due to a disease attack. Sudden death at the wheel due to a disease attack. Traffic Inj. Prev. **14**(2), 138–144 (2013)
7. Bella, A., Latif, R., Saddik, A.,Jamad, L.: Review and evaluation of heart rate monitoring based vital signs, a case study: Covid-19 pandemic. In: 2020 6th IEEE Congress on Information Science and Technology (CiSt), pp. 79–83 (2020)
8. Lee, Y., Park, J.Y., Choi, Y.W., et al.: A novel non-contact heart rate monitor using impulse-radio ultra-wideband (IR-UWB) radar technology. Sci. Rep. **8**, 13053 (2018)
9. Lv, W., He, W., Lin, X., Miao, J.: Non-contact monitoring of human vital signs using FMCW millimeter wave radar in the 120 GHz band. Sensors (Basel, Switzerland) **21**(8), 2732 (2021)
10. Ohata, T., Ishibashi, K., Sun, G.: Non-contact blood pressure measurement scheme using doppler radar. 2019 41st Annual International Conference of the IEEE Engineering in Medicine and Biology Society (EMBC), pp. 778–781 (2019)
11. Hassan, M., Alam, J.B., Datta, A., Mim, A.T., Islam, M.N.: Machine learning approach for predicting COVID-19 suspect using non-contact vital signs monitoring system by RGB camera. In: 6th International Congress on Information and Communication Technology, ICICT 2021, vol. 217, pp. 465–473 (2022)
12. Marchionni, P., Scalise, L., Ercoli, I., Tomasini, E.P.: An optical measurement method for the simultaneous assessment of respiration and heart rates in preterm infants. Rev. Sci. Instrum. **84**(12), 121705 (2013)
13. Gan, K., Yahyavi, E., Ismail, M.: Contactless respiration rate measurement using optical method and empirical mode decomposition. Technol. Health Care **24**(5), 761–768 (2016)
14. Wuerich, C., Wichum, F., Wiede, C., Grabmaier, A.: Contactless optical respiration rate measurement for a fast triage of SARS-CoV-2 patients in hospitals. IN: Proceedings of the International Conference on Image Processing and Vision Engineering (2021)
15. Moraes, J., Rocha, M., Vasconcelos, G., Vasconcelos Filho, J., de Albuquerque, V., Alexandria, A.: Advances in photopletysmography signal analysis for biomedical applications. Sensors **18**(6), 1894 (2018)
16. Castaneda1, D., Esparza1, A., Ghamari, M., Soltanpur, C., Nazeran, H.: A review on wearable photoplethysmography sensors and their potential future applications in health care. Int. J. Biosens. Bioelectron. **4**(4) (2018)
17. Loh, H.W., et al.: Application of photoplethysmography signals for healthcare systems: an in-depth review. Comput. Methods Programs Biomed. **216**, 106677 (2022)
18. Nguyen, S.H., Bourouina, R., Allin-Pfister, A.C.: Manuel d'anatomie et de physiologie. 2nd edn. Lamarre (2010)
19. Drummond, G.B., Fischer, D., Arvind, D.: Current clinical methods of measurement of respiratory rate give imprecise values. ERJ Open Res. **6**(3), 00023–02020 (2020)
20. Ysehak Abay, T., Shafqat, K., Kyriacou, P.A.: Perfusion changes at the forehead measured by photoplethysmography during a head-down tilt protocol. Biosensors **9**(2), 71 (2019)
21. Lu, D., Yan, L.: Face detection and recognition algorithm in digital image based on computer vision sensor. J. Sens. 1–16 (2021)

22. Tran, Q.-V., Su, S.-F., Tran, Q.-M., Truong, V.: Intelligent non-invasive vital signs estimation from image analysis. In: 2020 International Conference on System Science and Engineering (ICSSE), pp. 1–6 (2020)
23. Redmon, J., Farhadi, A.: Yolov3: an incremental improvement. arXiv preprint arXiv:1804.02767 (2018)
24. Medium Homepage: Face Mask Detector using Deep Learning (YOLOv3). https://medium.com/face-mask-detector-using-deep-learning-yolov3. Accessed 15 June 2022
25. Xie, Y., Huang, J., Willett, R.: Multiscale online tracking of manifolds. In: 2012 IEEE Statistical Signal Processing Workshop (SSP) (2012)
26. Prakash, S.K.A., Tucker, C.S.: Bounded Kalman filter method for motion-robust, non-contact heart rate estimation. Biomed. Opt. Express **9**(2), 873 (2018)
27. Verkruysse, W., Svaasand, L.O., Nelson, J.S.: Remote plethysmographic imaging using ambient light. Opt. Express **16**(26), 21434 (2008)
28. Lam, A., Kuno, Y.: Robust heart rate measurement from video using select random patches. In: 2015 IEEE International Conference on Computer Vision (ICCV) (2015)
29. Goudarzi, R.H., Somayyeh Mousavi, S., Charmi, M.: Using imaging photoplethysmography (iPPG) signal for blood pressure estimation. In: 2020 International Conference on Machine Vision and Image Processing (MVIP) (2020)
30. De Haan, G., Jeanne, V.: Robust pulse rate from chrominance-based rPPG. IEEE Trans. Biomed. Eng. **60**(10), 2878–2886 (2013)
31. Wang, W., den Brinker, A.C., Stuijk, S., de Haan, G.: Algorithmic principles of remote PPG. IEEE Trans. Biomed. Eng. **64**(7), 1479–1491 (2017)
32. Wang, W., Stuijk, S., de Haan, G.: A novel algorithm for remote photoplethysmography: spatial subspace rotation. IEEE Trans. Biomed. Eng. **63**(9), 1974–1984 (2016)
33. Patil, O.R., Wang, W., Gao, Y., Jin, Z.: A camera-based pulse transit time estimation approach towards non-intrusive blood pressure monitoring. In: 2019 IEEE International Conference on Healthcare Informatics (ICHI) (2019)
34. Fan, X., Ye, Q., Yang, X., Choudhury, S.: Robust blood pressure estimation using an RGB camera. J. Ambient Intell. Human. Comput. (2020)
35. Fujita, D., Suzuki, A., Ryu, K.: PPG-based systolic blood pressure estimation method using PLS and level-crossing feature. Appl. Sci. **9**(2), 304 (2019)
36. Goudarzi, R.H., Somayyeh Mousavi, S., Charmi, M.: Using imaging photoplethysmography (iPPG) signal for blood pressure estimation. In: 2020 International Conference on Machine Vision and Image Processing (MVIP), pp. 1–6 (2020)
37. Rouast, P.V., Adam, M.T.P., Chiong, R., Cornforth, D., Lux, E.: Remote heart rate measurement using low-cost RGB face video: a technical literature review. Front. Comput. Sci. **12**(5), 858–872 (2018). https://doi.org/10.1007/s11704-016-6243-6
38. Bensalah, K., Othmani, M., Kherallah, M.: Contactless heart rate estimation from facial video using skin detection and multi-resolution analysis. In: WSCG 2021 29. International Conference in Central Europe on Computer Graphics, Visualization and Computer Vision, pp. 283–292 (2021)
39. Massaroni, C., Lopes, D.S., lo Presti, D., Schena, E., Silvestri, S.: Contactless monitoring of breathing patterns and respiratory rate at the pit of the neck: a single camera approach. J. Sens. 1–13 (2018)
40. Hassan, M.A., Malik, A.S., Fofi, D., Saad, N., Meriaudeau, F.: Novel health monitoring method using an RGB camera. Biomed. Opt. Express **8**(11), 4838 (2017)

Arabic Handwritten Characters Recognition by Combining PHOG Descriptor with Ensemble Methods

M. Dahbali(✉), Noureddine Aboutabit, and N. Lamghari

LIPIM laboratory, National School of Applied Sciences, Sultan Moulay Slimane University, Beni Mellal, Morocco
dahbali.mohamed@gmail.com, {n.aboutabit,n.lamghari}@usms.ma

Abstract. Handwritten character recognition is a system that is widely used in the modern world for a variety of applications in various fields, but it is still difficult in the case of arabic language. The arabic alphabet is distinguished by the fact that many characters have similar shapes but differ in where the dots are placed in relation to the main body of the character. Furthermore, some people handwrite these dots as dashes, complicating the task of a recognition system. Over the last three decades, this field of study has received a lot of attention. However, it still faces several challenges, such as the variation in human handwriting and its cursive nature. In this paper, we created a new system with a PHOG descriptor inspired by the pyramid representation and the Histogram of Orientation Gradients. The performance of most machine learning models depends on the quality, quantity, and relevance of the data. However, insufficient data is one of the most common challenges in machine learning implementations, as collecting this data can be expensive and time consuming in many cases, for this purpose we tested various parameters of this descriptor and used a data augmentation method with Gaussian Noise, which significantly increased accuracy.

For the classification phase, we used five-fold cross-validation, which is a statistical method used to estimate the skill of machine learning models. K-nearest neighbors, decision trees, random forests and naive bayes are the four machine learning techniques we used to train and evaluate our system.

We used ensemble methods mainly bagging and stacking to evaluate the impact of combining multiple models. We tested our system using the AlexU Isolated Alphabet (AIA9K) Dataset, which contains 8,737 characters. When compared to other systems using the same database, the results of the suggested system are promising. The experiments revealed that using classification with stacking algorithm outperforms classification with individual classifiers and bagging algorithm when naive bayes, k-nearest neighbors and decision tree are used as Meta classifiers. We were able to achieve a high classification rate of 97.22% with random forest.

Keywords: Arabic handwritten characters recognition · PHOG descriptor · Data augmentation · Ensemble methods

N. Aboutabit et al. (Eds.): ICMICSA 2022, LNNS 656, pp. 141–153, 2023.
https://doi.org/10.1007/978-3-031-29313-9_13

1 Introduction

The field of optical character recognition (OCR) is critical, particularly for offline handwriting recognition systems. Offline handwritten recognition systems are different from online handwritten recognition systems [1], the capacity to handle vast amounts of handwritten script data is invaluable in some situations. One of these situations is the automation of script copying in old documents, which takes into consideration the difficult and irregular writing style [2]. In comparison to other languages, Arabic optical text recognition is progressing slowly [3]. The cursive nature of the characters causes the primary problem in arabic script recognition. Furthermore, depending on their position in the word, certain characters take two to four forms in the word and a number of characters are linked to complementary parts above, below, or within them. The Arabic alphabet has similar shapes, but the dots are placed differently in relation to the main character. Besides that, these dots are sometimes handwritten as dashes, which complicates the task of a recognition system. The Arabic alphabet is made up of 28 letters and the majority are written in cursive. Nevertheless, many people in other countries use the arabic alphabet, including all arab nations, as well as in Persian, Urdu, and Jawi [4]. It would be excellent to utilize arabic handwritten character recognition (AHCR) to convert a large number of documents into a digital format that may be observed electronically. Some benefits of a handwritten recognition system include reading postal addresses from envelops, digitizing documents, language translation and assisting the blind in reading. The method of feature extraction chosen in automated optical character recognition systems may be the most important factor in achieving high recognition accuracy. In this paper, we propose using the Pyramid Histogram of Oriented Gradients (PHOG). The literature has revealed that gradient orientation-based feature descriptors for example SIFT [5] and HOG [6] descriptors are local histogram-based feature descriptors. These techniques are incapable of capturing small changes in local regions, resulting in poor performance in object recognition systems. Bosch et al. [7] proposed PHOG, the spatial shape feature descriptor based on histograms of orientation gradients of spatial cells at all pyramid levels. There are two classification techniques: manual classification and automatic classification. Manual classification is based on rule-based classification, whereas automatic classification is based on machine learning algorithms. In this study, we used an automatic classification approach with four different algorithms: random forest, decision tree, naive bayes and k-nearest neighbors. To improve the accuracy of theses classifiers, ensemble methods were used which are bagging and stacking algorithms. The basic purpose of these algorithms is to turn a bad learning algorithm into a good one. The progress of the ensemble approach is determined by the variety of individual classifiers in terms of misclassified instances [8]. We used the data augmentation method to improve our method which aims to increase the amount of data by adding slightly modified copies of existing images. In most machine learning algorithms, the results show that the ensemble algorithms outperforms individual classifiers.

The remainder of the paper is organized as follows. Section 2 provides a quick overview of related work. Section 3 explains the proposed technique. The details of the experiments are shown in Sect. 4. The final section concludes the paper and presents future work.

2 Related Works

The recognition of arabic characters is a crucial issue because it may be necessary for the more challenging recognition of arabic words or sentences [9]. Despite advances in recognition algorithms and because the wide variety of handwritten characters, arabic handwritten character recognition techniques are still less effective than those created to recognize printed characters. Various approaches for arabic handwritten character recognition have been presented and reported. However, in this section, we will discuss some related work for tackling the AHCR problem. In [10], a convolutional neural network (CNN) model with regularization parameters like batch normalization to stop overfitting was introduced. Experiments were carried out with the AIA9k and AHCD databases, and the classification accuracies for the two datasets were 94.8% and 97.6%, respectively. Recent work [11] suggests using a two-stage hybrid classifier with Support Vector Machine (SVM) and a neural network (NN) with discrete cosine transform (DCT) as the method for obtaining features. The first step is a two-class SVM classifier that categorizes a character as containing dot(s) or not. The result of this phase is used to prolong the feature vector of the character by the class value, granting it an additional unique feature. In order to classify the character, the extend feature vector is fed into a multi-class neural network model. Using the AIA9K database, the proposed method achieved an average recognition accuracy of 91.84%. The authors of this paper [12] presented a project centered on developing a deep learning architecture model using a convolutional neural network (CNN) and a multilayer perceptron. This research examines the performance of a public database which is Arabic Handwritten Characters Dataset (AHCD). The authors found that there are two ways to overcome the overfitting problem in the case of MLP: increase dropout or reduce network size. The MLP was unable to accurately and without overfitting recognize the database. In the case of CNN, experiments were carried out using various parameters such as batch size, filter size and dropout value. Training this database with the CNN model resulted in a test accuracy of 95.27%, while training it with the MLP model resulted in a test accuracy of 72.08%. A new technique was developed in [13] that is based on an artificial immune system. To deal with complex classification problems, the system incorporates natural immunity concepts. The purpose of this study is to look into the applicability of this system in offline isolated handwritten arabic characters. The system was trained and tested using ten-fold cross-validation on a database constructed from well-known IFN/ENIT. A grid-search algorithm with leave-one-out cross-validation was used to tune the parameters. This experiment achieved an accuracy rate of 93.25%. This research [14] provides an efficient deep convolutional neural network architecture for extracting and classifying arabic handwritten characters, as well as a dropout support vector machine for classifying and recognizing missing features that DCNN does not accurately classify. Based on K-means clustering, the suggested system divides the multi-stroke with similar arabic characters into 13 clusters. On the AHCD database, the suggested approach achieves 95.07% correct classification accuracy.

3 Methods

This section describes the features extraction method we employed as well as the classification step.

3.1 Feature Extraction

The PHOG descriptor is an extension of the histogram of oriented gradients (HOG) descriptor [6], which describes the edge orientation distribution that is resistant to illumination, pose, and local geometric variation. The spatial layout of the shape and its local shape are presented in PHOG descriptor. The distribution over edge orientation within a region captures local features, and spatial layout is captured by tiling the image into non-overlapping regions of varying resolutions. In the PHOG descriptor, all features are extracted in the same way that they are in the HOG descriptor, but in a hierarchical structure called the pyramid. The number of levels of a pyramid is used to determine the global PHOG descriptor of an image.

We should extract edge contours to extract PHOG features. Edge contours are useful in describing any shape since they are independent of illumination and pose changes. We used canny edge detector to extract the edge contours. Thereafter, the image is divided into cells at various pyramid levels. The number of levels chosen has an impact on image recognition since increasing the number of levels increases the PHOG descriptor dimension.

The following step is to compute HOG for each grid at each pyramid resolution level. A histogram of edge orientations inside an image subregion was quantized into K bins. Each bin in the histogram represents the number of edges with orientations that fall within a given angular range. The orientations gradients along edge contours were obtained using central differences in the interior and first differences at boundaries. Finally, the PHOG descriptor for the image is a concatenation of all the HOG vectors at each pyramid resolution. L2 normalization is then performed to the final vector. The PHOG descriptor for the entire image is a vector with a dimensionality of $K \sum_{l \in L} 4^l$. For levels up to $l = 2$ and $K = 10$ bins it is a 210-vector.

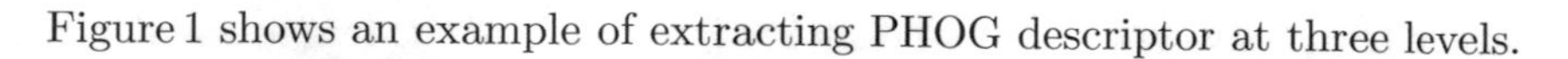

Figure 1 shows an example of extracting PHOG descriptor at three levels.

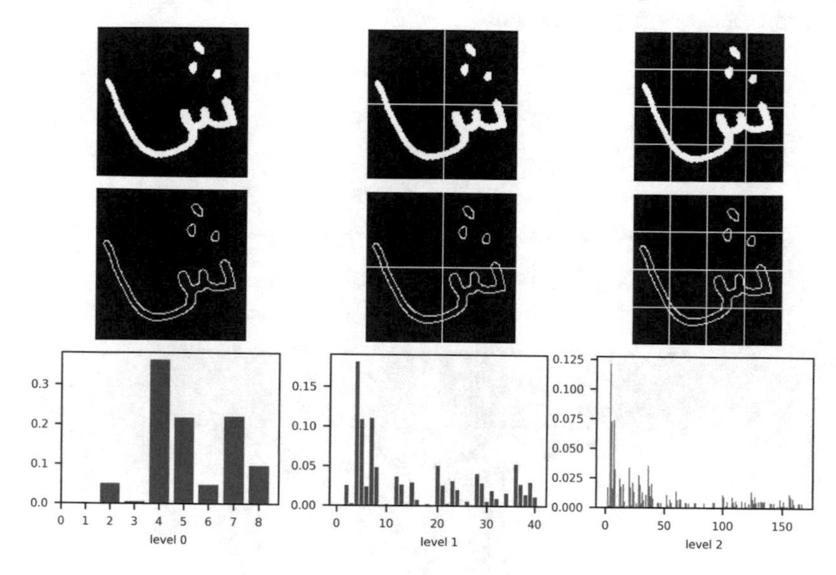

Fig. 1. Extraction of PHOG features from image at three pyramid resolution

3.2 Classification

To increase the robustness of machine learning algorithms, we applied data augmentation to the entire dataset to increase the size of training dataset by producing changed versions of images in the dataset, allowing us to construct more skilled models. The augmentation strategies can generate image variations that increase the capability of models to generalize what they have learned. On small datasets, data augmentation has also been widely employed to prevent overfitting. In image processing field, the simplest approach for data augmentation is adding noise and applying affine transformations (translation, zoom, flips, shear, mirror and color perturbation) [15]. In this study, we used data augmentation method with gaussian noise and we investigated the effectiveness of four machine learning methods on recognition rate: decision tree, naive bayes, k-nearest neighbors and random forest.

Bagging Algorithm: we used bagging algorithm to better grasp the power of using several classifiers in recognition rate. The bagging algorithm is based on the idea that bootstrap samples of the original training set will show a slight difference from the original training set, but enough variability to produce various classifiers. Each ensemble member is trained on a different training set, and the predictions are combined through averaging or voting. The different datasets are generated by sampling from the original set, choosing N items uniformly at random with replacement [16]. Figure 2 presents a simple example of how a bagging algorithm works. In this algorithm, we used three classifiers which are:

decision tree, naive bayes and k-nearest neighbors. The random forest algorithm is regarded as a bagging algorithm.

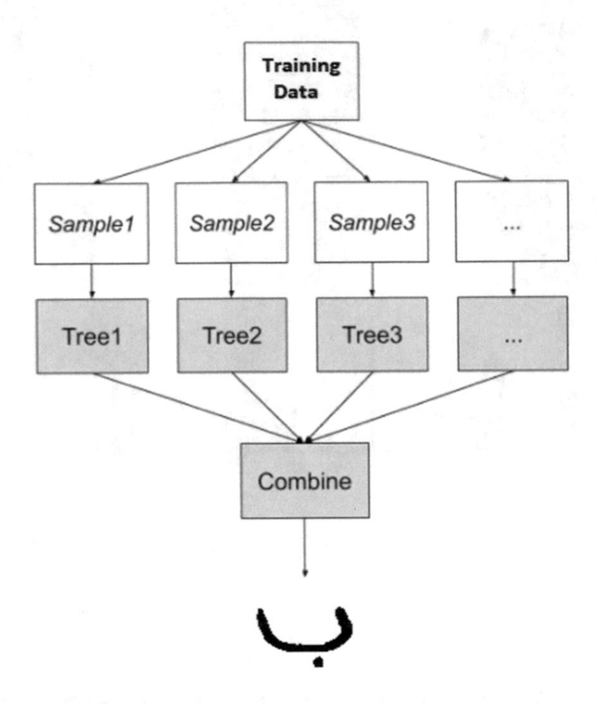

Fig. 2. Working process of a bagging classifier

Stacking Algorithm: stacking algorithm is the second method we used to improve the accuracy of arabic handwritten characters recognition. Stacking is a different method of combining multiple classifiers. In contrast to bagging, the models are distinct and fit to the same dataset. A stacking model architecture is made up of two or more base models (also known as level-0 models) and a meta-model (also known as a level-1 model) that combines the predictions of the base models. Improving performance is dependent on the problem complexity and if it is effectively represented by the training data and complicated enough that there is more to learn by combining predictions. It is also reliant on the selection of base models. We put this method to the test on decision tree, naive bayes, k-nearest neighbors and random forest classifiers. As shown in Figure.3, predictions made by models are fed into the meta-classifier to produce the final prediction.

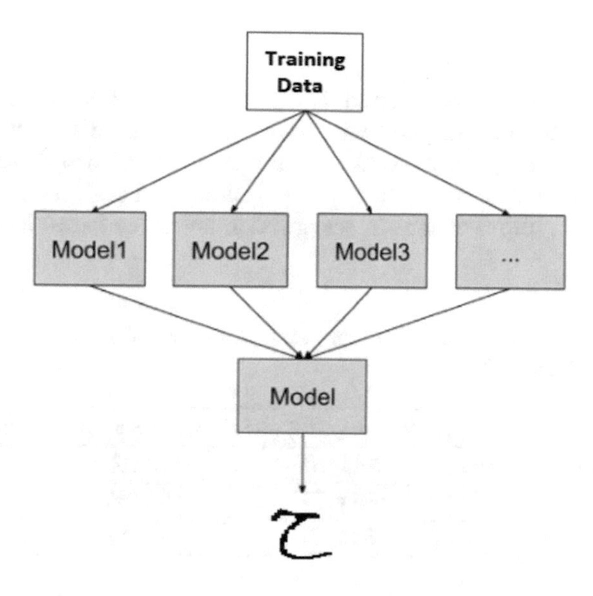

Fig. 3. Working process of a stacking classifier

4 Experiments and Results

To assess the effectiveness of the proposed arabic character recognition system, the experiments were carried out using Python 3.8 with the following hardware specifications: Windows 10, 64 bits Processor Intel (R) core (TM) i7-7600U and CPU @ 2.80 GHz 2.90 GHz, RAM (8 GB). AlexU Isolated Alphabet dataset (AIA9K) [17] was used in the experiments. There are 8,737 valid samples of the 28 arabic characters in the database. The obtained images of arabic handwritten characters were generated by 107 volunteer arabic writers from the students in the Faculty of Engineering at Alexandria University. We employed an image data augmentation technique to enhance the size of the full dataset with gaussian noise with mean = 0 and variance = 1. We tested the suggested system by doing experiments on individual classifiers and applying bagging and stacking methods to the same classifiers. It is worth noting that the random forest classifier is regarded as a bagging algorithm. In stacking algorithm, we used two base classifiers which are random forest and decision tree. We used a five-fold cross validation technique to train and test our system on four classifiers: random forest, decision tree, naive bayes and k-nearest neighbors. The dataset is divided randomly into 5 blocks, each block is held out once and the classifier is trained on the remaining. To evaluate the effect of PHOG parameters, we limited the number of bins to five and assessed recognition accuracy at three different levels. The same procedure is done with the number of bins in ten. We noticed that utilizing angle range in [0–360] was more effective than using any other angle range, so we utilize it in all experiments.

4.1 Experiment 1

The first experiment was conducted by using a bin size equal to five and three levels. We evaluated our method with two approaches: with and without data augmentation. By comparing the recognition accuracy of two strategies as shown in Table 1 and Table 2, data augmentation is efficient in the case of random forest, and we can see an improvement in recognition accuracy as we increase the level value in Table 2.

Table 1. Results when setting the number of bins at 5 for Random forest classifier without Data augmentation

Classifier	level	Bagging classifier	Stacking classifier
Random forest	1	74.05%	73.7%
	2	85.46%	85.3%
	3	84.74%	85.0%

Table 2. Results when setting the number of bins at 5 for Random forest classifier with Data augmentation

Classifier	level	Bagging classifier	Stacking classifier
Random forest	1	94.26%	94.17%
	2	96.53%	96.68%
	3	96.53%	96.69%

The data augmentation technique enhanced recognition accuracy for k-nearest neighbors, naive bayes, and decision tree classifiers, as seen in Table 3 and Table 4. In terms of recognition accuracy, Table 4 indicates that the stacking method surpasses both the individual classifiers and the bagging method.

Table 3. Results when setting the number of bins at 5 for remaining classifiers without Data augmentation

Classifier	level	individual classifier	Bagging classifier	Stacking classifier
K-nearest neighbors	1	68.43%	69.02%	67.5%
	2	79.85%	80.42%	80.0%
	3	80.91%	81.32%	79.4%
Naive bayes	1	61.67%	61.78%	45.3%
	2	70.02%	70.25%	57.9%
	3	67.89%	68.89%	56.1%
Decision tree	1	54.9%	66.8%	63.4%
	2	64.51%	77.88%	78.9%
	3	62.78%	77.59%	78.7%

Table 4. Results when setting the number of bins at 5 for remaining classifiers with Data augmentation

Classifier	level	individual classifier	Bagging classifier	Stacking classifier
K-nearest neighbors	1	75.21%	75.90%	91.28%
	2	84.73%	84.70%	94.32%
	3	85.85%	85.87%	93.87%
Naive bayes	1	61.87%	61.92%	76.82%
	2	71.44%	71.25%	86.66%
	3	70.72%	71.46%	87.45%
Decision tree	1	83.86%	90.47%	91.30%
	2	87.38%	93.84%	94.87%
	3	86.29%	93.51%	94.71%

As illustrated in Table 2 and Table 4, the random forest algorithm had the highest recognition accuracy (96.69%) in the third level using stacking algorithm, followed by stacking algorithm for decision tree classifier (94.87%) in the second level, k-nearest neighbors (94.32%) in the second level, and naive bayes (87.45%) in the second level.

4.2 Experiment 2

The second experiment was carried out with a bin size of ten and three levels. In this experiment, like in the first, we assessed our method using two approaches: with and without data augmentation. Data augmentation is effective in the case of random forest, according to a comparison of the recognition accuracy of two strategies, as shown in Table 5 and Table 6.

Table 5. Results when setting the number of bins at 10 for Random forest classifier without Data augmentation

Classifier	level	Bagging classifier	Stacking classifier
Random forest	1	76.34%	76.3%
	2	85.32%	85.6%
	3	84.83%	85.0%

Table 6. Results when setting the number of bins at 10 for Random forest classifier with Data augmentation

Classifier	level	Bagging classifier	Stacking classifier
Random forest	1	95.09%	94.94%
	2	97.06%	97.22%
	3	96.80%	96.76%

As shown in Table 7 and Table 8, the data augmentation approach improved recognition accuracy for k-nearest neighbors, naive bayes, and decision tree classifiers. In terms of recognition accuracy, Table 8 shows that the stacking method outperforms both the individual classifiers and the bagging method.

Table 7. Results when setting the number of bins at 10 for remaining classifiers without Data augmentation

Classifier	level	individual classifier	Bagging classifier	Stacking classifier
K-nearest neighbor	1	68.14%	68.96%	69.9%
	2	77.65%	77.81%	79.3%
	3	78.33%	78.9%	78.8%
Naive bayes	1	64.30%	64.28%	45.5%
	2	73.02%	73.31%	56.9%
	3	71.29%	72.39%	57.7%
Decision tree	1	51.91%	66.74%	66.6%
	2	60.58%	77.28%	79.7%
	3	60.12%	77.34%	77.9%

Table 8. Results when setting the number of bins at 10 for remaining classifiers with Data augmentation

Classifier	level	individual classifier	Bagging classifier	Stacking classifier
K-nearest neighbor	1	75.85%	76.62%	92.28%
	2	83.92%	84.02%	94.02%
	3	84.39%	84.55%	93.77%
Naive bayes	1	65.88%	66.06%	82.57%
	2	75.60%	75.80%	86.88%
	3	78.19%	78.68%	85.85%
Decision tree	1	81.29%	90.04%	92.19%
	2	83.96%	93.34%	95.22%
	3	82.80%	92.67%	94.76%

As indicated in Table 6 and Table 8, the random forest method surpasses all remaining classifiers with the highest recognition accuracy (97.22%) in the second level with stacking algorithm, followed by the same order of classifiers as in the first experiment. Stacking algorithm for decision tree classifier (95.22%) in the second level, k-nearest neighbors (94.02%) in the second level, and naive bayes (86.88%) in the second level.

4.3 Discussion of the Results and Comparative Study

Regular bagging and random forests are both bagging algorithms. In the bagging approach, each model uses a bootstrapped data set and the full set of features, and the predictions of the models are aggregated. The main difference between regular bagging method and random forest classifier is that random forest uses a subset of features to train individual trees. When compared to regular bagging, the trees are more independent of each other due to the random feature selection. We can see in two experiments that random forest outperforms the bagging algorithm on all classifiers and stacking algorithm surpasses bagging algorithm and individual classifiers when data augmentation is used. Stacking models, in contrast to bagging, are different and fit on the same dataset instead of samples from the training dataset.

The comparative results with prior methods are reported in Table 9 and shows the results of recognition rates obtained by the proposed system compared with those of other arabic handwritten character recognition systems that use the same database. The AIA9K is accessible to the general public for research purposes, making system comparisons meaningful. Our system achieved a high accuracy rate of 97.22% using random forest classifier. It is apparent from Table 9 that the performance of the proposed system is competitive and outperforms some recent research.

Table 9. Comparison between proposed approach and other approaches on the same dataset

Authors	Database	Models/Techniques	Recognition Rate
Al-Jourishi and Omari [11]	AIA9K	Discrete Cosine Transform (DCT) as the feature extraction and two stages using SVM and NN classifiers	91.84%
Torki et al. [17]	AIA9K	Different window-based descriptors and LR, ANN and SVM classifiers	94,28%
Younis and Khaled [10]	AIA9K and AHCD	Deep CNN classifier	94.8% and 97.6%
Proposed system	AIA9K	PHOG descriptor as the feature extraction and classification using individual and ensemble methods classifiers	97.22%

5 Conclusion

This research suggested an isolated arabic handwritten characters recognition system. In the proposed approach, we applied data augmentation technique to investigate the recognition accuracy in greater depth. The PHOG descriptor is used as a feature extraction and four machine learning algorithms were proposed which are random forest, decision tree, naive bayes and k-nearest neighbors.

The best classifier given high accuracy was random forest when data augmentation is used. In this paper, ensemble methods (bagging and stacking) were used, and we can notice an improvement in recognition accuracy. This work may be expanded in the future by employing different feature extraction methods and other ensemble methods.

References

1. Plamondon, R., Srihari, S.N.: Online and off-line handwriting recognition: a comprehensive survey. IEEE Trans. Pattern Anal. Mach. Intell. **22**(1), 63–84 (2000)
2. Belaïd, A., Ouwayed, N.: Segmentation of ancient Arabic documents. In: Märgner, V., El Abed, H. (eds.) Guide to OCR for Arabic Scripts, pp. 103–122. Springer, London (2012). https://doi.org/10.1007/978-1-4471-4072-6_5
3. Abandah, G.A., Younis, K.S., Khedher, M.Z.: Handwritten Arabic character recognition using multiple classifiers based on letter form. In: Proceedings of the 5th International Conference on Signal Processing, Pattern Recognition, and Applications (SPPRA), pp. 128–133 (2008)
4. Naz, S., Umar, A.I., Ahmed, R., Razzak, M.I., Rashid, S.F., Shafait, F.: Urdu Nasta'liq text recognition using implicit segmentation based on multi-dimensional long short term memory neural networks. Springerplus **5**(1), 1–16 (2016)
5. Lowe, D.G.: Distinctive image features from scale-invariant keypoints. Int. J. Comput. Vision **60**(2), 91–110 (2004)
6. Dalal, N., Triggs, B.: Histograms of oriented gradients for human detection. In: 2005 IEEE Computer Society Conference on Computer Vision and Pattern Recognition (CVPR 2005), vol. 1, pp. 886–893. IEEE (2005)

7. Bosch, A., Zisserman, A., Munoz, X.: Representing shape with a spatial pyramid kernel. In: Proceedings of the 6th ACM International Conference on Image and Video Retrieval, pp. 401–408 (2007)
8. Lee, K.C., Cho, H.: Performance of ensemble classifier for location prediction task: emphasis on Markov blanket perspective. Int. J. u- e-Serv. Sci. Technol. **3**(3), 2010 (2010)
9. Elleuch, M., Tagougui, N., Kherallah, M.: Arabic handwritten characters recognition using deep belief neural networks. In: 2015 IEEE 12th International Multi-Conference on Systems, Signals and Devices (SSD15). pp. 1–5. IEEE (2015)
10. Younis, K.S.: Arabic handwritten character recognition based on deep convolutional neural networks. Jordanian J. Comput. Inf. Technol. (JJCIT) **3**(3), 186–200 (2017)
11. Al-Jourishi, A.A., Omari, M.: Handwritten Arabic characters recognition using a hybrid two-stage classifier. Int. J. Adv. Comput. Sci. Appl. **11**(6) (2020). http://dx.doi.org/10.14569/IJACSA.2020.0110619
12. Almansari, O.A., Hashim, N.N.W.N.: Recognition of isolated handwritten arabic characters. In: 2019 7th International Conference on Mechatronics Engineering (ICOM), pp. 1–5. IEEE (2019)
13. Boufenar, C., Batouche, M., Schoenauer, M.: An artificial immune system for offline isolated handwritten Arabic character recognition. Evol. Syst. **9**(1), 25–41 (2018)
14. Shams, M., Elsonbaty, A.A., ElSawy, W.Z.: Arabic handwritten character recognition based on convolution neural networks and support vector machine. Int. J. Adv. Comput. Sci. Appl. **11**(8) (2020). http://dx.doi.org/10.14569/IJACSA.2020.0110819
15. Flusser, J., Suk, T.: Pattern recognition by affine moment invariants. Pattern Recogn. **26**(1), 167–174 (1993)
16. Rokach, L.: Pattern Classification Using Ensemble Methods, vol. 75. World Scientific, Singapore (2010)
17. Torki, M., Hussein, M.E., Elsallamy, A., Fayyaz, M., Yaser, S.: Window-based descriptors for Arabic handwritten alphabet recognition: a comparative study on a novel dataset. arXiv preprint arXiv:1411.3519 (2014)

Emotion Recognition Techniques

Maryam Knouzi(✉), Fatima Zohra Ennaji, and Imad Hafidi

Laboratory of Process Engineering, Computer Science, and Mathematics, Department of Mathematics and Informatics, National School of Applied Sciences, University Sultan Moulay Slimane, Beni-Mellal, Morocco
maryamknouzi@gmail.com

Abstract. This article evaluates techniques for obtaining facial images from convolutional neural networks to identify emotion expressions. The primary goal of the paper is to discuss the most often used approaches to analyzing and recognizing emotional expressions on the face. The works that were reviewed might be divided into two major trends: the traditional techniques and those that were especially developed using neural networks.The analysis revealed that using CNNs improved level of performance.

Keywords: Emotion recognition · CNN · Facial expressions

1 Introduction

It is simple for a person to tell or predict when someone is happy or sad, disgusted or confused, the expression of facial emotion is a task that humans perform naturally and effortlessly, but the challenge of recognizing emotions by machines and computers is extremely fascinating and difficult [1].

The development of algorithmic systems that can recognize facial expressions from human faces brings up new possibilities for the domain of human-computer interaction, such as in robotics and digital marketing [2]. The sooner we create these mechanisms for recognition, the more we can improve our knowledge of psychology, neurology, and other natural learning domains [3].

This topic is studied by the fields of image processing, computer vision, pattern recognition, and artificial intelligence, and there are strong correlations between each. Interesting approaches are documented in the literature. Algorithm performance can be enhanced in general, and in particular because of recent advancements in computer power and new high-performance computing architectures.

Many techniques have been updated for their parallel version with near real-time execution time with the introduction of graphics processing units (GPUs) [4]. With the development of new neural network types, such as convolutional neural networks, a significant achievement has been made in recent years with regard to the difficulty of recognizing emotions from images (CNN). The evolution of classical models, their use, and the emergence of convolutional neural

N. Aboutabit et al. (Eds.): ICMICSA 2022, LNNS 656, pp. 154–163, 2023.
https://doi.org/10.1007/978-3-031-29313-9_14

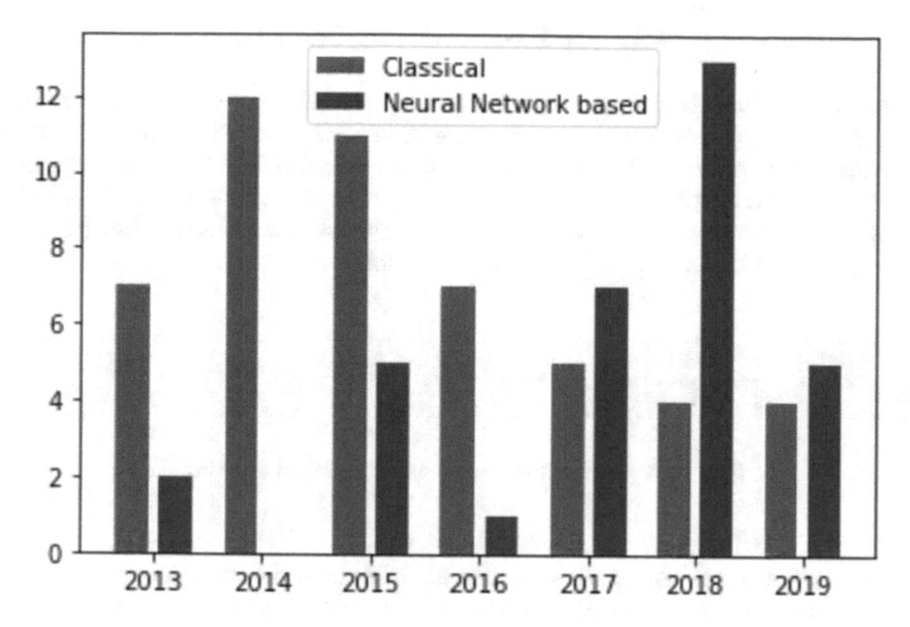

Fig. 1. Traditional image processing or computer vision methods (Classic) and neural network-based techniques were used to classify the techniques (NNB). Only methods that clearly stated the classification algorithm, dataset and results were counted. This illustration is from the article [5] of Felipe Zago et al.

networks are shown in the image below. Convolutional neural network provides robust mechanisms for extracting important features like corners and texture. Convolutional neural networks are frequently used to solve challenging issues, where an input signal is deconvolved into a set of invariant features. The classifier can interpret the image very effectively and accurately after a learning phase. CNNs are therefore increasingly used in a variety of industries, particularly those involving image processing, pattern recognition, and the classification of medical images.

Facial expression recognition is a challenging task and its reliability depends entirely on the parameters chosen, citing image-related parameters such as illumination factors, occlusion, namely obstruction on the face such as the hand, age and glasses. In order to achieve high accuracy, researchers in this field take these aspects into account when developing their models of emotion recognition. Here are a few components that are crucial for emotion recognition: Factor of illumination, the ability to recognize an expression is highly influenced by its level of intensity. Therefore, it is crucial to understand that working with CNNs requires a massive database in order to achieve very high accuracy (Fig. 1).

2 A General Flow for Recognizing Facial Emotions

Facial expression recognition systems can be categorized into two groups: the first group is based on static images, or, more specifically, the FER(Facial Emotion Recognition) only considers the location information of the feature representation point of a single image. The second group, or the so-called dynamic FER, however, also considers the temporal information with continuous images [6,7]. The phases of the static FER process are as follows (Fig. 2):

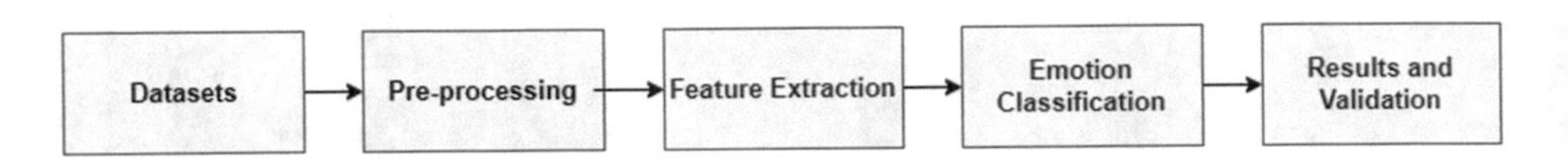

Fig. 2. A general flow for recognizing facial emotions

Datasets or Data Acquisition: The algorithms and techniques used for the FER process must be tested, trained on, and validated using a dataset with clearly defined emotion labels for facial expressions. These datasets include a set of images with various feelings and emotions. To train various models for real use, a diverse variety of datasets have been examined.

Pre-processing: This crucial step, which directly influences the behavior of the model, pre-processes the dataset by reducing noise levels, reducing the dimensionality of spatial resolution, or converting color information to grayscale. The following are the various data pre-processing steps: **(i) Face detection** is the ability to locate a face in an image or frame. **(ii) dimension reduction** is used to reduce the variables by a set of principal variables [8]. it becomes more difficult to visualize the training set and work on it if the number of features is higher. In this case, PCA (Principal Component Analysis) and LDA (Linear Discriminant Analysis) can be used to handle the above situation. **Normalization:** It is also known as feature scaling. After the dimension reduction step, the reduced features are normalized without distorting the differences in the range of feature values. There are different normalization methods, namely Z-normalization, Min-Max normalization and unit vector normalization, which improve numerical stability and speed up model learning.

Selection/Extraction of Features: This is the way features that are relevant to the FER are extracted. This leads to smaller and richer attribute sets that include characteristics like facial contours, corners, and diagonals as well as other crucial details like the separation between two eyes and the distance between the lips, which facilitates quick learning of trained data.

Emotion Classification: The features obtained in the previous steps of the FER process are used in this classification model step, that further utilizes different algorithms to categorize the emotions based on the extracted features and can generate a similarity or dissimilarity score for the facial emotions.

Results and Validation: This phase is generally required since it determines if the model is valid or not; nevertheless, because many models are based on a learning process, it is essential to complete this phase first. Then, a validation method is conducted to determine the accuracy of the adopted strategy. This final phase often involves comparing the proposed recognition model to baseline data provided by the dataset used.

3 Face Detection, Dimension Reduction and Normalization

The many face detection, dimension reduction, and normalization methods that are frequently applied in FER models are covered in this section. We also highlight the face detection methods presented by several researchers.

Pre-processing is a process that can be used to improve the performance of the FER system, and it can be performed before the feature extraction process. Image pre-processing includes different types of processes such as image brightness and scaling, contrast adjustment, and other enhancement processes to improve the expression frames [9]. The face was cropped and scaled, with the nose serving as the focal point and the other significant facial features being physically included.

Normalization is the pre-processing method which can be designed to reduce the illumination and variations of face images with median filter [10] and to get an enhanced image, the normalization method is also used for extracting the eye position, which provides more clarity to the input images.

Localization is a pre-processing method and it uses the Viola-Jones algorithm [11–13], the authors of [14–17] also suggested Viola-Jones as a face and landmark detection algorithm. While [18], in addition to the algorithm, to decrease the number of redundant features between various classes, they merged PCA and LDA. Then, [19] proposed an alternative preprocessing technique that exploited haar features split into different cascades for face detection. They also suggested a DNN (Deep Neural Network) approach to detect faces that were based on a single plane detector and ResNet.

The histogram equalization approach is applied to address fluctuations in illumination [11,13,20]. The primary purposes of this technique are to enhance image contrast and lighting as well as the ability to discern between intensities. Although FER uses a variety of pre-processing techniques, the ROI (region of interest) segmentation approach is better suitable since it accurately recognizes the facial features, which are primarily used for expression recognition. The equalization histogram, which enhances image differentiation, is a further crucial pre-processing method for FER. Table 1 displays the different techniques discussed in face detection, dimension reduction, and normalization section.

Table 1. Face detection techniques

Authors	Year	Used methods
Nievola J.C. [13]	2016	Viola-Jones, Histogram equalization
J.Jayalekshmi [14]	2017	Viola-Jones
S.K Zhou [16]	2017	Viola-Jones and Intraface
G.Ramrakhiani [17]	2018	Viola-Jones
kar et al. [18]	2019	Viola-Jones and PCA
Shah et al. [19]	2019	Haar Cascades, DNN
Priyanka, S. [20]	2019	Histogram equalization

4 Feature Extraction

The FER system then proceeds to the feature extraction method. Finding and displaying the advantageous features of an image for further processing is a process known as feature extraction. The technique of feature extraction in image processing is crucial because it signifies the transition from graphical representation to implicit data representation [21]. These data representations can then be a classification input.

The feature extraction techniques are divided into five categories: geometric feature-based, patch-based, global and local feature-based, texture-based, and edge-based [21]. These are the descriptors used to extract texture-based features: The Gabor filter is a texture descriptor that includes magnitude and phase information for feature extraction, what is presented by the authors of [22–29]. The local binary pattern (LBP), which can be applied for feature extraction, is also a texture descriptor. The thresholding between the center pixel and its neighboring pixels can be used to get LBP features, which are typically produced with the binary code [13]. This is the same concept that the authors of [14,30–32] articulated.

The transformation of the scale-invariant feature comes next (SIFT). It transforms the input image data into scale-invariant coordinates with regard to geographic features, which are subsequently stored in a database [8]. SIFT features can match a single characteristic from a database with a high probability because they are highly distinctive. Additionally, SIFT features are scale and rotation invariant, which means that they are unaffected by changes to the scale or rotation of an image. It is helpful when a rotated image is used as the input for FER, the authors of [33] show that if we rotate the image, the features are likewise kept and we may effectively extract them.

In addition to, the Histogram of oriented gradients (HOG) is feature descriptor that uses the gradient filter. The edge information of the captured facial images is the basis for the features that were extracted. It extracts distinctive features, such as the fact that a smiling face indicates curved eyes [34]. HOG was applied by the authors of [34,35] to extract characteristics from an image.

Furthermore, The descriptors that extract features using the global and local approaches are outlined as follows: for feature extraction, the Principal Component Analysis (PCA) method is used according to [18], it is used to extract low-dimensional and global characteristics. Consequently, Independent Component Analysis (ICA) is another feature extraction technique that uses multi-channel data to extract local characteristics [9]. Table 2 shows the different techniques mentioned in the feature extraction section.

Table 2. Various feature extraction techniques

Authors	Year	Used methods
Lui et al. [24]	2010	Gabor Filter
Luo et al. [30]	2013	Local Binary Pattern
Dahmane, M. [34]	2014	HOG
Kravets et al. [33]	2015	SIFT
Biswas et al. [31]	2015	LBP
Bonarini, A. [28]	2015	Gabor Filter
Hegde et al. [29]	2016	Gabor Filter
Mehta et al. [25]	2016	Gabor Filter
Konika, V. [27]	2017	Gabor Filter
Jayalekshmi et al. [14]	2017	LPB, Zernik moment
Sajjad et al. [35]	2018	HOG
Lopes et al. [22]	2018	Gabor Filter
Islam et al. [23]	2018	Gabor Filter
Kar et al. [18]	2019	PCA
Ramos et al. [26]	2020	Gabor Filter
Ravi et al. [32]	2020	LPB

5 Emotion Classification

Classification is the final stage of the FER system in which the classifier categorizes expressions such as smile, sadness, surprise, anger, fear, disgust, and neutral. This section covers the various classification methods that are applied to classify the facial expressions of human emotions.

One of the classification methods that uses two different sorts of approaches is the Support Vector Machine (SVM). Those methods are one-to-one and one-to-all. A sample is generated for each class in a one-to-all classification method, and a class is formed for each class of interest in a one-to-one classification method. For complex dimensionality situations, SVM is one of the most effective classification techniques, according to [34]. SVM is a supervised classification method that optimizes performance by using four different kernel types. They are sigmoid, radial basis function (RBF), polynomial, and linear.

The most popular architecture in both machine learning and computer vision is CNN. To improve its ability to address complicated challenges, training must be conducted on massive amounts of data. Instead of the traditional fully connected deep neural network, CNN uses convolution, min-max pooling, and fully connected as layers. A deep CNN for FER is proposed by Mollahosseini et al. [36] using a variety of readily accessible databases, the images were resized to 48×48 pixels after the facial landmarks were extracted from the dataset, the data augmentation technique was then used. They suggest the use of the network-in-network technique, which lowers over-fitting and boosts local performance by using convolution layers that are applied locally. They obtained an accuracy of 93.2% using the CK+ [41] dataset.

K. Jain [37] applied a deep convolutional neural network which is a combination of a convolutional neural network coupled with deep residual blocks. They have trained their model on two datasets CK+ [41] and JAFFE [44], and they achieved **95.24%** accuracy. Agrawal and Mittal [38] conduct a study on the impact of the variation of CNN parameters on the recognition rate using FER2013 [39] database, they achieved an accuracy of 65%. G. Wen [40] devised a method to circumvent the constraints of CNN using CK+ [41], and the accuracy of the model was 76.05%. Furthermore, S. Jadhav [42] implemented a general CNN for emotion recognition with a Haar classifier using the FER2013 dataset, the accuracy of this model was about 63%. E. Pranav [43] implemented a Deep CNN to accurately classify emotions with their own dataset, they achieved an accuracy of 78.04%. While Ravi et al. [32] used a CNN for emotion recognition using the CK+, JAFFE, RAFD, and YaleFace datasets, they had 89.62% accuracy for CK+, and 73.81% for JAFFE [44].

The flow chart in the next figure (Fig. 3) provides an overview of the emotion recognition methods covered in this article.

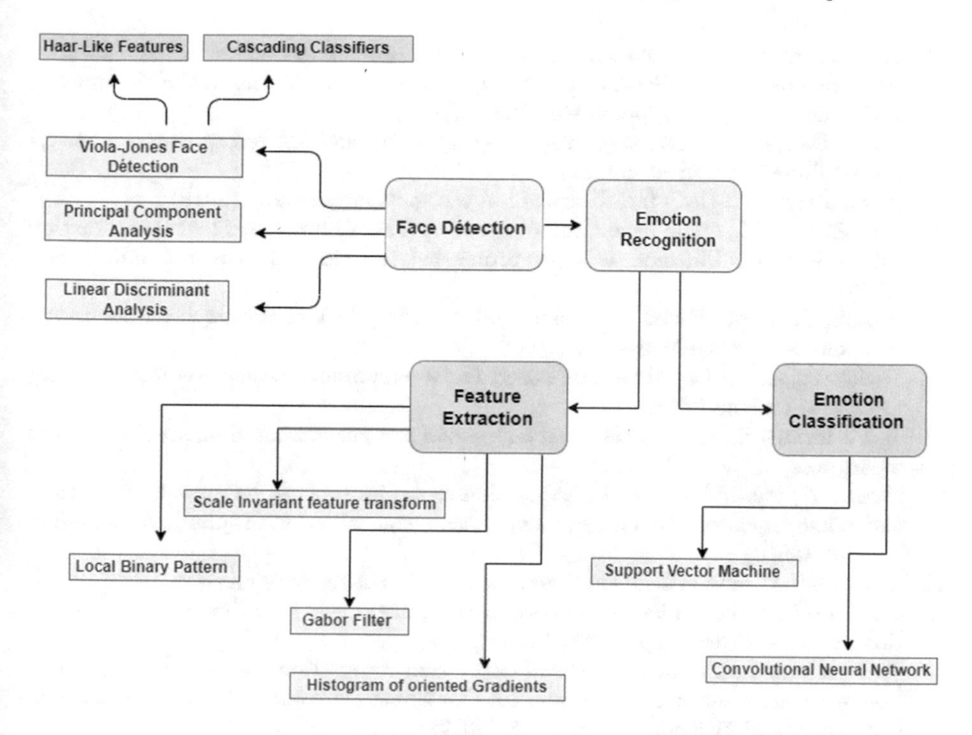

Fig. 3. A comprehensive flow chart of emotion recognition techniques

6 Conclusion

This work mentions facial expressions recognition, a challenging issue that is currently becoming more significant because of its numerous applications. The emotion or state of a person can be determined by their facial expressions. We have discussed in this paper methods for recognizing human emotions, starting with image pre-processing and feature extraction techniques, as well as strategies for classifying emotions using either classical or convolutional neural network approaches. We have observed that using CNNs enhances the performance of the model by up to **95,24%**.

References

1. Biswas, S., Sil, J.: An efficient expression recognition method using contourlet transform, In: Proceedings of the 2nd International Conference on Perception and Machine Intelligence, pp. 167–174. ACM (2015)
2. Lopes, A.T., et al.: A facial expression recognition system using convolutional networks. In: 28th SIBGRAPI Conference on Graphics, Patterns and Images, pp. 273–280. IEEE (2015)
3. Puthanidam, R.V., Moh, T.S.: A hybrid approach for facial expression recognition. In: Proceedings of the 12th International Conference on Ubiquitous Information Management and Communication, p. 60. ACM (2018)

4. Turabzadeh, S., et al.: Real-time emotional state detection from facial expression on embedded devices. In: Seventh International Conference on Innovative Computing Technology (INTECH), pp. 46–51. IEEE (2017)
5. Canal, F.Z., et al.: A survey on facial emotion recognition techniques: a state-of-the-art literature review. Inf. Sci. 593–617 (2021)
6. Li, S., Deng, W.: Deep facial expression recognition: a survey (2018)
7. Abhishree, T.M., et al.: Face recognition using Gabor filter based feature extraction with anisotropic diffusion as a pre-processing technique. Procedia Comput. Sci. (2015)
8. Rajesh, G., et al.: Facial sentiment analysis using AI techniques: state-of-the-art, taxonomies, and challenges. **25** (2020)
9. Taylor, P., et al.: Depth camera-based facial expression recognition system using multilayer scheme (2014)
10. Ji, Y., Idrissi, K.: Automatic facial expression recognition based on spatiotemporal descriptors, 1373–1380 (2012)
11. Demir, Y., et al., A new facial expression recognition based on curvelet transform and online sequential extreme learning machine initialized with spherical clustering, 131–142 (2014)
12. Salmam, F.Z., et al.: Facial expression recognition using decision trees. IEEE (2016)
13. Cossetin, M.J., et al.: Facial expression recognition using a pairwise feature selection and classification approach. IEEE (2016)
14. Jayalekshmi, J., Mathew, T.: Facial expression recognition and emotion classification system for sentiment analysis. In: Conference on Networks and Advances in Computational Technologies, pp. 1–8 (2017)
15. Li, H., et al.: A convolutional neural network cascade for face detection. In: IEEE Conference on Computer Vision and Pattern Recognition (2015)
16. Ding, H., et al.: FaceNet2ExpNet: regularizing a deep face recognition net for expression recognition, pp. 118–126 (2017)
17. Chaudhari, M.N., et al.: Face detection using viola jones algorithm and neural networks. In: Conference on Computing Communication Control and Automation, ICCUBEA, pp. 1–6 (2018)
18. Kar, N.B., et al.: Face expression recognition system based on ripplet transform type II and least square SVM. **78** (2019)
19. Shah, H.M., et al.: Analysis of facial landmark features to determine the best subset for finding face orientation. In: ICCIDS, pp. 1–4 (2019)
20. Dhara Mungra et al., PRATIT: a CNN-based emotion recognition system using histogram equalization and data augmentation. Multimed. Tools Appl. **79**, 2285–2307 (2019). Springer
21. Revina, I.M., Sam Emmanuel, W.R.: A survey on human face expression recognition techniques. J. King Saud Univ. - Comput. Inf. Sci. **10** (2018)
22. Lopes, N., et al.: Facial emotion recognition in the elderly using a SVM classifier. In: Proceedings of 2nd International Conference on Technology and Innovation in Sports, Health and Wellbeing, pp. 1–5 (2018)
23. Islam, B., et al.: Facial expression region segmentation based approach to emotion recognition using 2D Gabor filter and multiclass support vector machine, pp. 1–6 (2018)
24. Liu, S.-S., Tian, Y.-T.: Facial expression recognition method based on Gabor wavelet features and fractional power polynomial kernel PCA, pp. 144–151 (2010)
25. Mehta, N., Jadhav, S.: Facial emotion recognition using log Gabor filter and PCA, pp. 1–5 (2016)

26. Ramos, A.L.A., et al.: Classifying emotion based on facial expression analysis using Gabor filter: a basis for adaptive effective teaching strategy, pp. 469–479 (2020)
27. Verma, K., Khuntetatitle, A.: Facial expression recognition using Gabor filter and multi-layer artificial neural network, pp. 1–5 (2017)
28. Matamoros, H., et al.: A facial expression recognition with automatic segmentation of face regions, pp. 529–540 (2015)
29. Hegde, G.P., et al.: Kernel locality preserving symmetrical weighted fisher discriminant analysis based subspace approach for expression recognition (2016)
30. Luo, Y., et al.: Facial expression recognition based on fusion feature of PCA and LBP with SVM, pp. 2767–2770 (2013)
31. Biswas, S., Sil, J.: Facial expression recognition using modified local binary pattern, pp. 595–604 (2015)
32. Ravi, R., et al.: A face expression recognition using CNN & LBP. In: Proceedings of the Fourth International Conference on Computing Methodologies and Communication, pp. 684–689 (2020)
33. Chickerur, S., et al.: Parallel scale invariant feature transform based approach for facial expression recognition, creativity in intelligent technologies and data science, pp. 621–636 (2015)
34. Dahmane, M., Meunier, J.: Prototype-based modeling for facial expression analysis. IEEE (2014)
35. Sajjad, M., et al.: Facial appearance and texture feature-based robust facial expression recognition framework for sentiment knowledge discovery (2015)
36. Mollahosseini, A., et al.: Going deeper in facial expression recognition using deep neural networks. In: Winter Conference on Applications of Computer Vision (2016)
37. Jain, D.K., et al.: Extended deep neural network for facial emotion recognition (2019)
38. Agrawal, A., Mittal, N.: Using CNN for facial expression recognition: a study of the effects of kernel size and number of filters on accuracy (2019)
39. Goodfellow, I.J., et al.: Challenges in representation learning: a report on three machine learning contests, pp. 59–63 (2015)
40. Wen, G., et al.: Ensemble of deep neural networks with probability-based fusion for facial expression recognition, pp. 597–610 (2017)
41. Lucey, P., et al.: The extended Cohn-Kanade dataset (CK+): a complete dataset for action unit and emotion-specified expression, pp. 94–101. IEEE (2010)
42. Jadhav, R.S., Ghadekar, P.: Content based facial emotion recognition model using machine learning algorithm. In: ICACAT, pp. 1–5 (2018)
43. Pranav, E., et al.: Facial emotion recognition using deep convolutional neural network, pp. 317–320 (2020)
44. Zheng, W., et al.: Facial expression recognition using kernel canonical correlation analysis (KCCA), pp. 233–238 (2006)

A Deep Convolutional Neural Networks Approach for Word-Level Handwritten Script Identification Using a Large Dataset

Siham El Bahy(✉), Noureddine Aboutabit, and Hind Ait Mait

IPIM Laboratory, ENSA Khouribga, Sultan Moulay Slimane University,
PO Box 523, 23000 Beni Mellal, Morocco
siham.elbahy@gmail.com, n.aboutabit@usms.ma

Abstract. In this work, we propose a convolutional neural network (CNN) architecture to identify six word-level handwritten scripts involving Arabic, Latin, Chinese, Bangla, Devanagari and Telugu. A large dataset of 14k word images per script was constructed based on several public handwritten datasets. Then, three architectures are proposed and compared based on standard metrics performance and time execution. Experiments conducted on both test and validation classification show high performances that outperform the state-of-art techniques. Indeed, the best result was provided by CNN model with three-convolutional-polling pairs layers that achieved an average script identification accuracy of 97.67% and ran in a sufficiently fast time of 2 ms per frame during the test phase.

Keywords: Handwritten script identification · Deep learning · Convolutional Neural Networks · Word-level handwritten dataset

1 Introduction

Optical Character Recognition (OCR) is the translation of printed or handwritten documents into electronic encoded text. This conversion generally depends on the script in which the document was written. Consequently, identifying the script is an essential step towards OCR process. It becomes more important when a single document includes multiple scripts. Studies dealing with the identification of scripts in documents did indeed begin early with the invention of computers in the 1960s. Early works were focused essentially on printed documents and many systems were developed [19]. Further works have been carried out on the handwritten documents that have received increased attention in recent decades. Indeed, unlike printed texts, handwriting presented, and is still, several challenges due to its large variability between the written styles of scripts. So, for the same language, people write different and a single writer can provide different shapes for the same character at different times. The complexity

N. Aboutabit et al. (Eds.): ICMICSA 2022, LNNS 656, pp. 164–174, 2023.
https://doi.org/10.1007/978-3-031-29313-9_15

of processing handwritten documents also arises from other issues that can be addressed in pre-processing. Such as issues are ruling line, skewness, noise, low contrast that caused character fragmentation ...etc.

To identify handwritten scripts from document image, many propositions were reported in the literature where each depends on several specifications. In fact, regarding the real time aspect, such identification might be online or offline. Input data were captured point by point in the case of online, whereas in offline it was represented in the image form. Also, each proposed work performed identification script on a selected number of scripts/languages. Furthermore, scripts identification systems can perform at many different levels depending on the expected application: page, block, line, word or character. Several published works deal with handwritten script identification using these text levels even if it appears that the word level is the most used [20]. Normally, the visual appearance of a script is more distinguishable at the word level than at the character level.

Otherwise, the deep learning based concept was mostly used in various fields of pattern recognition for feature extraction or/and classification. While it has been used several times in recent years for OCR [1,17,18,21], there are a few works based on this concept for script identification [1]. It is necessary to note that the deep learning techniques outperform the classical machine learning ones when it is applied to a big size data.

In this context, developing an accurate and efficient method to identify scripts becomes a need in order to propose a full system for digitizing the large amount of handwritten documents in a multilingual country. In Morocco, the most handwritten documents are written in Arabic and/or French languages where each has a different script style (Arabic or Latin). Accordingly, in this paper, we propose a word-level approach based on Convolutional Neural Networks (CNN) to identify the script of handwritten texts of Arabic, Latin, Chinese, and three indic scripts: Bangla, Devanagari and Telugu. In addition to Arabic and Latin, we added Chinese and 3 other Indic scripts to evaluate the robustness of our technique. We have been careful to build a database with a high number of samples for each script. We analyse the proposed CNN architecture and compare three variants through the obtained accuracy and time execution.

The rest of the paper is organized as follows. In Sect. 2 the state-of-art of handwritten scripts identification is discussed. The proposed approach is described in Sect. 3. In Sect. 4 the experiments set-up and the obtained results are presented and discussed.

2 Related Works

Script identification can be performed according to different specifications. It can be online or offline, applied to printed or handwritten documents, and using various text levels. In this section we will focus on works carried out on offline identification of handwritten scripts including Arabic and Latin/Roman. Based on this, we reported various proposed systems to identify handwritten scripts using page, block, line and word levels while character level does not attract much attention until now.

In most cases of these systems each input data was processed to extract a feature set. Then, a classification step was performed. The processing of input data may vary depending on the text level. If in the page level, the full image document is processed, for the other levels an operation of segmentation is done first to extract the expected entities: block, line or word.

[3] proposed a document level system to identify six handwritten scripts included Roman, Arabic, Cyrillic, Chinese, Japanese and Devanagari. In this system, a set of feature were extracted based on a summary statistics taken across the document's black connected components. Then, a linear discriminant analysis were applied to scripts classification. An average correct identification rate of 88% was obtained in this study.

[2] proposed a hybrid method for Arabic and Latin script identification. In this method, morphological analysis of the text block level and geometrical analysis of the line and the connected component level were combined and experimented on two datasets. Using a K-Nearest Neighbor (KNN) classifier, the authors obtained an identification rate of 88% and 98% for handwritten Arabic and Latin respectively.

To discriminate the same scripts, [4] developed a scheme based on fractal analysis to extract 12 features. Tested on various styles and sizes of the analyzed scripts, this scheme provided 96.64% in average as correct script identification rate when using a KNN classifier and 98.72% with Radial Basic Function (RBF) based classifier.

[5] designed a system based on steerable pyramid transform for Arabic and Latin script identification considering word level. They proposed to extract features from pyramid sub bands and used KNN for classification. The proposed approach was evaluated on a dataset of 400 word images and the authors provided an identification accuracy of 97% for handwritten Arabic and 96% for Latin.

Recently, [6] used a Hidden Markov Model (HMM) in a identification system to recognize the writing type (handwritten or machine-printed) and the script nature (Arabic or Latin) of an input image in both word and line levels. They considered features based on Histogram of oriented Gradient (HoG) extracted in two direction from right to left and from left to right. Experiments made by the authors had been shown a high word identification accuracy for handwritten Arabic of 98,35% and a low rate for Latin of 78%. In case of line level identification, the reported accuracies were 99,8% and 89% for Arabic and Latin respectively.

Besides, only very limited contributions based on deep learning have been made for Arabic and Latin script identification. In a study that can be related to our case, [1] proposed a CNN modeling to automatically extract relevant features to classify 11 Indian handwritten scripts. Two types of CNN input images were considered with different sizes: raw images and wavelet transformed images. Also, the considered CNNs had 3 or 2 pairs of convolution-max polling layers to extract the features. For all these CNNs models the classification was performed by a neural network with 3 fully connected layers. The authors reported a best average identification accuracy of 94.73%.

3 The Proposed Approach

Our approach is to design a CNN that could both extract relevant features and classify them to identify scripts.

3.1 Dataset

To evaluate the performance of our deep learning identification system in the best condition, a large amount of dataset is necessary. Thus, in the lack of a standard handwritten word dataset for all the involved scripts, we have collected our corpus from different public word datasets. For the Arabic script, the complex KHATT [7] database was used in association with the IFN/ENIT database [8]. The first is a database of line-level handwritten Arabic text written by 623 different writers. Further, the space to scale technique proposed by [22] was applied to segment lines of text into words. Ditto, for Chinese script, we didn't find a word-level dataset. So, we have segmented two handwritten line-level datasets HWDB2.0-2.2 [10] and HIT-MW text-line level [11] based on a vertical projection process. In this latter, we count the horizontal number of pixels for each part which is considered a word if this number is greater than 80 pixels. Otherwise, it is merged with the next part. For the other analysed scripts, we have considered existing public word-level datasets as shown in Table 1.

Finally, the collected dataset has 84000 images for the six concerned scripts; So, 14000 word images for each one. It was divided on three subsets: a training subset with 72900 images, a validation subset with 8100 images and a test subset with 3000 images.

Table 1. Sources of our word-level datasets

Script	Sources	Size (number of words)
Arabic	KHATT [7] + IFN/ENIT [8]	14000
Latin	IAM [13]	14000
Chinese	HWDB2.0-2.2 [10] + HIT-MW text-line level [11]	14000
Bangla	CMATERdb 2.1.2 [9]	14000
Devanagari	IIIT-HW-Dev v1 [12]	14000
Telugu	IIIT-HW-Telugu v1 [14]	14000

3.2 Preprocessing

The images in our database were pre-processed before being input into the CNN network. As these images were collected from different databases, they were in gray-level format and had different dimensions. So, we first pruned the extra white pixels that are present in the word images to keep only the region of interest containing the writing. This operation was applied in order to reduce execution time. Secondly, we resized the whole images into a fixed dimension of 128 × 64 pixels. Finally, the images were binarized using global thresholding approach. Figures 1 and 2 illustrate examples of images used in our experiment before and after the pre-processing step.

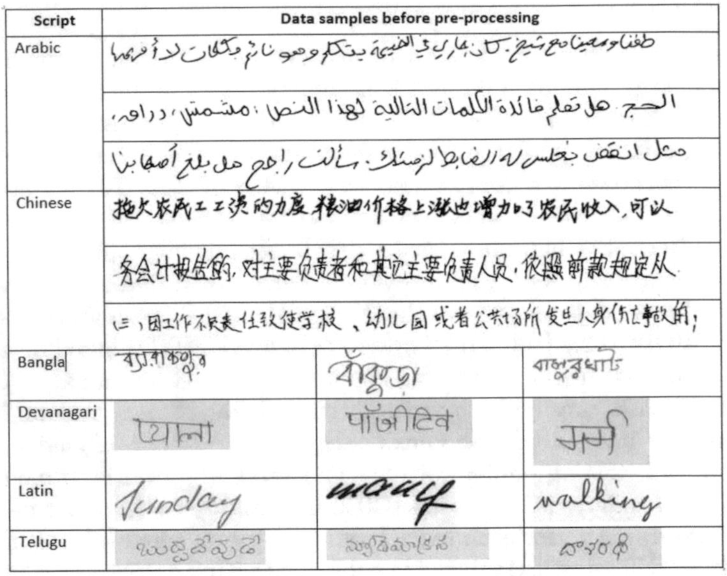

Script	Data samples before pre-processing		
Arabic	[illegible]		
	الحج. هل تعلم فائدة الكلمات التالية لهذا النص: مشمش، دراهم،		
	[illegible]		
Chinese	扩大农民工工资的力度，粮油价格上涨也增加了农民收入，可以		
	务会计报告的，对主要负责者和其它主要负责人员，依照前款规定从		
	（三）因工作不负责任致使学校、幼儿园或者公共场所发生人身伤亡事故的；		
Bangla	[illegible]	[illegible]	[illegible]
Devanagari	[illegible]	[illegible]	[illegible]
Latin	Sunday	[illegible]	walking
Telugu	[illegible]	[illegible]	[illegible]

Fig. 1. Examples of collected data

3.3 Description of CNN Architecture

CNN was developed in the 1980s and 1990s [15]. However, in this period, such technique has not been used for real-world applications because of slow CPUs and no existing GPU [16]. In the last decade, CNN were revived due to the improvement of the computing power with faster CPUs and the availability of GPU as multipurpose computing tool. In addition, public databases have become increasingly available and accessible [18].

For image recognition task, using the original images directly provides low performances. Instead that, feature maps are provided using a feature extractor. In our approach, we used a sequential CNN architecture to automatically extract and select relevant feature. Precisely, we are setting up a first architecture and we are trying to modify some of their control parameters to improve its accuracy and execution time. The basic architecture consists of a two 2D-convolution layers followed each one by a max-pooling layer. The convolutional layer converts the input by a convolution based filtering and the output passes through an activation function. The max-pooling layer reduce the dimension of the input based on replacing a block of neighbouring pixels by the maximum value one. The output from the feature extractor are transformed into a one-dimensional

Script	Data samples after pre-processing				
Arabic	[illegible]	[illegible]	[illegible]	[illegible]	[illegible]
Bangla	[illegible]	[illegible]	[illegible]	[illegible]	[illegible]
Chinese	[illegible]	[illegible]	[illegible]	[illegible]	[illegible]
Devanagari	[illegible]	[illegible]	[illegible]	[illegible]	[illegible]
Latin	subject	paying	remarkable	because	smiled
Telugu	[illegible]	[illegible]	[illegible]	[illegible]	[illegible]

Fig. 2. Examples of data after the pre-processing step

vector and then enter into the classifier network that identify the label of the expected script. Our classifier is a fully connected neural network consisting of two layer. The Fig. 3 illustrates an overview of our first architecture.

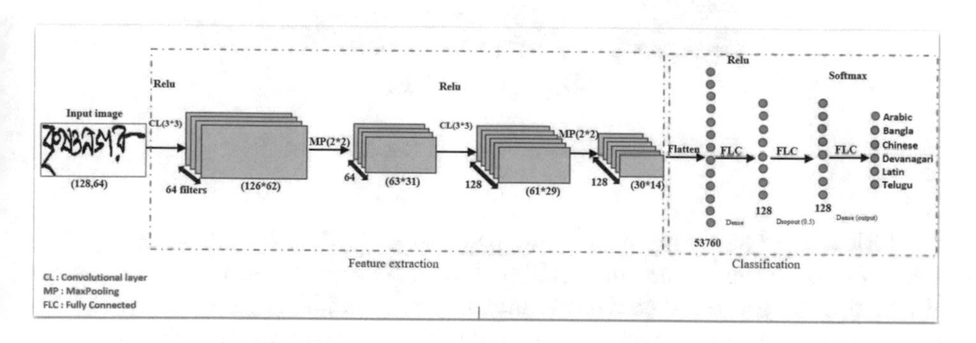

Fig. 3. Overview of the first CNN architecture

In this first architecture, input convolution layer processes the input images, which are considered to be two-dimensional arrays of dimensions (64,128,1). The number of nodes in this layer is 64 with convolution filters of 3×3 size. Its output of 64 feature maps with (126×62) size is the input of a max pooling layer where the image stack is reduced to a smaller size of 63×31. The third layer is another convolution layer, with 128 filters of size 3×3. Its output is of 128 feature maps with (61×29) dimension feeds a fourth max Pooling layer which outputs 128 images of (30×14) size.

The Flatten layer is required after the convolutional layers so that we can use fully connected layers. The latter have no local limitation like convolutional layers (which only observe a local part of an image using convolutional filters).

The Flatten layer outputs 53760 ($30 \times 14 \times 128$) nodes which are fully connected to 128 nodes through a Dense layer used with a Relu activation function. In order to regularize this dense layer, we use a Dropout (0.5) that sets to zero the output of each neuron which is lower than 0.5. This would drop such neuron type to participate in the back-propagation and thus avoid over-learning. A second dense layer is then used to reduce the number of output nodes to 6, one for each possible result (Arabic, Bangla, Chinese, Devanagari, Latin, Telugu). We use for this last layer a Softmax activation to allow the prediction of the script class. Figure 4 describes the activation maps for our CNN. The above CNN model takes $128 * 64$ size images as input. So, in a second version, we maintain the same structure and modify some settings parameters. Indeed, we scale the input images to a size $128 * 32$ in order to reduce the execution time and we modify the filter size used in the first layer to $5 * 5$ and in the second layer to $3 * 3$.

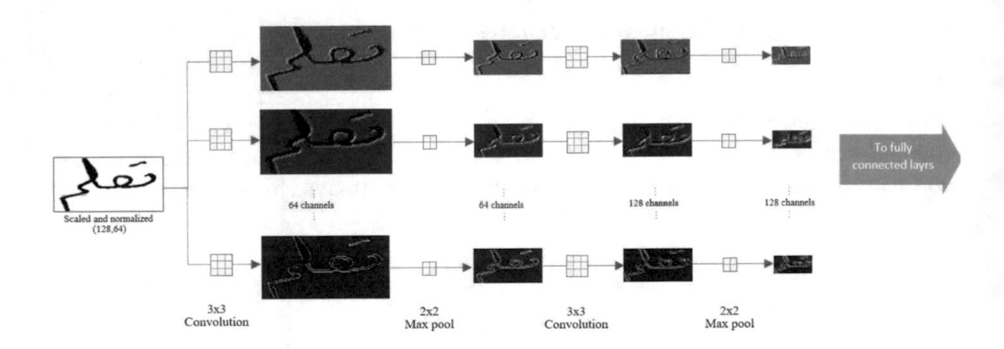

Fig. 4. Activation maps of the first model

With a $128 * 32$ as input size, we propose a third CNN model using three pairs of convolutional and max polling layers. The used convolution filters are of 3*3 size for the three convolutionnal layers. In addition, a Dropout layer is added to convolutional part of the model before attacking the classifier. The Fig. 5 shows the structure of this model.

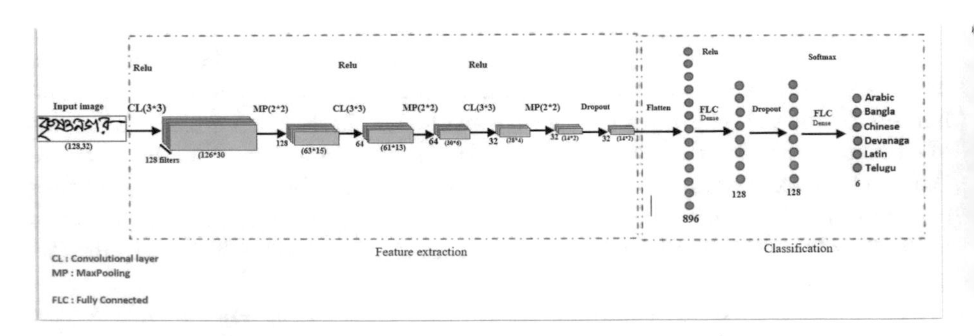

Fig. 5. Overview of the third CNN architecture

4 Experiments and Results

The above CNN models were implemented and carried out on the collected word-level dataset. As mentioned earlier, this dataset were split on three subsets for training, validation and test. To fit the model, we formed batches of 20 images from the training subset for model 1 and 512 images for models 2 and 3. We had chosen an Adam optimization with the parameters: $learningrate = 0.001$, $\beta 1 = 0.9$, $\beta 2 = 0.999$ and $\epsilon = 1e^{-7}$. We had also chosen the most common loss function "categorical crossentropy" for our multi-classes prediction. In addition, we used the "accuracy" metric on the validation set when we formed the model. The training phase was conducted over 20 iterations.

The proposed approach is implemented using Python language on i7-6500U CPU 2.50 GHz 2.59 GHz with 12 GB RAM. To evaluate and compare the three models, we consider the following conventional evaluation metrics: accuracy, precision, recall, f-score and runtime per frame. The values of all these metrics were computed as averages over the 6 classes to predict.

After constructing and validating our CNN models, we obtain the results summarized in Table 2. This latter shows a high rates obtained for all models that all metric values achieve more than 96% on the different subsets training, validation and test. We can note from these results that the reduction in input image resolution from $128 * 64$ to $128 * 32$ does not affect the accuracy of models 2 and 3. It seems, however, that the third model, with more convolutional layers, provides better rates than the others. It performs the highest accuracy of 97,67% when applied on test subset. This would mean that 2,33% of the test words are not correctly classified. In this case, the confusion matrix in Fig. 6 indicates the different misclassifications among the six analyzed scripts. We could note that all scripts were well classified with approximatively the same rate. Thus, as regards our case of interest, with the few errors made, Arabic and Latin are still slightly confused.

In addition to the preceding evaluation metrics, we noted the execution time during the validation and test phases (Table 3). Evidently, the runtime depends on the dimension of the input image. That is why the first CNN model using

Table 2. Results provided by our three CNN models based on the used evaluation metrics

Metric		CNN model 1	CNN model 2	CNN model 3
Accuracy	training	99.99	100	97.17
	validation	97.31	97.3	98
	test	96.43	97	***97.67***
Precision	training	100	100	99.16
	validation	97.5	97.66	98.16
	test	96.66	97.16	***97.83***
Recall	training	100	100	99.33
	validation	97.33	97.33	97.83
	test	96.33	96.83	***97.5***
F-score	training	100	100	99.17
	validation	97.33	97.16	98.16
	test	96.33	97.16	***97.66***

128 * 64 input image ran in a high time than the 2 others models. Nevertheless, the average execution time for all models is still fast.

Table 3. Average execution time per frame taken by each CNN model on milliseconds (ms)

		CNN model 1	CNN model 2	CNN model 3
runtime per frame	validation	9 ms	1 ms	2 ms
	test	6 ms	0.932 ms	2 ms

With respect to the state-of-art, it is delicate to compare our results to those found in the literature works due to many reasons. In fact, the number and types of scripts to classify are not often the same. Furthermore, our approach is performed on a large collected database different in amount and sources from the previous studies. However, we could state that our results are more accurate when contrasted to those reported by [1] who used deep learning approach for 11 handwritten scripts identification at the word-level. [1] provided in the best case a maximum script identification accuracy of 94,73%. Further, Our identification accuracy is significantly higher than the average accuracy of 88% obtained by [6] for the identification of handwritten Arabic and Latin scripts (98% for Arabic and 78% for Latin).

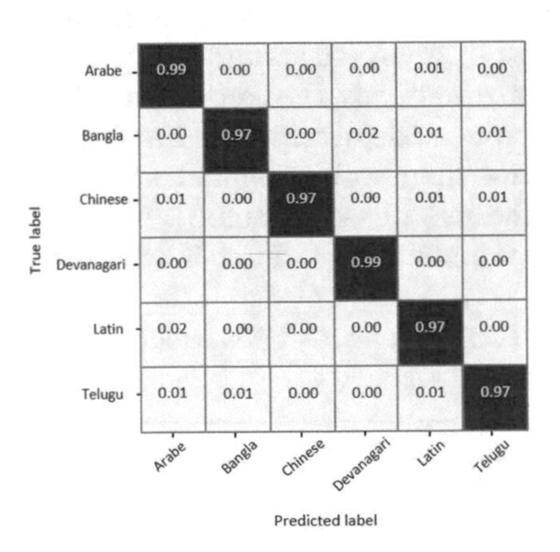

Fig. 6. Confusion matrix of the third CNN model when applied on test subset

5 Conclusion

In this paper, we propose an automatic identification system of handwritten scripts using CNN approach. The proposed identification is made with considering word level text for six scripts: Arabic, Latin, Chinese, Bangla, Devanagari and Telugu. It consists mainly of a feature extractor based on a succession of convolutional max-polling layers and fully connected layers as classifier. Three CNN architectures are investigated and evaluated using standard usual metrics. Experiments are carried out on a dataset with high size and more complexity in which we have collected over than 84000 word images, where 14000 words are used for each script. We consider that our dataset is the biggest one that is collected for word level. Hence, we have achieved a highest average accuracy of 97,67%, which supersedes the results achieved by other state-of-art studies.

This study is a preliminary step to develop a handwriting recognition system for documents with mixed Arabic and Latin scripts. The significant results achieved in this study lead the way to develop such a handwriting recognition system based on the proposed optimal CNN architecture.

References

1. Ukil, S., Ghosh, S., Obaidullah, S.M., Santosh, K.C., Roy, K., Das, N.: Deep learning for word-level handwritten indic script identification. In: Santosh, K.C., Gawali, B. (eds.) RTIP2R 2020. CCIS, vol. 1380, pp. 499–510. Springer, Singapore (2021). https://doi.org/10.1007/978-981-16-0507-9_42
2. Kanoun, S., et al.: Script and nature differentiation for Arabic and Latin text images. In: Proceedings Eighth International Workshop on Frontiers in Handwriting Recognition. IEEE (2002)
3. Hochberg, J., et al.: Script and language identification for handwritten document images. Int. J. Doc. Anal. Recogn. **2**(2) (1999)
4. Moussa, S.B., et al.: Fractal-based system for Arabic/Latin, printed/handwritten script identification. In: 2008 19th International Conference on Pattern Recognition. IEEE (2008)
5. Benjelil, M., et al.: Arabic and Latin script identification in printed and handwritten types based on steerable pyramid features. In: 2009 10th International Conference on Document Analysis and Recognition. IEEE (2009)
6. Cheikh Rouhou, A., Abdelhedi, Z., Kessentini, Y.: A HMM-based Arabic/Latin handwritten/printed identification system. In: Abraham, A., Haqiq, A., Alimi, A.M., Mezzour, G., Rokbani, N., Muda, A.K. (eds.) HIS 2016. AISC, vol. 552, pp. 298–307. Springer, Cham (2017). https://doi.org/10.1007/978-3-319-52941-7_30
7. Mahmoud, S.A., et al.: Online-khatt: an open-vocabulary database for Arabic online-text processing. Open Cybern. Syst. J. **12**(1) (2018)
8. Pechwitz, M., et al.: IFN/ENIT-database of handwritten Arabic words. In: Proceedings of CIFED, vol. 2. Citeseer (2002)
9. Sarkar, R., et al.: CMATERdb1: a database of unconstrained handwritten Bangla and Bangla-English mixed script document image. Int. J. Doc. Anal. Recogn. (IJDAR) **15**(1) (2012)
10. Liu, C.-L., et al.: Online and offline handwritten Chinese character recognition: benchmarking on new databases. Pattern Recogn. **46**(1) (2013)
11. Su, T., Zhang, T., Guan, D.: HIT-MW dataset for offline Chinese handwritten text recognition. In: Tenth International Workshop on Frontiers in Handwriting Recognition, Suvisoft (2006)
12. Dutta, K., et al.: Offline handwriting recognition on Devanagari using a new benchmark dataset. In: 2018 13th IAPR International Workshop on Document Analysis Systems (DAS). IEEE (2018)

13. Liwicki, M., Bunke, H.: IAM-OnDB-an on-line English sentence database acquired from handwritten text on a whiteboard. In: Eighth International Conference on Document Analysis and Recognition (ICDAR 2005). IEEE (2005)
14. Dutta, K., et al.: Towards spotting and recognition of handwritten words in Indic scripts. In: 2018 16th International Conference on Frontiers in Handwriting Recognition (ICFHR). IEEE (2018)
15. LeCun, Y., et al.: Gradient-based learning applied to document recognition. Proc. IEEE **86**(11) (1998)
16. Culurciello, E.: Neural Network Architectures. Synthesis Lectures on Artificial Intelligence and Machine Learning, San Francisco (2017)
17. Cireşan, D.C., et al.: Deep, big, simple neural nets for handwritten digit recognition. Neural Comput. **22**(12) (2010)
18. Ciregan, D., Meier, U., Schmidhuber, J.: Multi-column deep neural networks for image classification. IN: 2012 IEEE Conference on Computer Vision and Pattern Recognition. IEEE (2012)
19. Mantas, J.: An overview of character recognition methodologies. Pattern Recogn. **19**(6) (1986)
20. Obaidullah, S.Md., et al.: Handwritten Indic script identification in multi-script document images: a survey. Int. J. Pattern Recogn. Artif. Intell. **32**(10) (2018)
21. Jaderberg, M., et al.: Deep structured output learning for unconstrained text recognition. arXiv preprint arXiv:1412.5903 (2014)
22. Manmatha, R., Srimal, N.: Scale space technique for word segmentation in handwritten documents. In: Nielsen, M., Johansen, P., Olsen, O.F., Weickert, J. (eds.) Scale-Space 1999. LNCS, vol. 1682, pp. 22–33. Springer, Heidelberg (1999). https://doi.org/10.1007/3-540-48236-9_3

Arabic Sign Language Analysis and Recognition

Ihssane Bouhanou(✉) and Noureddine Aboutabit

Department of Computer System and Technology, ENSA, 25000 Khouribga, Morocco
ihssan.bouhanou@usms.ac.ma

Abstract. For those who are deaf or hard of hearing, sign language continues to be the preferred form of communication. As technology has advanced, systems that can automatically distinguish between spoken language and vision-based sign language have been created. This paper examines and identifies Arabic alphabet sign language (ArSLR). The Convolutional Neural Network (CNN) model is used by the system to visually recognize motions from the input sequence of hand photographs. We employed two datasets: the isolated words dataset for dynamic gestures given across multiple frames and the alphabet dataset for static gestures presented throughout a single frame. The suggested systems combine model training, feature extraction, and prediction. To boost performance, we concentrated on hyperparameter validation. The system's accuracy has exceeded 98% after testing and comparing various metrics, which is comparative to other works utilizing the similar dataset.

Keywords: Arabic sign language recognition (ArSLR) · continuous sign recognition · image-based · isolated word · recognition · CNN

1 Introduction

Sign language (SL) is the primary form of communication between deaf and hard-hearing people that represent More than 360 million individuals worldwide, or 5% of the total population, and primarily live in low to middle-income countries [1]. In contrast to spoken languages, sign language is a visual language in which signers communicate using hand gestures together with facial emotions and body postures.

There are special context and grammar rules that sustain the expression of sign languages [2]. In general, there are 3 main categories in sign language recognition: alphabet and numbers, isolated sign recognition (a single sign gesture), and continuous sign recognition (a sign gesture sentence). They consist of static and dynamic gestures: Static gestures, which do not involve movement, are employed for most of ArSL alphabets and numbers (Out-Of-Vocabulary words, such as proper names, are usually finger spelled [3]), dynamic gestures, on the other hand, entail hand motion while making signs [4] and are used for isolated and continuous signs. While dynamic movements require multiple frames to depict gesture motion, static gestures can be represented by a single frame or image.

N. Aboutabit et al. (Eds.): ICMICSA 2022, LNNS 656, pp. 175–185, 2023.
https://doi.org/10.1007/978-3-031-29313-9_16

Sign language varies from one country to another and sometimes within the same country. In Morocco, several sign languages are used such as Arabic, French, and Spanish. To unify them, an effort has been made by the Arab League Educational, Cultural and Scientific Organization (ALECSO) in 1999 to standardize the Arabic Sign Language (ArSL) which resulted in a dictionary consisting of about 3200 words published in two parts [1].

There are few hearing people who acknowledge sign language. Therefore, it is necessary to clear the way communication between hearing and deaf communities through a sign language recognition system, technologies such as gesture recognition are getting a lot of attention lately, and several attempts have been made to develop recognition systems for ArSL.

Our motivation for working on sign language gesture recognition lies in creating a system capable of classifying and predicting the Arabic sign language gesture represented by a human being. Likewise, we used two databases to create an ArSL recognition (ArSLR) system. We suggest two CNN models based on sequences of body postures and static hand forms, deep learning image classification models by transforming the video of the sign into single images.

The following is how the paper is set up: the literature on current sign language systems is briefly presented in Sect. 2, and an overview of the databases used in the experiments is provided in Sect. 3. The proposed recognition systems are provided in Sect. 4, and Sect. 5 discusses the outcomes of the experimental evaluation. Finally, Sect. 6 presents the conclusions.

2 Related Work

Sign language includes not only static hand gestures, finger spellings, and hand movements, but also facial expressions and body language. Each sign language has its structure, grammar, syntax, semantics, pragmatics, morphology, and phonology. As such, techniques should be developed to decode each sign language independently, including ArSL, as is presently in progress.

2.1 Alphabet Recognition

In this scenario, the signer performs each letter separately. The letters are represented by a static posture and the size of the vocabulary is limited. The Arabic alphabet consists of 28 letters.

The suggested system in [5] visually recognizes motions from input of hand images. The Hidden Markov Model (HMM) is used by the Arabic alphabet sign language recognition algorithm to classify the shapes of the letters based on hand geometry and the variations in the shape of a hand in each sign. The experiment's findings demonstrate that expanding the number of locations by segmenting the rectangle enclosing the hand is necessary to increase the rate of gesture identification.

In [6] they developed a sign language letter recognition system that starts by locating the images' histograms. The KNN classifier receives the profiles that were retrieved from these histograms as input. The system achieved an accuracy of 50% for the naked hand, 75% for the hand wearing a red glove, 65% for the hand wearing a black glove, and 80% for the hand wearing a white glove.

Hassanien and Hemayed [7] discussed an Arabic sign language alphabet recognition system that transforms signs into voices. The approach is much closer to a real configuration, yet, the recognition is not executed in real-time. The system concentrates on static and simple mobile gestures. The inputs are color images of the gestures. YCbCr space is used to extract the skin blobs and Prewitt edge detector is utilized to extract the shape of the hand. To convert the image area into feature vectors, main component analysis (PCA) is used with KNN algorithm in the classification step.

A polynomial classifier was employed by Assaleh and Al-Rousan [8] to identify the alphabetic signs. Five distinct colored gloves were utilized for the fingertips and one glove for the wrist area. Features were created using various geometric metrics including lengths and angles. On a dataset containing more than 200 samplings representing 42 gestures, a recognition rate of 93.4% was attained.

Arabic alphabet sign recognition is a uncomplicated problem, the simplest of all image-based ArSLR methods, because the vocabulary size is fixed, and the signs are represented with mostly static images. Such approaches accomplish high recognition rates. Alphabet signs are not commonly used in day-to-day practice. Their usage is limited to the fingerspelling of words without precise signs. For these motivations, much of the current study effort has proceeded into generating systems that concentrate on isolated words or even continuous sign recognition.

2.2 Recognition of Isolated Words

Unlike alphabet sign recognition, word sign recognition techniques analyze a sequence of images representing the entire sign.

In [9], Mohandes and Deriche used an HMM (Hidden Markov Model) to identify isolated Arabic signs from images. They used a dataset consisting of 500 samples representing 50 signs. A Gaussian skin color model was used to find the signer's face which was then taken as a reference for hand movements. Two colored gloves (orange and yellow) were used for the right and left hands to facilitate segmentation of the hand region. A simple region growth technique was used for hand segmentation.

In [10], Elons and al. used PCNN (Pulse Coupled Neural Network) for the generation of image features from two different viewing angles. The characteristics were evaluated using a fitness function to obtain a weighting factor for each camera. Features derived from both images were used to achieve 3-D optimized features. The dataset used in the experiment contains 50 isolated words.

In [11], Al Mashagba and al. developed an automatic isolated word recognition system using two different colored gloves and an additional colored reference mark on the head. After extracting the three colored regions, five geometric features are extracted from a given video sequence. These characteristics are: the speed of the angle of the hand, the horizontal speed of the hand, the vertical speed of the hand, the horizontal position of the hand at the center of the head and the vertical position of the hand at the center of the head. A time-delayed neural network is used in the recognition phase, achieving a recognition accuracy of 77.4.

In [12], Shanableh and al built a database of 23 selected signs, each sign was repeated 50 times by 3 signers, the videos are segmented to keep each sign in an individual video. They presented various spatio-temporal feature extraction techniques with applications to online and offline gesture recognition isolated from Arabic Sign Language (ArSL). These techniques compress the motion information in a video segment into a single representative image. This was done based on the concept of temporal prediction and accumulated differences. The representative image is then transformed into the frequency domain and parameterized into a precise and concise feature set. This process allowed the use of simple classification techniques that are normally used with time-independent feature sets. To establish baselines for comparison with other classical techniques, they conducted a series of experiments using the classical method of classifying data with temporal dependencies. Namely, Hidden Markov Models (HMM). The experimental results revealed that the proposed feature extraction scheme combined with a simple KNN or Bayesian classification gives comparable results to the classical HMM-based scheme. Experimental results showed classification performance of 97% recognition rate.

In [13], they presented two approaches for feature extraction applied to video-based Arabic sign language recognition: motion representation via motion estimation and motion representation via movement residues: In the first case, motion estimation is used to calculate the motion vectors of a video-based deaf sign or gesture. In the pre-processing step for feature extraction, the horizontal and vertical components of these vectors are rearranged into intensity images and transformed into the frequency domain. In the second approach, motion is represented by motion residuals. The residuals are then thresholded and transformed into the frequency domain. Since in both approaches the time dimension of the video-based gesture needs to be preserved, hidden Markov models are used for classification tasks. Additionally, this paper proposes to project motion information into the time domain through either a telescopic motion vector composition or accumulated polar differences of motion residuals. The feature vectors are then extracted from the projected motion information. The model parameters can then be assessed using straightforward classifiers like Fisher's linear discriminant. It has been shown that motion residuals-based feature extraction outperforms motion estimation in terms of reducing computing complexity and achieving greater classification rates for sign language.

In [14], the authors investigated the use of different transformation techniques for feature extraction and description from an accumulation of sign frames in a single image. They assessed using different machine learning tools the performance of three transformation techniques, these transformations including Fourier, Hartley and Log-Gabor, were applied on the whole and slices of the image of the accumulated sign. Different classification schemes are evaluated and compared, the results obtained show that the Hartley transformation is effective for the recognition of Arabic sign language with an accuracy of 98.8% using the SVM classifier. When the sign image is segmented into 3×3 fragments, this accuracy increases to above 99%.

2.3 Continuous Sign Language Recognition

Sign language Continuous recognition is more difficult than alphabet recognition and isolated signs. Such systems would be more representative of real signing situations for deaf people. The main challenge lies in detecting and modeling the additional movement resulting from the transition between the end of a certain sign and the start of the next.

In [15], Assaleh and al. developed a continuous user-dependent ArSLR system for a 40-sentence database consisting of 80 commonly used words. Spatio-temporal features, based on the DCT (Discrete Cosine Transform), were used with an HMM classifier. The classifier has been optimized regarding the number of features, the number of states and the number of Gaussian mixtures. Albelwi and Alginahi extended their initial alphabet recognition system to a continuous real-time ArSLR system [16]. After segmenting the hand regions, they used Fourier descriptors to model the outer profile of each hand. For classification, they used a simple KNN algorithm and achieved a recognition accuracy of 90.6% on a limited database size.

In [10], Tolba and al. used PCNN and graph matching for continuous ArSLR. The experiments focused on sentences made up of three to four words. The signs are broken down into basic elements and static postures before applying the graph matching technique. The recognition accuracy obtained was greater than 70% for 30 continuous sentences composed of 100 gestures.

Research on continuous ArSLR is still limited compared to the recognition of the alphabet and isolated signs. However, it has been observed that interest in such systems has increased. Continuous ArSLR systems are more suited to the practical situations of deaf people. An ideal continuous sign language recognition system should have a real-time response with a low error rate.

3 Methodology

CNN is a trainable fusion of multi-stage feature extraction/classification deep neural network architectures. It contains alternating layers of convolution and pooling. Given an input image, the convolution layer extracts and combines the local features by performing discrete convolutions with multiple filters, to form the feature maps. Each convolution layer is parameterized by the size of the filters and the number of feature maps. All neurons in a given feature map share their weights to improve learning efficiency by reducing the number of learned parameters. As a result, it enables CNNs to achieve better generalization of visual recognition tasks. The clustering layer learns higher-order features by subsampling feature maps. Eventually, these higher order features are converted to a 1-D vector as the target output at the last layer (Fig. 1).

We created a simple CNN classification model with two CNN layers. We did use Single frame CNN prediction function that produces the singular prediction

```
Layer (type)                 Output Shape              Param #
=================================================================
conv2d_6 (Conv2D)            (None, 48, 48, 32)        896
_________________________________________________________________
conv2d_7 (Conv2D)            (None, 46, 46, 64)        18496
_________________________________________________________________
batch_normalization_6 (Batch (None, 46, 46, 64)        256
_________________________________________________________________
max_pooling2d_3 (MaxPooling2 (None, 23, 23, 64)        0
_________________________________________________________________
global_average_pooling2d_3 ( (None, 64)                0
_________________________________________________________________
dense_6 (Dense)              (None, 128)               8320
_________________________________________________________________
batch_normalization_7 (Batch (None, 128)               512
_________________________________________________________________
dense_7 (Dense)              (None, 23)                2967
=================================================================
Total params: 31,447
Trainable params: 31,063
Non-trainable params: 384
```

Fig. 1. Structure of CNN Model

for a complete video. It takes images from the entire video and makes predictions. Ultimately, it averages the predictions from those frames to give us the final activity class for that video. This function is useful for predicting the activity name and score from a video containing an activity.

4 Experiments

4.1 Alphabet Dataset: Static Signs

The dataset named ArSL2018 consists of 47,937 images for the 28 Arabic alphabets signs collected from 40 participants of different age groups. Different dimensions and different variations were present in the images. The ArSL2018 dataset is unique in that it is the first large comprehensive dataset for Arabic sign language as far as the authors know [12]. It consists of 47,937 gray scale images with $64 * 64$ dimensions saved in jpeg file format. Variations of images were introduced with different lighting and different background.

To tune and test the proposed system, ArSL2018 Arabic Alphabet Sign Language Recognition Database is generated as follows: The ArASLRDB corpus consists of 28 letters of the Arabic alphabet.

Number of signs	**28**
Number of images	**47,937**
Number of training images	**38,192**
Number of test images	**9,745**
Average Images per sign	**1712**
Average training per sign	**1363.96**
Average test images per sign	**348.03**
Percentage of training images	**79.67%**
Percentage of test images	**29.13%**

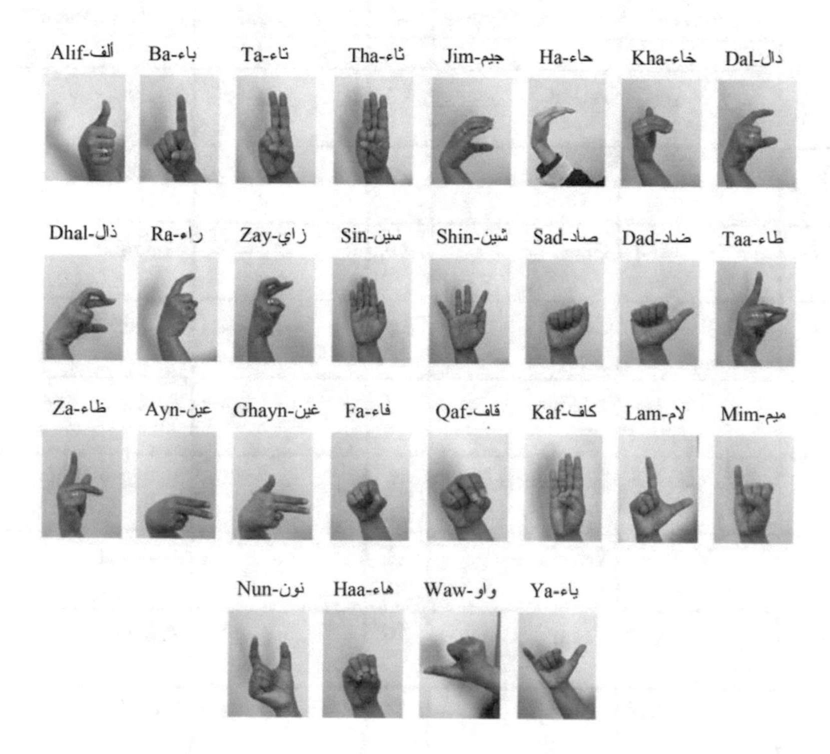

Fig. 2. Arabic sign language alphabet

4.2 Isolated Words Dataset: Single Sign Gestures

As the authors mentioned in [17], unlike other sign languages Arabic did not have a publicly available standard database. Therefore, In [17], Shanableh and al collected their own Isolated Sentence Arabic Sign Language database "ArSLR-Gestures-No-gloves", a database of 23 Arabic gesture words/phrases of 3 different signers (70% for training and 30% for testing). Each signer repeated each gesture 50 times over 3 different sessions, 150 repetitions of the 23 gestures (3450 video segments). Each of the three signers was asked to repeat each gesture 50 times during three different sessions, resulting in a total of 150 repetitions of the 23 gestures which correspond to 3450 video segments. They were recorded using an analog camcorder (Video camera, analog camcorders record information in analog form, like VCRs of the same type, on magnetic tape) without imposing any restrictions on clothing or image background. Video segments from each session were digitized and partitioned into short sequences representing each gesture individually with variability in clothing, use of right and left hands and distance between camera and signers is not constant (Figs. 3 and 4).

#	Arabic word	English Meaning	#	Arabic word	English Meaning
1	صديق	Friend	13	أكل	To Eat
2	جار	Neighbor	14	نام	To sleep
3	ضيف	Guest	15	يشرب	To Drink
4	هدية	Gift	16	يستيقظ	To wake up
5	عدو	Enemy	17	يسمع	To listen
6	السلام عليكم	Peace upon you	18	يسكت	To stop talking
7	اهلا وسهلا	Welcome	19	يشم	To smell
8	شكرا	Thank you	20	يساعد	To help
9	تفضل	Come in	21	أمس	Yesterday
10	عيب	Shame	22	ذهب	To go
11	بيت	House	23	أتى	To come
12	انا	I/me			

Fig. 3. Arabic Alphabet Training Set

Fig. 4. Video sequence of gesture 6 (Peace upon you)

5 Results and Discussion

After evaluating the algorithm and configuring the optimal hyperparameters, we were able to reach using Adamax optimizer and 115 epochs to the following results for isolated words (Fig. 5).

Training Loss	0.0008
Training accuracy	1
Validation loss	0.03
Validation accuracy	0.99
Prediction	98.39%

Fig. 5. Final result

We have 16 misclassified gestures out of 1035 with a success prediction percentage of 98.39%. The exhaustive test led to show the effectiveness of our approach used with a success rate of 98.39%:

5.1 Deductions

- **Activation functions:** The Relu function is efficient, it converges quickly, and its execution time is short. Yet Sigmoid is slow and expensive and Softmax is better to be used for the output layer.
- **Optimizers:** The Adam optimizer is stable, computationally efficient and fast learning, it has no significant decrease in accuracy. SGD is slowly converging its low learning rate.
- **Batch size:** The larger batch size results in faster progression of training, but does not always converge as quickly. Smaller batches train slower, but can converge faster.

5.2 Discussion

CNN is efficient in terms of calculation, it uses special convolution and pooling operations and performs parameter sharing. This CNN model is powerful and efficient, it performs automatic feature extraction to achieve high accuracy that we didn't need to manually extract features from the image. The network learns to extract features while training. The practical benefit is that having fewer parameters greatly improves training time and reduces the amount of data needed to train the model. In CNN, the number of parameters is independent of the size of the original image. We can run the same CNN on images of other sizes and the number of parameters will not change in the convolution layer.

6 Conclusion and Future Work

In this project, CNN is used to recognize Arabic sign language gestures. The proposed system is demonstrated experimentally. The phases of the proposed algorithm include feature extraction, model training and prediction. We worked on hyperparameter validation to improve performance. Experimental results show that the proposed algorithm achieves a recognition rate of 76.21% for alphabet recognition and 98.39% for isolated words using K-fold cross-validation.

As a perspective, we need to develop models to cover other Arabic Sign Language signs and move to continuous recognition instead of single words. An even bigger challenge will be to collect our own database where a larger number of signers will be used. As such, we will be able to focus on the user independence capability of the proposed system. Additionally, we plan to collect ArSL data for continuous gestures too with considering hybrid systems that combine not only several algorithms, but also non-homogeneous sensors such as cameras, sensors, etc. These systems are expected to translate ArSL in real time with the least restriction and with high accuracy.

References

1. Sidig, A.A.I., Luqman, H., Mahmoud, S., Mohandes, M.: KArSL: Arabic sign language database. King Fahd University of Petroleum and Minerals, Saudi Arabia (2022)
2. Lim, K.M., Tan, A.W.C., Lee, C.P., Tan, S.C.: Isolated sign language recognition using convolutional neural network hand modelling and hand energy image. Multimedia University Jalan Ayer Keroh Lama, Malaysia (2019)
3. Luqman, H., Mahmoud, S.A.: Automatic translation of Arabic text-to-Arabic sign language (2018)
4. Neiva, D.H., Zanchettin, C.: Gesture recognition: a review focusing on sign language in a mobile context (2018)
5. Abdo, M.Z., El-Rahman Salem, S.A.: Arabic alphabet and numbers sign language recognition. Faculty of Engineering, Helwan University, Egypt (2015)
6. Naoum, R., Owaied, H.H., Joudeh, S.: Development of a new Arabic sign language recognition using k-nearest neighbor algorithm. J. Emerg. Trends Comput. (2012)

7. Hemayed, E.E., Hassanien, A.S.: Edge-based recognizer for Arabic sign language alphabet (ArS2V-Arabic sign to voice) (2010)
8. Assaleh, K., Al-Rousan, M.: Recognition of Arabic sign language alphabet using polynomial classifiers (2005)
9. Mohandes, M., Deriche, M.: Image based Arabic sign language recognition (2005)
10. Elons, A.S., Abull-Ela, M., Tolba, M.F.: A proposed PCNN features quality optimization technique for pose-invariant 3D Arabic sign language recognition (2013)
11. Al Mashagba, F., Nassar, M.O.: Automatic isolated-word Arabic sign language recognition system based on time delay neural networks: new improvements (2013)
12. Shanableh, T., Assaleh, K., Al-Rousan, M.: Spatio-temporal feature-extraction techniques for isolated gesture recognition in Arabic sign language (2007)
13. Shanableh, T., Assaleh, K.: Telescopic vector composition and polar accumulated motion residuals for feature extraction in Arabic sign language recognition (2007)
14. Ala, A.I., Hamzah Luqman, A.: Mahmoud Rybach: transform-based Arabic sign language recognition. King Fahd University of Petroleum and Minerals, Saudi Arabia (2017)
15. Assaleh, K., Shanableh, T., Fanaswala, M., Amin, F., Bajaj, H.: Continuous Arabic sign language recognition in user dependent mode (2010)
16. Albelwi, N.R., Alginahi, Y.M.: Real-time Arabic sign language (ArSL) recognition (2012)
17. Mohandes, M., Deriche, M., Liu, J.: Image-based and sensor-based approaches to Arabic sign language recognition. King Fahd University, Saudi Arabia (2014)

Masked Facial Recognition Using Deep Metric Learning

S. Ahmam(✉), N. Lamghari, H. Khalfi, and A. Ourdou

LIPIM, ENSA Khouribga, USMS, Khouribga, Morocco
siham.ahmam@usms.ac.ma, n.lamghari@usms.ma

Abstract. Due to the Covid-19 disease, masked faces identification has become the current challenge. This type of identification is difficult because masks cover noses and mouths, obscuring important features for facial recognition. A deep learning-based model for recognizing masked faces is presented in this paper. We tested our system on a dataset of 2113 images collected from 179 people with and without masks. The obtained results are analysed using various metrics and appear to be motivating.

Keywords: Face recognition · Deep learning · Face mask · Convolutional neural network

1 Introduction

With the development of video surveillance, the popularity of digital cameras, smartphones and social networks that allow them to be shared, the number of digital images increased significantly, additionally more and more attention is paid to Computer Vision. In this vast field, we are interested in the branch of object recognition, more specifically facial recognition. Facial recognition is a method of identification that has become one of the most successful branches of computer vision. A face recognition system allows access control based on the identity verification of individuals. But before verifying the identity, it is necessary to detect the face and extract the facial components required for the recognition procedure.

Recently, the world is facing an elusive health crisis with the emergence of the COVID-19 disease of the coronavirus family. One of the modes of transmission of COVID-19 is the airborne transmission. This transmission occurs when humans breathe in droplets emitted by an infected person while breathing [1] talking, singing, coughing or sneezing. To fight against this disease, the government forced people to put on a face mask. The mask hides an important part of the face. Because of this, essential features of facial recognition are missing. While facial recognition is a task that humans perform naturally and effortlessly in their daily lives, equipping a computer with this capability is still a big challenge. With a face mask, most of facial features are covered. Consequently, face recognition of masked faces is extremely challenging. Our goal in this paper is to identify

N. Aboutabit et al. (Eds.): ICMICSA 2022, LNNS 656, pp. 186–196, 2023.
https://doi.org/10.1007/978-3-031-29313-9_17

masked faces. In this context, we have developed a model based on deep learning and tested it on a database of masked faces of 2113 images corresponding to 179 individuals. The results of our model are motivating.

This paper is organized in 3 sections, we will start by citing some works related to our paper in the first section. Then we will describe the methodology we used to build our system to recognize masked faces. In this section, we will describe the architecture of our model. Then, we will describe in detail the proposed process of masked face recognition. We will then discuss the results obtained during the testing of our model and compare it to five other pre-trained systems, using our own dataset and two other datasets namely COMASK20 and Real World Masked Face Recognition Dataset. Finally, a conclusion summarizing our work is given.

2 Related Work

There have been several cases of COVID-19 infection around the world. To combat the propagation of this epidemic, the government has forced people to wear masks. Therefore, facial recognition has become more and more difficult with the lack of features hidden under the face mask. During the coronavirus pandemic, one of the main objectives of researchers is to make recommendations for dealing with this problem with quick and effective solutions. Hariri et al. [2] proposed a reliable method based on eliminating masked parts to give a resolution to the recognition of masked faces. First, they eliminated the masked portion of the face. The unmasked portion is extracted by applying a cropping filter. Then, they applied VGG-16 pre-trained convolutional neural networks to extract the best features of the obtained regions. Vu et al. [3] proposed a method that leveraged the combination of deep learning and local binary pattern (LBP) features to recognize masked faces using RetinaFace, a multi-task learning face detector that can handle different scales of faces, as a fast but efficient encoder. In addition, they extracted local binary pattern features from the eye, forehead, and eyebrow areas of masked faces and combine them with the learned features of RetinaFace in a unitary framework for masked face recognition. In addition, they collected a dataset called COMASK20 of 2754 facial images labeled for 300 different identities. Anwar et al. [4] focused on developing a model able to extract key points of the masked part of the face to reconstruct masked faces. Another system was developed by Li et al. [5] which was more efficient and performing than most recent methods in the field of masked face recognition. In this model, the authors are interested in the unmasked part of the face. They devised a module called convolutional block attention module (CBAM) to pay more attention to the regions not masked by the face mask. Golwalkar et al. [6] have developed a system capable of identifying people with face masks in images and videos. To this end, the authors of this paper focus on the generation of 128-d encodings using their deep learning network called FaceMaskNet-21.

Another approach was proposed by Ullah et al. [12], the authors focused in this work not only on the recognition of masked faces, but also on creating a system named (MDMFR) that can detect masked faces using a dataset of their own creation which served to test their model.

Another vision has been implemented by Din et al. [15], they shifted their attention to the removal of the hidden part of the face and then the reconstruction of the face in the next part. They built a system based on Generative Adversarial Networks to reconstruct the masked parts. This system served multiple purposes as it can be seen in [18–21].

Ejaz et al. [16] used principal component analysis (PCA) to recognize masked and unmasked faces in their work. According to the findings, PCA provided motivating results in the recognition of masked faces. The existing networks such as the ResNet-50 [7,11], Vgg16 [8], MobileNetV2 [9,10], AlexNet [11] and Inception-V3 [11], were tested on databases of masked faces to see their ability and performance in recognizing masked faces. Indeed, the accuracy was increased and they demonstrated their performance in identifying masked faces.

We also find another work proposed in the detection of facial masks such as the approach of Loey et al. [14]. The authors of this paper have tried to combine two systems: the pre-trained model is called ResNet50 and Support Vector Machine (SVM). The first one was used for face feature extraction and the other one for image classification. The combination of these two models gave good results on the datasets used in this work. In Shaohui Lin's paper et al. [17], they also tried to detect masked faces by modifying a pre-trained system into a model capable of detecting masked faces called MLeNet. They tested their system on a dataset composed of 1140 images collected from videos.

3 Methodology

Our goal in this paper is to identify individuals with face masks, in fact, we propose a system that has yielded motivating results in the recognition of masked faces. To do this, we have developed a model based on deep learning, which was tuned to produce encodings of 128-d. In this work, we have implemented deep metric learning to guarantee an increase in accuracy and ensure the speed of a masked face. This technique focuses on generating a 128-d feature vector of real floating point number, contrary to the traditional deep learning technique estimates a decision function that correctly assigns to each input its appropriate class. Our network's training is done using a dataset called quadruplet (shown in Fig. 1). Indeed, we take four images of two people wearing a face mask. Each person has two doubles. Our system quantizes the input image of the face and constructs feature encodings of length 128.

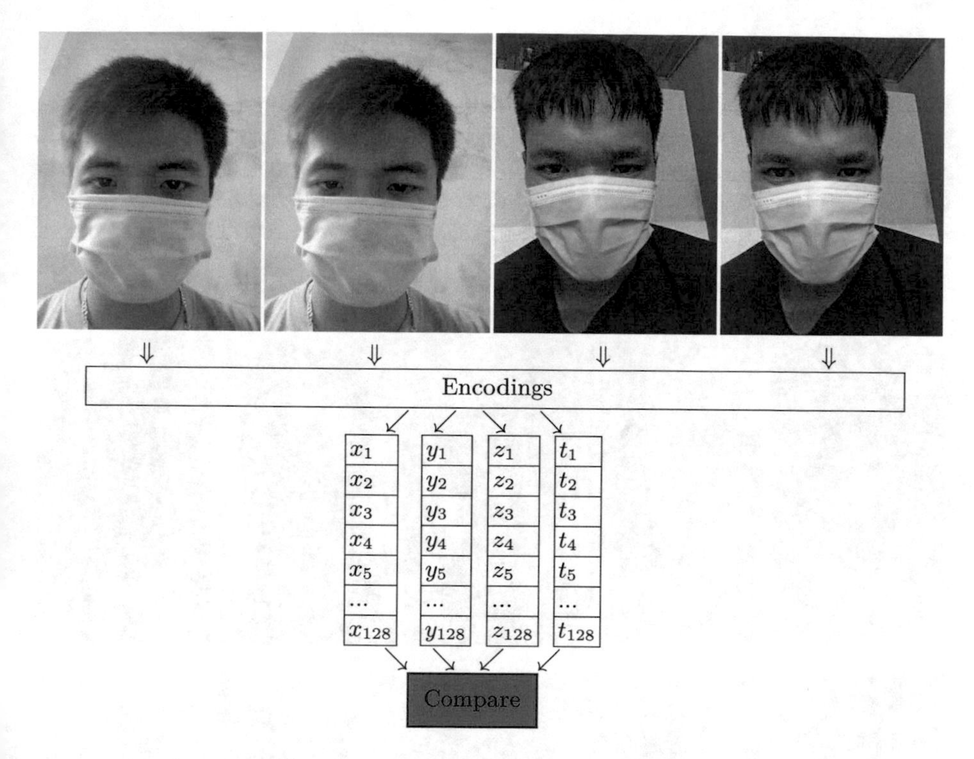

Fig. 1. The process of adjusting our network. The adjustment of our network is done using four images of two people (quadruplets) following two important steps. The first is the generation of 128-d encodings of each image. The $(x_1,...,x_{128})$, $(y_1,...,y_{128})$, $(z_1,...,z_{128})$ and $(t_1,...,t_{128})$ are the generated vectors of the four input images.

Our neural network continues to adjust the weights using the quadruplets images until it produces clearly distinguishable encodings for both images corresponding to different people meanwhile leaving the encodings of fairly close for images of the same person. Training the network with a quadruplet has improved its performance. In addition, using masked faces in the quadruplets allowed our system to gain robustness and successfully identify masked faces. Because face masks only allow us to see the areas around the eyes, recognizing masked faces remains difficult due to a lack of facial characteristics that can help us identify these people. We built our neural network to consider primarily unmasked facial features, for the masked face recognition process.

3.1 Dataset

The dataset on which we tested our model (shown in the Fig. 2, Fig. 3 and Fig. 4) is composed of 2113 images of the masked faces of 179 people. We collected our dataset from three datasets of masked face images. A part of it goes back to the COMASK20 [3,13] (shown in the Fig. 2) dataset. In this dataset, there are 300 people and each person has a number of masked and unmasked images grouped in a directory.

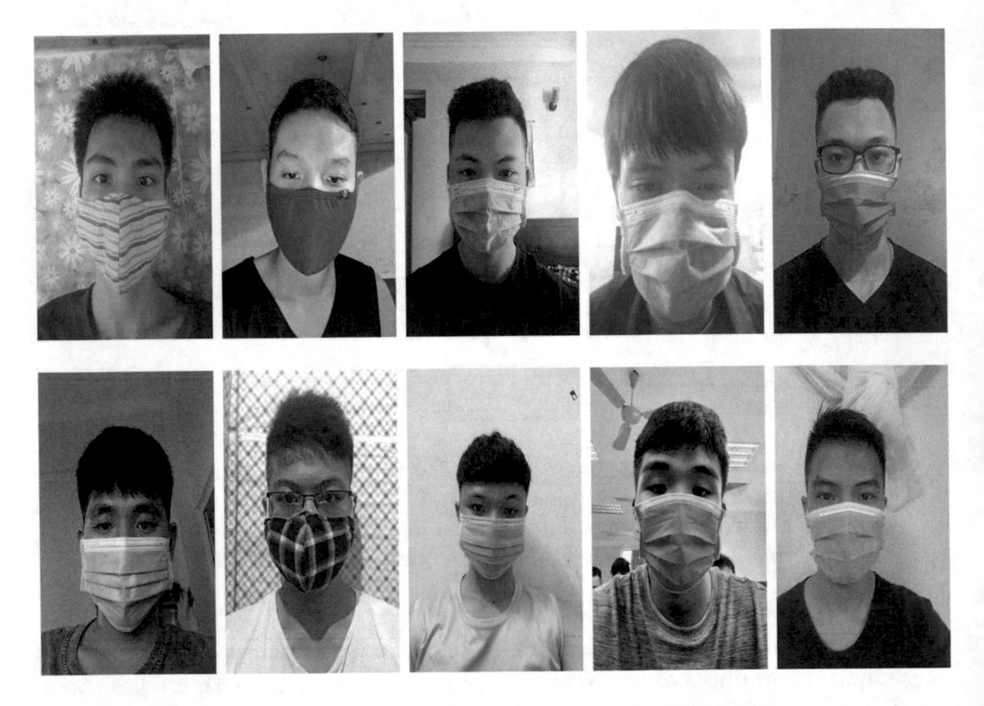

Fig. 2. A sample of dataset COMASK20 ([13])

Another part of our dataset is part of the large RMFRD dataset (shown in Fig. 3). We have taken only a part of the masked face images of which the images of each person are collected in a directory. This made it easier for us to group the images for each person.

Fig. 3. A sample of dataset RMFRD ([22])

The last part is part of the HFRD-MakedFaceDatasets [23] dataset (shown in Fig. 4) which is a dataset with images of masked faces, there are images that are in grayscale, anothers in RGB with different types and color of facial masks and anothers with a change of poses. We are interested in the images in RGB

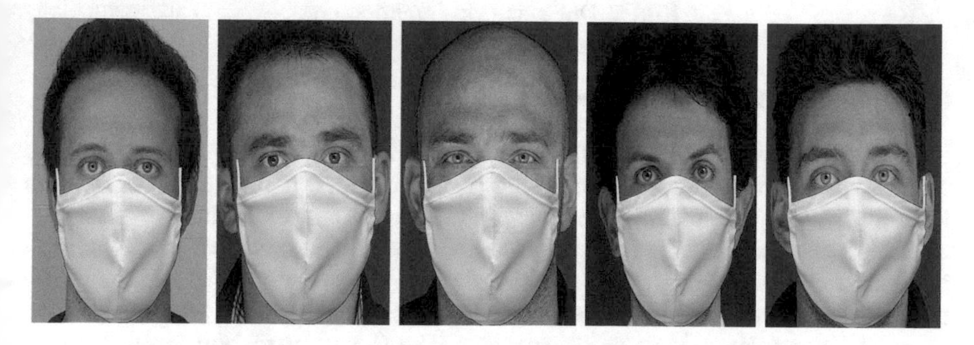

Fig. 4. A sample of dataset HFRD-MakedFaceDatasets ([23])

The number of images in our dataset is not the same for each person. We have resized all the images to size (227, 227) and we divided our dataset into 70% for training our system and 30% remained for the test.

3.2 Our Neural Network Architecture

Our developed network, as shown in Fig. 5, has an objective to generate 128-d encodings. The input images of our network are in RGB of dimension (227, 227). First, we normalized the different-shaped images to a unified dimension of (227, 227). Then, we started our network with a convolution layer of the dimensions of (55, 55) with 96 filters. The role of this layer is to collect image features and detect specific patterns by applying filters. After the convolution layer, we then used a max-pooling layer with dimensions (27, 27), consisting of 96 filters. We then applied a next convolution layer with the same dimensions as the max-pooling layer, but with 256 filters. The next is a second layer of max-pooling. Its role is to calculate the largest value of the part defined by the pool size of (3, 3) and by the dimensions (13, 13), with 256 filters. Then, we applied a last convolution layer as well with dimensions (13, 13) consisting of 384 filters. Before the Softmax layer, we added a fully connected layer. The latter has an output size of 128. At the end, we generated the 128-d output encodings.

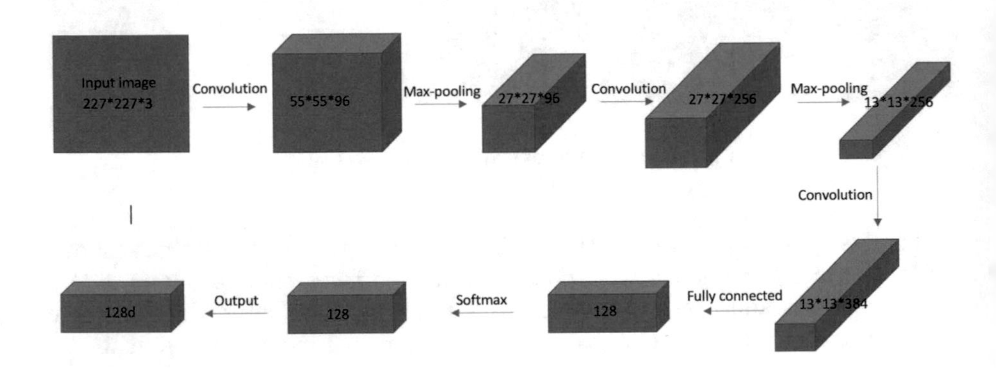

Fig. 5. Our network's architecture.

3.3 Face Recognition Process

First of all, we loaded the input image in RGB. Next, we initialized a list for storing the names of people after detecting their faces in the input image. The following step is the generation of encodings for each face of the image. After calculating the encodings for this image, we compared the newly extracted encodings with the encodings pre-computed during the training of the model. We applied an Euclidean distance between these encodings to see if these people are known by our system. If there is a match of the faces, the distance was minimal after comparison with all the existing 128-d encodings. Then we store the names of the matching faces in the pre-initialized list. After determining the name of the recognized face and extracting the coordinates from the face bounding box, the image of the recognized face with the name of the person is displayed.

4 Result and Discussion

We started our face recognition process by generating 128-d encodings of all the images in the dataset. After comparing the new encoding extracted from the input person's face with the pre-computed encodings, if there is a correspondence with such a known encoding, the corresponding face image is displayed, along with the name of the known person and a box detecting the face. If the person is not recognized, the corresponding image is displayed with the name "unknown". The precision during training of our model achieved with our dataset was 88.31%. This accuracy is motivating in the masked face recognition process.

Figure 6 shows the accuracy obtained by our developed network and 5 other existing models by training on our dataset. ResNet-50 achieved the highest accuracy of 88.7%. Our model achieved an accuracy of 88.31% and VGG-16 achieved 87.2% accuracy. The MobileNetV2 and Inception-V3 models achieved 80.6% and 75.9% accuracy. The AlexNet model achieved the lowest accuracy of 72.4%. Therefore, the proposed model achieved high accuracy compared to most other deep learning models.

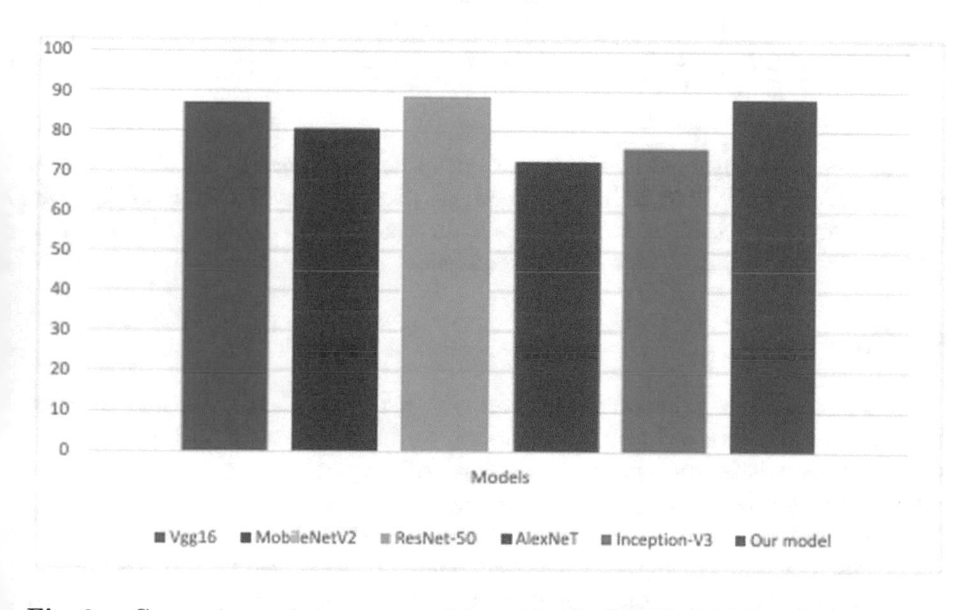

Fig. 6. Comparison between models AlexNet,VGG16,MobileNetV4,Inception-V3,ResNet-50 and our developed network according to the accuracy obtained during the test on our dataset.

We tested our model on two other databases COMASK20 [3,13],and Real World Masked Face Recognition Dataset (RMFRD) [24]. The latter contains 5000 masked images of 525 people. The accuracy obtained on the test on the COMASK20 database is 88.2% but on the RMFRD dataset, we got an accuracy of 80.5%.

We have made a comparison with other facial recognition systems, the results of this comparison are illustrated in the Table 1. This comparison is based on the accuracy and the ability of these systems to identify masked faces. In addition, the comparison includes the dataset and the technique used to identify the people. The recognition accuracy of these systems is high.

Table 1. A comparison table with the existing systems.

Method	Accuracy	Dataset	Number of image	Technique
Hriri et al. [2]	91.30%	RMFRD	90000	VGG-16 pre-trained model
Vu et al. [3]	97%	COMASK20	2754	LBP
Golwalkar et al. [6]	88.92%	User dataset	2000	FaceMaskNet-21
	82.22%	RMFRD	5000	
	88.186%	User dataset including images of children	1992	
Proposed method	88.31%	Our dataset	2113	Our model

The precision obtained by the system proposed by Hariri et al. [2] is 91.30% for face recognition while Golwalkar et al. [6] presented an accuracy of 88.92%. Training our model using our dataset gave an accuracy of 88.31%. Which means that our model is able to identify masked faces.

5 Conclusion

Our work recognizes a person's face even with a mask on and gives an accuracy of 88.31% on our dataset. Using our model, we have generated 128-d encodings of each faces existing on the input images. Afterwards, a comparison is made between these encodings with the pre-calculated encodings. Then, we used the Euclidean distance to see the correspondence between these encodings. Finally, the person's face was displayed with their name on the screen with a box detecting their face. We tested our model on two other databases COMASK20 and RMFRD and we obtained an accuracy of 88.2% on the first database and an accuracy of 80.5% on the second one.

References

1. Rahman, H.S., et al.: The transmission modes and sources of COVID-19: a systematic review, Int. J. Surg. Open **26**, 125–136 (2020)
2. Hariri, W.: Efficient masked face recognition method during the COVID-19 pandemic. IViP **16**, 605–612 (2022)

3. Vu, H.N., Nguyen, M.H., Pham, C.: Masked face recognition with convolutional neural networks and local binary patterns. Appl. Intell. **52**, 5497–5512 (2022)
4. Anwar, A., Raychowdhury, A.: Masked face recognition for secure authentication, CoRR, abs/2008.11104 (2020)
5. Li, Y., Guo, K., Lu, Y., Liu, L.: Cropping and attention based approach for masked face recognition. Appl. Intell. **51**, 3012–3025 (2021)
6. Golwalkar, R., Mehendale, N.: Masked-face recognition using deep metric learning and FaceMaskNet-21. Appl. Intell. **52**, 13268–13279 (2022)
7. Mandal, B., Okeukwu, A., Theis, Y.: Masked face recognition using ResNet-50, CoRR, abs/2104.08997 (2021)
8. Perdana, A.B., Prahara, A.: Face recognition using light-convolutional neural networks based on modified Vgg16 model. In: International Conference of Computer Science and Information Technology (ICoSNIKOM), pp. 1–4 (2019)
9. Patel, V., Patel, D.: Face mask recognition using MobileNetV2. Int. J. Sci. Res. Comput. Sci. Eng. Inf. Technol. 35–42 (2021)
10. Adhinata, F.D., Tanjung, N.A.F., Widayat, W., Pasfica, G.R., Satura, F.R.: Comparative study of VGG16 and MobileNetV2 for masked face. Universitas Ahmad Dahlan **7** (2021)
11. Nyarko, B.N.E., Bin, W., Zhou, J., Agordzo, G.K., Odoom, J., Koukoyi, E.: Comparative analysis of AlexNet, Resnet-50, and Inception-V3 models on masked face recognition. In: IEEE World AI IoT Congress (AIIoT), pp. 337–343 (2022)
12. Ullah, N., Javed, A., Ghazanfar, M.A., Alsufyani, A., Bourouis, S.: A novel Deep-MaskNet model for face mask detection and masked facial. J. King Saud Univ. - Comput. Inf. Sci. (2022)
13. https://github.com/tuminguyen/COMASK20
14. Loey, M., Manogaran, G., Taha, M.H.N., Khalifa, N.E.M.: A hybrid deep transfer learning model with machine learning methods for face mask detection in the era of the COVID-19 pandemic. Measurement **167**, 108288 (2020)
15. Din, N.U., Javed, K., Bae, S., Yi,J.: A novel GAN-based network for unmasking of masked face. IEEE Access **8**, 44276–44287 (2020)
16. Ejaz, M.S., Islam. M.R., Sifatullah, M., Sarker, A.: Implementation of principal component analysis on masked and non-masked face recognition. In: 1st International Conference on Advances in Science, Engineering and Robotics Technology (ICASERT), pp. 1–5 (2019)
17. Lin, S., Cai, L., Lin, X., Ji, R.: Masked face detection via a modified LeNet. Neurocomputing 197–202 (2016)
18. Zhao, J., et al.: Dual-agent GANs for photorealistic and identity preserving profile face synthesis. In: Advances in Neural Information Processing Systems NIPS (2017)
19. Yang, H., Huang, D., Wang, Y., Jain, A.K.: Learning face age progression: a pyramid architecture of GANs, CoRR (2018)
20. Na, I.S., Tran, C., Nguyen, D., Dinh, S.: Facial UV map completion for pose-invariant face recognition: a novel adversarial approach based on coupled attention residual UNets. Hum.-Centric Comput. Inf. Sci. (2020)
21. Li, L., Song, L., Wu, X., He, R., Tan, T.: Learning a bi-level adversarial network with global and local perception for makeup-invariant face verification. Pattern Recogn. (2019)

22. https://drive.google.com/le/d/1UlOk6EtiaXTHylRUx2mySgvJX9ycoeBp/view
23. Al-Khatib, R.M., Barhoush, M., Migdady, A., Al-Madi, M.: HFRD: human face recognition datasets on COVID-19 pandemic stage (v2020). Mendeley Data, V3 (2020). https://doi.org/10.17632/v992cb6bw7.3
24. https://github.com/X-zhangyang/Real-World-Masked-Face-Dataset

Intelligent Systems

Hardware Software Co-design Approch for ECG Signal Analysis

Bouchra Bendahane(✉), Wissam Jenkal, Mostafa Laaboubi, and Rachid Latif

Ibn Zohr University, National School of Applied Science, System Engineering and Information Technology Laboratory, Agadir, Morocco
bouchra.bendahanne@edu.uiz.ac.ma

Abstract. An essential tool for diagnosing heart diseases is the electrocardiogram (ECG) signal. The accuracy of the diagnosis is impacted by the noise that occurs while this signal is being acquired. Denoising turns into a foundational step in this context. DWT-ADTF is an effective ECG signal-denoising method. This work tries to provide an HW/SW co-design solution to this method. The signal is well denoised, according to the simulation results of the developed designs.

Keywords: ECG Signal · Denoising · HW/SW Co-design · FPGA

1 Introduction

The electrical activity of the heart produces biological signals called electrocardiogram signals (ECGs). These signals are used in biomedicine to diagnose several heart diseases as well as some physiological issues in the human body. An ECG signal analysis system involves a denoising stage, a feature detection and extraction stage, in particular the QRS feature, and a signal classification stage that allows the distinction between a normal state and a pathological one of the heart. This work aims to give an embedded system solution to this entire system. Researchers in signal processing have developed several algorithms at the level of each stage. As researchers in embedded systems, our role is to raise and study these algorithms in terms of performance and complexity to select the most performant and least complex one (s). In this paper, we are interested in the denoising stage; the remaining stages will be the subject of future works.

The signal processing literature has provided efficient methods for ECG signal analysis. The empirical mode decomposition (EMD) [1–3], digital filters [4,5], discrete wavelet transform (DWT) [6–9], and artificial intelligence algorithms (AI) [10,11] are the most well-known for the denoising stage. To enhance the denoising quality, researchers have developed hybrid methods based on the previous techniques, like the work presented in [12], which is based on EMD and DWT. an EMD-EWT-based work in [13] , and another work based on the EMD and the mean filter in [14]. In [7] a DWT-ADTF-based work is presented, which is

Supported by System Engineering and Information Technology Laboratory

N. Aboutabit et al. (Eds.): ICMICSA 2022, LNNS 656, pp. 199–210, 2023.
https://doi.org/10.1007/978-3-031-29313-9_18

the algorithm purpose of this embedded system design paper. The DWT-ADTF method gathers the performance of two denoising techniques: the DWT and the adaptive dual threshold filter (ADTF). The DWT is the most commonly used method in denoising tasks, and this is thanks to its ability to decompose signals and show their frequency composition according to the time at which these frequencies appear. The ADTF is a recently published ECG signal denoising algorithm [15]. The dual-threshold median filter used in image processing served as the inspiration for the basic idea behind the ADTF method [15].

Embedded systems consist of application-specific hardware parts and software programming parts [16]. HW/SW co-design is an intermediate solution for embedded system design that allows the system to be partitioned into a software and a hardware part, combining the benefits of both. The goal of this paper is to provide an HW/SW co-design solution for the DWT-ADTF algorithm, which was developed for ECG signal denoising. The present paper is organized as follows: Sect. 2 reviews the literature on ECG signal pre-processing as well as embedded systems HW/SW co-design. Section 3 describes the methodology followed to choose the DWT-ADTF as the appropriate ECG signal denoising algorithm for embedded system applications. This section also includes a description of the proposed architectural design for the chosen algorithm, in addition to the used tools and source. The results justifying the DWT-ADTF algorithm choice, its proposed HW/SW architecture simulation are presented in Sect. 4. Finally, Sect. 5 brings the paper to a close.

2 Literature

During its acquisition, the ECG signal is infected by several noises which hide the signal features desired for the diagnosis, especially the QRS complex. Therefore, denoising is an essential stage in the ECG signal analysis process.

An embedded system could be designed following a hardware or a software approach. The hardware approach allows higher performance. However, it is a resource and power-consuming solution. Unlike this, the software approach is characterized by a fixed area so power consumption is reduced. However, it is less performant compared to hardware. An intermediate solution is to gather the advantages of both approaches in an HW/SW co-design approach.

This section discusses the ECG signal denoising literature as well as that of the HW/SW co-design.

2.1 ECG Signal Denoising Literature

The denoising stage is a crucial step in the ECG signal analysis since it makes appear useful features for a correct diagnosis. Due to the importance of this stage, researchers have developed several technics. The most known and used are the empirical mode decomposition (EMD)based algorithms [1–3], the digital filters [4,5], the discrete wavelet transform (DWT)based works [6–9], and the artificial intelligence (AI) algorithms [10,11].

The discrete wavelet transform is a signal decomposition and reconstruction process using low-pass and high-pass filters given according to the chosen wavelet mother. The DWT process makes appear the frequency composition of the signal in the form of frequency intervals by decomposition level.

The empirical mode decomposition is an iterative calculation process of IMF functions from upper and lower envelopes resulting respectively from interpolating the local maximums and minimums of the input signal. This process is repeated while extremums exist in between the zero-crossing. The process continues on residues until the function becomes monotone or inexistence of extremums in between its zero-crossing.

The artificial intelligence (AI) models involve convolutional, recursive, and other neural networks. It is a set of interconnected layers of neurons where each neuron has a certain number of inputs and outputs and is activated with a determinate activation function.

To enhance the denoising quality, researchers have developed hybrid methods based on the previous techniques like that presented in [12] based on EMD and DWT. In [13] an EMD-EWT-based work, and in [14] another work based on the EMD and the mean filter. In [7] a DWT-ADTF-based work is presented which is the algorithm purpose of this embedded system design paper.

2.2 Embedded Systems HW/SW Co-design Literature

The HW/SW co-design approach allows for bringing together the advantages of both software and hardware. However, the most difficult challenge in HW/SW co-design is determining which system components should be implemented in hardware and which in software [16]. Traditionally, manual HW/SW partitioning has been used [16]. But as the system's complexity increased, partitioning automation became essential. Therefore, several methodologies and tools have been developed [16–18]. In [16], a high-level HW/SW partitioning methodology is presented. The paper discusses the partitioning problem. The partitioning problem is modeled with a cost function that is a weighted sum of the system constraints. Based on the particle swarm optimization algorithm, the function is resolved and several HW/SW partitioning solutions are given. Thus, the designer could decide which solution is suitable.

Field programmable gate arrays (FPGA) have become increasingly popular thanks to the faster growth of the transistor density and range diversity that cover several domains' needs and constraints. The FPGAs allow complete embedded system study from design, and verification to cost estimation, which allows for important post-fabrication error cost reduction. Furthermore, these SoC (System-on-Chip) FPGAs include one or more integrated processors, whether soft or hard core, promoting HW/SW co-design.

3 Methodology

The goal of this paper is to provide an embedded system solution for the ECG signal denoising stage. However, we are faced with multiple algorithms. Therefore, an appropriate methodology is required to choose the most performant and least complex algorithm. This section discusses the methodology used to select the ECG signal denoising algorithm, as well as the methodology used to design the corresponding HW and SW architecture.

3.1 ECG Signal Denoising Algorithm Choice

The ECG signal denoising literature involves numerous ge algorithms. As a result, a performance comparison study is carried out to determine the best algorithm for the ECG signal denoise stage. On the other hand, as we target embedded systems applications, the study has considered not only performance evaluation but also the algorithmic complexity since embedded systems are mainly constrained by resource limitations.

ECG signal denoising algorithms are grouped by category, and only the category of the less complex algorithms is guarded; others are ignored by the algorithm selection process. Afterward, the retained category of algorithms is assessed in terms of performance constraint via a comparative study based on the mean square error (MSE), the percentage root difference (PRD), the signal to noise ratio improvement (SNRimp), and the execution time metrics. From this category of algorithms, four of the most recent and most performant algorithms are selected for a performance comparison study. Consequently, among the four algorithms, the most performant will be chosen for the ECG signal denoising and then for embedded system design. A performant algorithm has the lowest MSE and PRD values, highest SNRimp, and shortest execution time.

3.2 HW/SW Co-design for the Chosen ECG Denoising Algorithm

Embedded systems consist of application-specific hardware parts and software programming parts [16]. Software parts are characterized by easier and faster development and modification compared to hardware parts [16] [17]. Therefore, in terms of development cost and time, the software parts are less expensive [16]). However, better performance is partly provided by hardware [16]. For these reasons, system design is constrained by the minimization of software delay, hardware area, and consumption [16,17]. HW/SW co-design is an intermediate solution for embedded system design, allowing the system to be partitioned into a software and hardware part and thus combining the benefits of both.

The DWT-ADTF is a simple computing algorithm. It consists of a two-level DWT decomposition and two threshold computations with simple mathematical formulas. Therefore, the partitioning is done manually. However, a complete ECG analysis algorithm co-design requires automated HW/SW partitioning since it involves other treatment stages as well as denoising.

The Cyclone FPGA devices have a general-purpose FPGA architecture with important logic elements (LE) and memory blocks. However, the Cyclone II family is a high-density architecture with advanced features, namely an important number of embedded multipliers, DSP, memories, and customized soft processor embedding possibilities, in addition to the large logic network. With these advanced features and at a price comparable to ASICs, Cyclone II devices can handle intricate digital systems on a single chip [19]. Other cyclone families are more advanced, have more embedded specific elements, and are designed for high-volume systems. Therefore, the Cyclone II family of devices is the preferred and appropriate platform for signal processing applications.

3.3 Verification Tools

This work involves algorithm verification at the algorithmic and architectural levels. Therefore, appropriate tools are required in addition to ECG signals and noise. The algorithmic validations are performed using the Matlab R2016a software, while the architectural validations are performed using the Modelsim simulation software. ECG signals are taken from the MIT-BIT Arrhythmia database, while the noise used is the additive white Gaussian noise (AWGN) generated by Matlab. For the test, ECG signals are corrupted with AWGN with a given SNR to simulate the noised version of the ECG signal, which represents the input signal of the algorithm or the architecture.

4 Results and Validation

After having discussed the methodology followed for choosing the appropriate ECG denoising algorithm as well as the design approach for embedded systems applications, we will talk about the results.

4.1 Algorithm Choice

The ECG signal denoising literature involves several algorithms that could be distributed into four main categories, namely, the EMD-based algorithms, the digital filters, the DWT-based algorithms, and the AI models. The AI models are the most complex algorithms, as they demand massive matrix calculations. In addition, these algorithms require pre-optimization to raise their implementability. The EMD involves an extremum calculation and interpolation process where the iteration number is unpredictable. Digital filters are simple to design and implement, but they demand huge adders and multipliers as well as a priori knowledge of the noise frequency to remove. The DWT involves a greater or lesser number of operations linked immediately to the coefficient number of the chosen wavelet mother and the decomposition level. So, the complexity is manageable in the case of this technique compared to previous methods. Following that, for the selection of the ECG denoising algorithm, we were only interested in DWT-based works, excluding those in which the DWT is combined with EMD or AI algorithms.

In the DWT-based denoising literature, the DWT is used with thresholding, adaptive thresholding, non-local mean, and other algorithms. We compared the denoising performances of four of literature's most powerful and recent works.

These algorithms are the DWT-Thresholding, which is the conventional DWT-based denoising use; the DWT-NLM, which combines the DWT with the non-local mean (NLM) algorithm; the DWT-VMD-NLM, which is the DWT-NLM with the variational mode decomposition (VMD) method; and the DWT-ADTF, which involves the ADTF filter. The comparison is carried out with an input SNR of 10 dB for different ECG signals taken from the MIT-BIH Arrhythmia database of Physionet. The results are summarized in the Table 1. The denoising performance comparison showed efficient results given by the DWT-ADTF algorithm, especially at the level of the MSE, SNRimp, and execution time(Table 1). From Table 1 we can conclude that the lowest MSE values and the highest SNRimp values are given by the DWT-ADTF algorithm, while acceptable values are achieved by this algorithm at the level of the PRD metric. On the other hand, in terms of the execution time metric, the DWT-ADTF achieved the minimum value (Table 2). According to these comparison results, the DWT-ADTF is the chosen algorithm for this embedded system application paper.

Table 1. Denoising Performance Comparison of Four DWT-based Methods

	DWT-Thresholding [6]			DWT-NLM [8]			DWT-VMD-NLM [6]			DWT-ADTF [7]		
Signal N°	MSE	PRD	SNRimp	MSE	PRD	SNRimp	MSE	PRD	SNRimp	MSE	PRD	SNRimp
100	0.0011	9.25	4.03	0.00039	5.50	8.58	0.0003	5.28	8.92	0.0005	6.46	13.90
103	0.003	13.93	4.99	0.00089	7.75	8.58	0.0008	7.63	8.56	0.0030	13.33	7.04
104	0.0028	14.62	4.22	0.00097	8.52	8.38	0.0009	8.46	8.44	0.0035	16.90	5.55
105	0.0024	12.82	5.67	0.0011	8.87	8.16	0.018	8.69	8.35	0.0013	9.10	10.54
106	0.0049	16.37	4.74	0.0012	9.71	8.22	0.0010	9.41	8.49	0.0038	15.63	6.59
115	0.003	9.68	4.60	0.0011	8.48	8.62	0.0010	8.01	9.12	0.0027	8.88	10.51
215	0.0018	15.19	4.85	0.0011	8.90	8.55	0.0010	8.30	9.15	0.0008	10.77	9.72

Table 2. Execution Time Comparison of Four DWT-based Methods

	DWT-Thresholding [6]	DWT-NLM [8]	DWT-VMD-NLM [6]	DWT-ADTF [7]
Execution Time (s)	0.1189	0.9986	0.5480	0.030

4.2 Chosen Algorithm Description

The DWT-ADTF method gathers the performance of two denoising techniques, the DWT and the adaptive dual threshold filter (ADTF). The DWT has been used as a decomposition technique in many hybridization works, where the results were very interesting compared to those of conventional DWT use (threshold algorithms). One of these works is the DWT-ADTF filter, which combines filtering performance with algorithmic simplicity.

The DWT Method is widely used in ECG signal analysis for signal denoising, feature extraction, and signal compression. It is a linear transform that allows the decomposition of the signal into a set of bands of frequencies up to a given level. Signal analysis and synthesis are accomplished with DWT by employing two filters: a low-pass and a high-pass associated with the scale and wavelet functions, respectively [6]. There are numerous filter coefficients available for signal analysis (e.g., debucies, coiflets, and symlets, among others) [7]. The decomposition and reconstruction processes of DWT are shown in Fig. 1, while the decomposition Eqs. ((1) and (2)) and reconstruction Eq. (3) are as follows [8,9]:

$$A[n] = \sum_{k=-\infty}^{k=+\infty} g(k)x(2n-k) \tag{1}$$

$$D[n] = \sum_{k=-\infty}^{k=+\infty} h(k)x(2n-k) \tag{2}$$

$$x[n] = \sum_{k=-\infty}^{k=+\infty} [D(k)h(2k-n) + A(k)g(2k-n)] \tag{3}$$

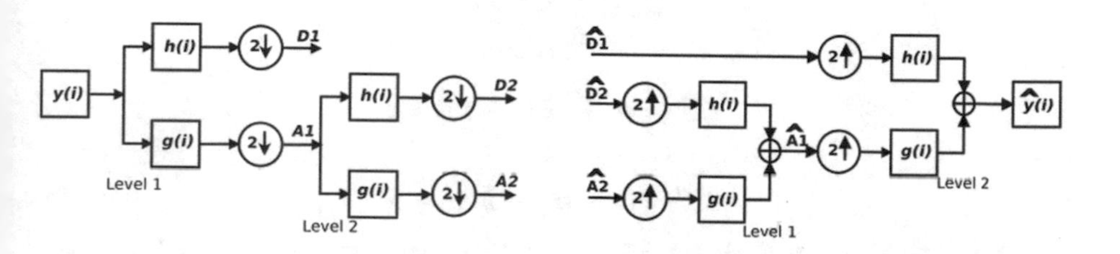

Fig. 1. DWT Decomposition and Reconstruction Processes.

A and D are the approximation and detail coefficients of a given level, respectively. A and D are the outputs of the low filter (g) and the high filter (h), respectively. Figure 2 shows the two-level decomposition results of the record n°103 using the debauchee 4 with an input SNR of 15 dB. A good analysis using DWT is based on the correct choice of the mother wavelet as well as the decomposition level. In general, the wavelet is chosen to be as similar as possible to the morphology of the studied signal.

The ADTF Method is a recent algorithm for ECG signal denoising. It is an adaptive method inspired by the dual-threshold mean filter used in image processing [15]. Two thresholding values are calculated for each window as the basis of the ADTF application [15]. As a result, noise is eliminated, and all signal-useful values are balanced between the two thresholds. The window's mean,

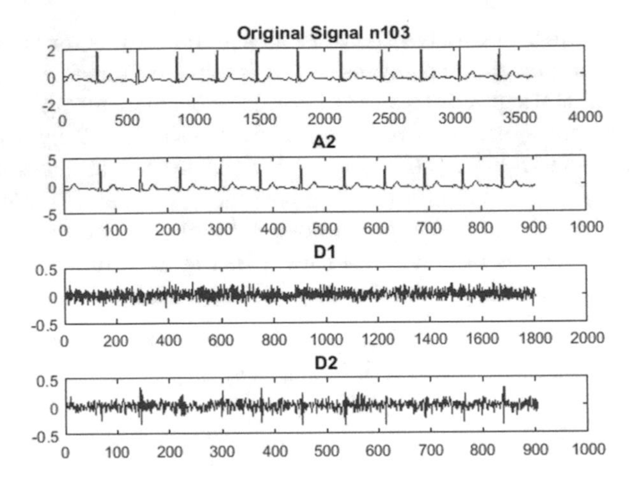

Fig. 2. Two-Level DWT Decomposition of Signal n°103 with input SNR of 15 dB.

upper, and lower thresholds—the ADTF parameters—are computed as follows [15]:

$$m = \frac{1}{L} \sum_{i=n}^{n+L} S(i) \tag{4}$$

$$HT = m + (Mx - m) * \beta \tag{5}$$

$$LT = m - (m - Mn) * \beta \tag{6}$$

where S is the noised signal, Mx and Mn are respectively the maximum and minimum values of the window. The thresholding coefficient, β, varies between zero and one [15], and L is the window length. Figure 3 shows the denoising result using the ADTF filter for the record n°103 that is corrupted with an input SNR of 15 dB.

The DWT-ADTF Method involves three main steps: the DWT application, which consists of the elimination of the two first details, D1 and D2 (Fig. 2), the ADTF application with a ten-sample moving window, and finally the high-peak correction. The algorithm is validated using the record n°103 and an input SNR of 15 dB (Fig. 4).

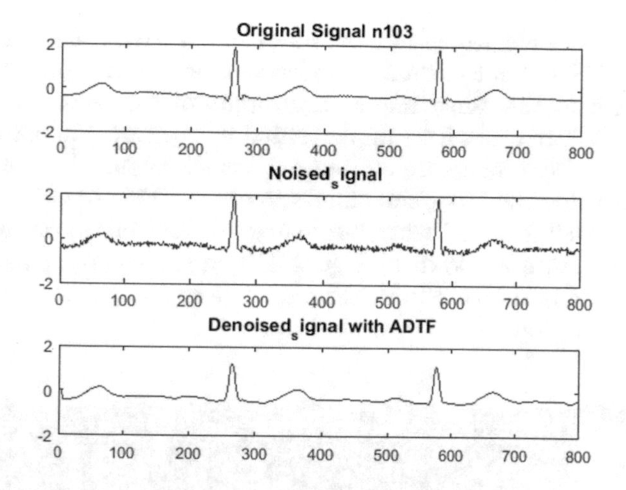

Fig. 3. Signal n°103 Denoising Results Using ADTF with input SNR of 15 dB.

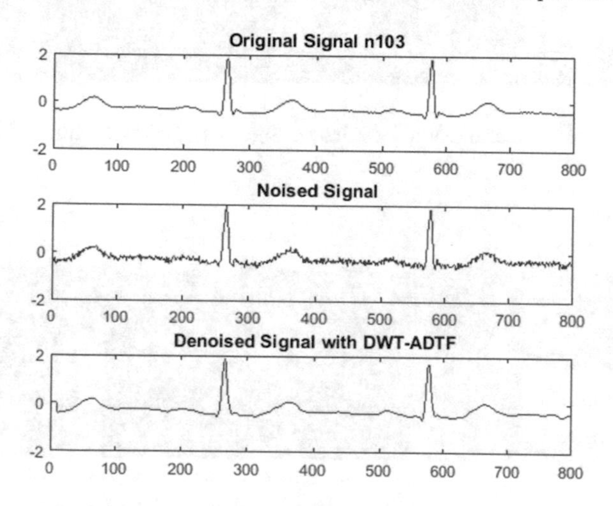

Fig. 4. Signal n°103 Denoising Results Using DWT-ADTF with input SNR of 15 dB.

4.3 Chosen Algorithm Design

The DWT-ADTF method is composed of three components: the DWT, the ADTF, and the peak correction, as was previously mentioned. The DWT is a sequential process and is the most complicated part of the entire algorithm since it is an FIR-based decomposition and reconstruction filter bank. Thus, huge multipliers and adders are required as the level increases. Therefore, DWT computing is dedicated to software implementation using C code. The peak correction involves a simple comparison between DWT and ADTF values to correct the attenuated R-peaks, but it is a critical stage that influences the QRS complex region. Accordingly, more performance is required here. Therefore, the peak correction part is implemented in hardware using VHDL code. Similarly to the DWT, the ADTF is implemented in software since both types of information must be present at the same time at the inputs of the peak correction stage; therefore, we preferred that it be implemented in software. Simulation results of software and hardware parts are carried out for the signal n°103 with an input SNR of 15 dB using the Modelsim simulation tool (Figs. 5 and 6). The signal is qualitatively well-denoised when compared to the Matlab result. The entire proposed architecture is shown in Fig. 7. The implementation platform is the Altera Intel Cyclone II DE1 FPGA (Fig. 8). The entire hardware synthesis cost of this implementation will be presented later.

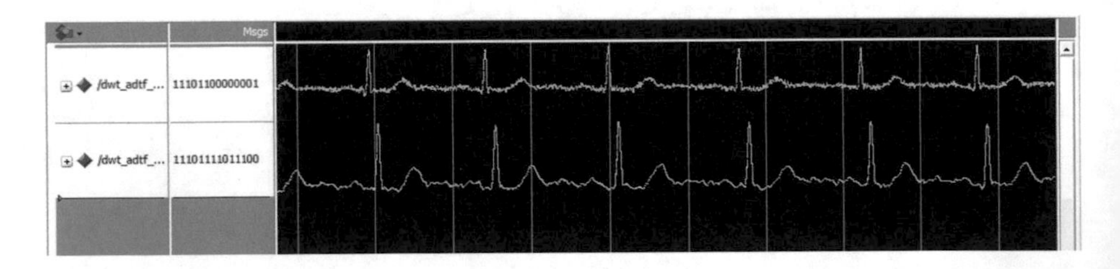

Fig. 5. Software Part Simulation Results of Signal n°103 with input SNR of 15 dB.

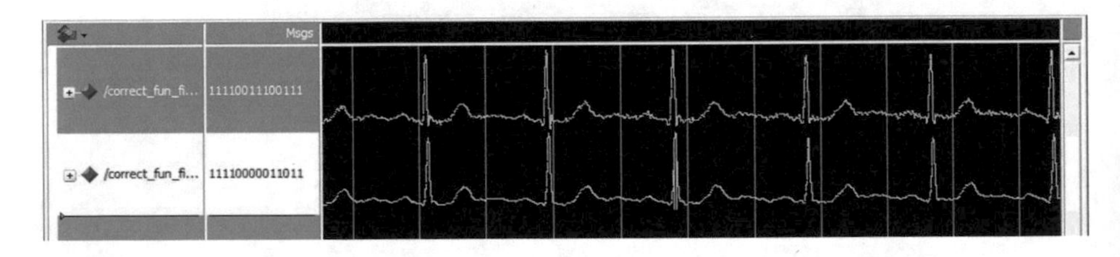

Fig. 6. Hardware Part Simulation Results of Signal n°103 with input SNR of 15 dB.

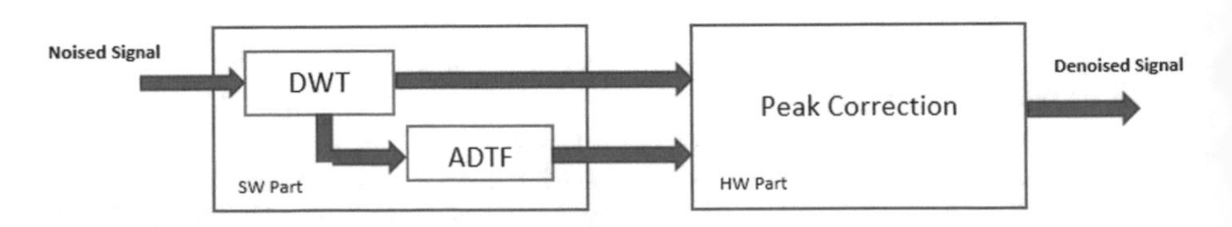

Fig. 7. HW/SW Architecture of DWT-ADTF Algorithm.

Fig. 8. Altera-Intel Cyclone II DE1 FPGA Implementation Platform.

5 Conclusion

This paper was the subject of a HW/SW co-design of the DWT-ADTF algorithm that is used for filtering the ECG signal. The HW/SW partitioning is carried out manually. The simulation results are obtained using the Modelsim simulation tool. The simulation results show that the signal is well-denoised on a qualitative level. The synthesis costs of the entire architecture will be presented later. The DWT-ADTF algorithm has been implemented in the hardware-only architecture in [20]. Future studies will compare the results of the HW-only and HW/SW designs to judge the effectiveness of each architecture. We will also propose an SW only architecture that will be included in the comparative study.

References

1. Colominas, M.A., Schlotthauer, G., Torres, M.E.: Improved complete ensemble EMD: a suitable tool for biomedical signal processing. Biomed. Signal Process. Control. **14**, 19-29 (2014). https://doi.org/10.1016/j.bspc.2014.06.009
2. Huang, W., Cai, N., Xie, W., Ye, Q., Yang, Z.: ECG baseline wander correction based on ensemble empirical mode decomposition with complementary adaptive noise. J. Med. Imaging Heal. Informatics. **5**, 1796-1799 (2015). https://doi.org/10.1166/jmihi.2015.1647
3. Yao, L., Pan, Z.: A new method based CEEMDAN for removal of baseline wander and powerline interference in ECG signals. Optik (Stuttg). **223**, 165566 (2020). https://doi.org/10.1016/j.ijleo.2020.165566
4. Cuomo, S., De Pietro, G., Farina, R., Galletti, A., Sannino, G.: A revised scheme for real time ECG Signal denoising based on recursive filtering. Biomed. Signal Process. Control. **27**, 134-144 (2016). https://doi.org/10.1016/j.bspc.2016.02.007
5. Liu, M., Hao, H.Q., Xiong, P., Lin, F., Hou, Z.G., Liu, X.: Constructing a guided filter by exploiting the butterworth filter for ECG signal enhancement. J. Med. Biol. Eng. **38**, 980-992 (2018). https://doi.org/10.1007/s40846-017-0350-1

6. Singh, P., Pradhan, G.: Variational mode decomposition based ECG denoising using non-local means and wavelet domain filtering. Australas. Phys. Eng. Sci. Med. **41**, 891-904 (2018). https://doi.org/10.1007/s13246-018-0685-0
7. Jenkal, W., Latif, R., Toumanari, A., Dliou, A., El, O., Maoulainine, F.M.R.: ScienceDirect an efficient algorithm of ECG signal denoising using the adaptive dual threshold filter and the discrete wavelet transform. Integr. Med. Res. **36**, 499-508 (2016). https://doi.org/10.1016/j.bbe.2016.04.001
8. Singh, P., Pradhan, G., Shahnawazuddin, S.: Denoising of ECG signal by non-local estimation of approximation coefficients in DWT. Biocybern. Biomed. Eng. **37**, 599-610 (2017). https://doi.org/10.1016/j.bbe.2017.06.001
9. Vargas, R.N., Veiga, A.C.P.: Electrocardiogram signal denoising by clustering and soft thresholding. IET Signal Process. **12**, 1165-1171 (2018). https://doi.org/10.1049/iet-spr.2018.5162
10. Xiong, P., Wang, H., Liu, M., Lin, F., Hou, Z., Liu, X.: A stacked contractive denoising auto-encoder for ECG signal denoising. Physiol. Meas. **37**, 2214-2230 (2016). https://doi.org/10.1088/0967-3334/37/12/2214
11. Wang, G., Yang, L., Liu, M., Yuan, X., Xiong, P., Lin, F., Liu, X.: ECG signal denoising based on deep factor analysis. Biomed. Signal Process. Control. **57**, 101824 (2020). https://doi.org/10.1016/j.bspc.2019.101824
12. Kabir, M.A., Shahnaz, C.: Denoising of ECG signals based on noise reduction algorithms in EMD and wavelet domains. Biomed. Signal Process. Control. **7**, 481-489 (2012). https://doi.org/10.1016/j.bspc.2011.11.003
13. Boda, S., Mahadevappa, M., Kumar, P.: Biomedical signal processing and control a hybrid method for removal of power line interference and baseline wander in ECG signals using EMD and EWT. Biomed. Signal Process. Control. **67**, 102466 (2021). https://doi.org/10.1016/j.bspc.2021.102466
14. Rakshit, M., Das, S.: Biomedical signal processing and control an efficient ecg denoising methodology using empirical mode decomposition and adaptive switching mean filter. Biomed. Signal Process. Control. **40**, 140-148 (2018). https://doi.org/10.1016/j.bspc.2017.09.020
15. Jenkal, W., Latif, R., Toumanari, A., Elouardi, A., Hatim, A., El'bcharri, O.: Real-time hardware architecture of the adaptive dual threshold filter based ECG signal denoising. J. Theor. Appl. Inf. Technol. **96**, 4649–4659 (2018)
16. Abdelhalim, M.B., Habib, S.E.D.: An integrated high-level hardware/software partitioning methodology. Des. Autom. Embed. Syst. **15**, 19-50 (2011). https://doi.org/10.1007/s10617-010-9068-9
17. Shannon, L., Chow, P.: Leveraging reconfigurability in the hardware/software codesign process. ACM Trans. Reconfigurable Technol. Syst. **4**, 2000840 (2011). https://doi.org/10.1145/2000832.2000840
18. Muck, T.R., Frohlich, A.A.: Toward unified design of hardware and software components using C++. IEEE Trans. Comput. **63**, 2880-2893 (2014). https://doi.org/10.1109/TC.2013.159
19. Corporation, A.: Cyclone II Device Handbook , Volume 1 Preliminary Information. Innovation 1 (2008)
20. Mejhoudi, S., Latif, R., Jenkal, W., Saddik, A., El Ouardi, A.: Hardware architecture for adaptive dual threshold filter and discrete wavelet transform based ECG signal denoising. Int. J. Adv. Comput. Sci. Appl. **12**, 45-54 (2021). https://doi.org/10.14569/IJACSA.2021.0121106

Dynamic Output Feedback Controller Design for a Class of Takagi-Sugeno Models: Application to One-Link Flexible Joint Robot

Boutayna Bentahra[1,2(✉)], Karim Bouassem[1,2], Abdellatif El Assoudi[1,2], and El Hassane El Yaagoubi[1,2]

[1] Laboratory of High Energy Physics and Condensed Matter, Faculty of Science Hassan II University of Casablanca, Casablanca, Maarif 5366, Morocco
bentahraboutayna@gmail.com

[2] ECPI, Department of Electrical Engineering, ENSEM, University Hassan II of Casablanca, Casablanca, Oasis 8118, Morocco

Abstract. This paper investigates the design problem of dynamic output feedback controller for a class of continuous time nonlinear systems described by Takagi-Sugeno (T-S) fuzzy model with measurable premise variables. First, the feedback controller design and the observer design of the considered class of T-S models are presented. Next, we show that state feedback stabilizing controller and observer make an output feedback stabilizing controller. The theory is studied by using the Lyapunov approach and the stability conditions of global system is given in term of linear matrix inequalities (LMIs). Finally, an illustrative application of a one link flexible joint robot is given to show the effectiveness of the exposed method.

Keywords: Takagi-Sugeno model · Dynamic output feedback controller · Lyapunov theory · Linear matrix inequalities (LMIs)

1 Introduction

In recent decades, industry sector have seen an unprecedented rapidity of change, driven by the 4th industrial revolution. At the same time, this situation poses number of new challenges in terms cost optimization, design complexity, process performance and reliability.

In this context, more attention is given to the output feedback control problem and particularly for Dynamic Output Feedback (DOF) [1,2]. Indeed, the presence of non-linearities in the model makes the controller design more complex. For that reason, in the field of nonlinear systems modelling and control a significant place is given to Takagi-Sugeno approach [3], which offers a universal approximator and allows to represent nonlinear systems through a combination of linear sub-models [4].

N. Aboutabit et al. (Eds.): ICMICSA 2022, LNNS 656, pp. 211–222, 2023.
https://doi.org/10.1007/978-3-031-29313-9_19

In this work, we are focusing on the separation principal approach design for nonlinear system described by T-S models [5–8]. The stability and the stabilization are studied through Lyapunov method [9–12]. As regards the separation principle, is defining as a theory that combines the couple (observer and controller). This theory is also called dynamic output feedback controller or observer-based controller. For nonlinear systems described by T-S models, the problems of control and observation design are wildly studied in the literature; see for instance [13–16]. Notice that, generally an interesting way to address the problems raised previously (control and observation) is to set the convergence conditions on the LMI form [17].

The main contribution in this paper is the development of an observer-based controller for a class of continuous-time T-S models with measurable premise variables. The stability of the closed-loop augmented system is studied using the Lyapunov theory and a judicious use of the Young's lemma allows to set the stability conditions in terms of LMIs.

The structure of this paper is as follows. In Sect. 2 we present the class of the studied systems. The feedback controller design and the observer design of the considered class of T-S models are presented respectively in Sects. 3 and 4. The main result about observer-based controller design for a class of continuous-time T-S models with measurable premise variables is exposed in Sect. 5. While in Sect. 6, an illustrative application of a flexible joint robot is given to illustrate the effectiveness of the exposed method, and the conclusion will be in Sect. 7.

2 System Description

In this work, the following class of T-S explicit systems is considered:

$$\begin{cases} \dot{x} = \sum_{i=1}^{q} \mu_i(\xi)(A_i x + B_i u) \\ y = \sum_{i=1}^{q} \mu_i(\xi) C_i x \end{cases} \tag{1}$$

where $x \in \mathbb{R}^r$ is the state vector, $u \in \mathbb{R}^m$ is the control input, $y \in \mathbb{R}^p$ is the measured output. $A_i \in \mathbb{R}^{n \times n}$, $B_i \in \mathbb{R}^{n \times m}$, $C_i \in \mathbb{R}^{p \times n}$ are real known constant matrices. The premise variable ξ is supposed to be real-time accessible. q is the number of sub-models. The $\mu_i(\xi)$ $(i = 1, \ \ldots, \ q)$ are the weighting functions that ensure the transition between the contribution of each sub model:

$$\begin{cases} \dot{x} = A_i x + B_i u \\ y = C_i x \end{cases} \tag{2}$$

They verify the so-called convex sum properties:

$$\begin{cases} \sum_{i=1}^{q} \mu_i(\xi) = 1 \\ 0 \leq \mu_i(\xi) \leq 1 \ , \ i = 1, \ldots, q \end{cases} \tag{3}$$

The aim of this work is to design an observer-based controller for system (1) which assures globally asymptotically stability of closed-loop system.

3 Feedback Controller Design

In this section, the goal is to give the gains K_i of the following control law:

$$u(x) = -\sum_{i=1}^{q} \mu_i(\xi) K_i x \tag{4}$$

Notice that the local linear feedback controller related with each sub-model (2) is given by: $u(x) = -K_i x$
By substituting (4) into (1), the following closed-loop fuzzy system can be represented as:

$$\begin{cases} \dot{x} = \sum_{i,j=1}^{q} \mu_i(\xi)\mu_j(\xi)\Gamma_{ij} x \\ y = \sum_{i=1}^{q} \mu_i(\xi) C_i x \end{cases} \tag{5}$$

where

$$\Gamma_{ij} = A_i - B_i K_j \tag{6}$$

Theorem 1. *[14] : The closed-loop system described by (5) is globally exponentially stable if there exist matrices $Y_1 > 0$, U_i, $i = 1, \ldots, q$ verifying the following LMIs:*

$$\begin{cases} A_i Y_1 + Y_1^T A_i^T - B_i U_i - U_i^T B_i^T < 0 \qquad i = 1, \ldots, q \\ A_i Y_1 + Y_1^T A_i^T + A_j Y_1 + Y_1^T A_j^T - B_i U_j - U_j^T B_i^T - B_j U_i - U_i^T B_j^T < 0 \\ i < j \ s.t. \ \mu_i \cap \mu_j \neq \emptyset \end{cases} \tag{7}$$

The fuzzy local feedback gains K_i, $i = 1, \ldots, q$ are given by:

$$K_i = U_i Y_1^{-1} \tag{8}$$

4 Observer Design

In this section, we present an observer design for a class of T-S explicit models (1). It takes the following form:

$$\begin{cases} \dot{\hat{x}} = \sum_{i=1}^{q} \mu_i(\xi)(A_i \hat{x} + B_i u - L_i(\hat{y} - y)) \\ \hat{y} = \sum_{i=1}^{q} \mu_i(\xi) C_i \hat{x} \end{cases} \tag{9}$$

where $\hat{x}$ and $\hat{y}$ denote the estimated state vector and output vector respectively. The activation functions $\mu_i(\xi)$ are the same as those used in the T-S model (1). L_i are the gains of observer which are to be determined such that the state estimation error $e = \hat{x} - x$ converges towards 0.

It follows from (1) and (9) that the observer error dynamic is given by the differential equation:

$$\dot{e} = \sum_{i,j=1}^{q} \mu_i(\xi)\mu_j(\xi)\Omega_{ij}e \tag{10}$$

where

$$\Omega_{ij} = A_i - L_i C_j \tag{11}$$

Theorem 2. *[14] : There exists an observer (9) for (1) if there exist matrices $P_2 > 0$, W_i, $i = 1, \ldots, q$ verifying the following LMIs:*

$$\begin{cases} A_i^T P_2 + P_2 A_i - C_i^T W_i^T - W_i C_i < 0 \qquad i = 1, \ldots, q \\ A_i^T P_2 + A_j^T P_2 + P_2 A_i + P_2 A_j - C_j^T W_i^T - C_i^T W_j^T - W_i C_j - W_j C_i < 0 \\ i < j \ s.t. \ \mu_i \cap \mu_j \neq \emptyset \end{cases} \tag{12}$$

The fuzzy local observer gains L_i, $i = 1, \ldots, q$ are given by:

$$L_i = P_2^{-1} W_i \tag{13}$$

5 Main Result

In this section, a fuzzy dynamic output feedback controller is designed for system (1). First the feedback control is given by:

$$u(\hat{x}) = -\sum_{i=1}^{q} \mu_i(\xi) K_i \hat{x} \tag{14}$$

where K_i can be determined by Theorem 3 below.

Then, we consider the augmented state: $x_a = \begin{pmatrix} \hat{x} \\ e \end{pmatrix}$

Combining (9), (10) and (14), we obtain the following augmented closed-loop system:

$$\dot{x}_a = \sum_{i,j=1}^{q} \mu_i(\xi)\mu_j(\xi)\Sigma_{ij} x_a \tag{15}$$

where

$$\Sigma_{ij} = \begin{pmatrix} \Gamma_{ij} & -L_i C_j \\ 0 & \Omega_{ij} \end{pmatrix} \tag{16}$$

Γ_{ij} and Ω_{ij} are defined in (6) and (11) respectively. The observer gain L_i are determined by Theorem 3.

The following theorem provides the main result of this work.

Theorem 3. *The closed-loop TS model described by (15)–(16) is globally asymptotically stable if for a fixed scalar $\varepsilon > 0$, there exist matrices $Y_1 > 0$, $Y_2 > 0$, U_i, W_i, $i = 1, \ldots, q$ verifying the following LMIs:*

$$\begin{cases} \begin{pmatrix} \Lambda_{ii}^{11} & 0 & I & 0 \\ 0 & \Lambda_{ii}^{22} & 0 & C_i^T W_i^T \\ I & 0 & -\varepsilon Y_2 & 0 \\ 0 & W_i C_i & 0 & -\varepsilon^{-1} Y_2 \end{pmatrix} < 0 \qquad i = 1, \ldots, q \\ \begin{pmatrix} \Lambda_{ij}^{11} + \Lambda_{ji}^{11} & 0 & I & 0 \\ 0 & \Lambda_{ij}^{22} + \Lambda_{ji}^{22} & 0 & C_j^T W_i^T + C_i^T W_j^T \\ I & 0 & -\varepsilon Y_2 & 0 \\ 0 & W_i C_j + W_j C_i & 0 & -\varepsilon^{-1} Y_2 \end{pmatrix} < 0 \\ i < j \ s.t. \ \mu_i \cap \mu_j \neq \emptyset \end{cases} \tag{17}$$

where

$$\begin{cases} \Lambda_{ii}^{11} = Y_1^T A_i^T - U_i^T B_i^T + A_i Y_1 - B_i U_i \\ \Lambda_{ii}^{22} = A_i^T Y_2 - C_i^T W_i^T + Y_2 A_i - W_i C_i \\ \Lambda_{ij}^{11} = Y_1^T A_i^T - U_j^T B_i^T + A_i Y_1 - B_i U_j \\ \Lambda_{ji}^{11} = Y_1^T A_j^T - U_i^T B_j^T + A_j Y_1 - B_j U_i \\ \Lambda_{ij}^{22} = A_i^T Y_2 - C_j^T W_i^T + Y_2 A_i - W_i C_j \\ \Lambda_{ji}^{22} = A_j^T Y_2 - C_i^T W_j^T + Y_2 A_j - W_j C_i \end{cases} \tag{18}$$

The dynamic controller gains K_i and L_i are given by:

$$\begin{cases} K_i = U_i Y_1^{-1} \\ L_i = Y_2^{-1} W_i \end{cases} \tag{19}$$

Proof. Let us consider the following candidate Lyapunov function:

$$V(x_a) = x_a^T P x_a \ , P = P^T > 0 \text{ with } P = \begin{pmatrix} P_1 & 0 \\ 0 & P_2 \end{pmatrix} \tag{20}$$

The time derivative of the Lyapunov function along the trajectories of the system (15) is obtained as:

$$\dot{V}(x_a) = \sum_{i,j=1}^{q} \mu_i(\xi)\mu_j(\xi) x_a^T (\Sigma_{ij}^T P + P \Sigma_{ij}) x_a \tag{21}$$

Therefore, we have the following stability conditions:

$$\begin{cases} \Sigma_{ii}^T P + P \Sigma_{ii} < 0 \qquad i = 1, \ldots, q \\ (\dfrac{\Sigma_{ij} + \Sigma_{ji}}{2})^T P + P(\dfrac{\Sigma_{ij} + \Sigma_{ji}}{2}) < 0 \quad i < j \ s.t. \ \mu_i \cap \mu_j \neq \emptyset \end{cases} \tag{22}$$

which becomes from (16) and (22):

$$\begin{cases} \begin{pmatrix} \Gamma_{ii}^T P_1 + P_1 \Gamma_{ii} & -P_1 L_i C_i \\ -(L_i C_i)^T P_1 & \Omega_{ii}^T P_2 + P_2 \Omega_{ii} \end{pmatrix} < 0 \qquad i = 1, \ldots, q \\ \begin{pmatrix} (\Gamma_{ij} + \Gamma_{ji})^T P_1 + P_1(\Gamma_{ij} + \Gamma_{ji}) & -P_1(L_i C_j + L_j C_i) \\ -(L_i C_j + L_j C_i)^T P_1 & (\Omega_{ij} + \Omega_{ji})^T P_2 + P_2(\Omega_{ij} + \Omega_{ji}) \end{pmatrix} < 0 \\ i < j \ s.t. \ \mu_i \cap \mu_j \neq \emptyset \end{cases} \tag{23}$$

Now multiplying the two inequalities given in (23) on the left and right by $diag([(P_1^{-1})^T, 0])$ and $diag([(P_1^{-1})^T, 0])$ respectively and defining new variables $Y_1 = P_1^{-1}$, $Y_2 = P_2$, we obtain:

$$\begin{cases} \Psi_1 = \begin{pmatrix} Y_1^T \Gamma_{ii}^T + \Gamma_{ii} Y_1 & -L_i C_i \\ -(L_i C_i)^T & \Omega_{ii}^T Y_2 + Y_2 \Omega_{ii} \end{pmatrix} < 0 \qquad i = 1, \dots, q \\ \Psi_2 = \begin{pmatrix} Y_1^T (\Gamma_{ij} + \Gamma_{ji})^T + (\Gamma_{ij} + \Gamma_{ji}) Y_1 & -L_i C_j + L_j C_i) \\ -(L_i C_j + L_j C_i)^T & (\Omega_{ij} + \Omega_{ji})^T Y_2 + Y_2 (\Omega_{ij} + \Omega_{ji}) \end{pmatrix} < 0 \\ i < j \ s.t. \ \mu_i \cap \mu_j \neq \emptyset \end{cases} \tag{24}$$

By considering the following:

$$\begin{cases} \Phi_1 = \begin{pmatrix} Y_1^T \Gamma_{ii}^T + \Gamma_{ii} Y_1 & 0 \\ 0 & \Omega_{ii}^T Y_2 + Y_2 \Omega_{ii} \end{pmatrix} \\ \Phi_2 = \begin{pmatrix} Y_1^T (\Gamma_{ij} + \Gamma_{ji})^T + (\Gamma_{ij} + \Gamma_{ji}) Y_1 & 0 \\ 0 & (\Omega_{ij} + \Omega_{ji})^T Y_2 + Y_2 (\Omega_{ij} + \Omega_{ji}) \end{pmatrix} \\ Z_1 = \begin{pmatrix} 0 \ L_i C_i \end{pmatrix} \\ Z_2 = \begin{pmatrix} 0 \ L_i C_j + L_j C_i \end{pmatrix} \\ Z = \begin{pmatrix} -I \ 0 \end{pmatrix} \end{cases} \tag{25}$$

Matrices Λ_1 and Λ_2 can be rewritten again as:

$$\begin{cases} \Psi_1 = \Phi_1 + Z_1^T Z + Z^T Z_1 \\ \Psi_2 = \Phi_2 + Z_2^T Z + Z^T Z_2 \end{cases} \tag{26}$$

Lemma 1 : Young's Inequality. For any matrices $\tilde{X}$ and $\tilde{Z}$ with appropriate dimensions, the following property holds for any invertible matrix J and scalar $\varepsilon > 0$:

$$\tilde{X}^T \tilde{Z} + \tilde{Z}^T \tilde{X} \leq \varepsilon \tilde{X}^T J \tilde{X} + \varepsilon^{-1} \tilde{Z}^T J^{-1} \tilde{Z} \tag{27}$$

Applying Lemma 1 and taking $J = Y_2$, (26) becomes:

$$\begin{cases} \Psi_1 \leq \Phi_1 - \begin{pmatrix} I & 0 \\ 0 & (L_i C_i)^T Y_2 \end{pmatrix} \begin{pmatrix} -\varepsilon Y_2 & 0 \\ 0 & -\varepsilon^{-1} Y_2 \end{pmatrix}^{-1} \begin{pmatrix} I & 0 \\ 0 & Y_2 L_i C_i \end{pmatrix} \\ \Psi_2 \leq \Phi_2 - \begin{pmatrix} I & 0 \\ 0 & (L_i C_j + L_j C_i)^T Y_2 \end{pmatrix} \begin{pmatrix} -\varepsilon Y_2 & 0 \\ 0 & -\varepsilon^{-1} Y_2 \end{pmatrix}^{-1} \begin{pmatrix} I & 0 \\ 0 & Y_2 (L_i C_j + L_j C_i) \end{pmatrix} \end{cases} \tag{28}$$

Hence, using the Schur complement [17], the inequalities $\Psi_1 < 0$ and $\Psi_2 < 0$ hold if the following two matrix inequalities are satisfied.

$$\begin{cases} \begin{pmatrix} Y_1^T \Gamma_{ii}^T + \Gamma_{ii} Y_1 & 0 & I & 0 \\ 0 & \Omega_{ii}^T Y_2 + Y_2 \Omega_{ii} & 0 & (L_i C_i)^T Y_2 \\ I & 0 & -\varepsilon Y_2 & 0 \\ 0 & Y_2 L_i C_i & 0 & -\varepsilon^{-1} Y_2 \end{pmatrix} < 0 \qquad i = 1, \dots, q \\ \begin{pmatrix} \Pi_{ij}^{11} & 0 & I & 0 \\ 0 & \Pi_{ij}^{22} & 0 & (L_i C_j + L_j C_i)^T Y_2 \\ I & 0 & -\varepsilon Y_2 & 0 \\ 0 & Y_2 (L_i C_j + L_j C_i) & 0 & -\varepsilon^{-1} Y_2 \end{pmatrix} < 0 \\ i < j \ s.t. \ \mu_i \cap \mu_j \neq \emptyset \end{cases} \tag{29}$$

where

$$\begin{cases} \Pi_{ij}^{11} = Y_1^T(\Gamma_{ij} + \Gamma_{ji})^T + (\Gamma_{ij} + \Gamma_{ji})Y_1 \\ \Pi_{ij}^{22} = (\Omega_{ij} + \Omega_{ji})^T Y_2 + Y_2(\Omega_{ij} + \Omega_{ji}) \end{cases} \tag{30}$$

Then, from (6), (11) and the use of the changes of variables:

$$\begin{cases} U_i = K_i Y_1 \\ W_i = Y_2 L_i \end{cases} \tag{31}$$

we establish the LMI conditions (17) given in Theorem 3.

6 Application to One-Link Flexible Joint Robot

To illustrate the effectiveness of the suggested theory, we consider, in this section, a one-link flexible joint robot given in [18]. The following equations represent this process :

$$\begin{cases} \dot{x} = A(x)x + Bu \\ y = Cx \end{cases} \tag{32}$$

where x, u and y are the state vector, the applied input force and the variable of output measurement respectively.
The state vector is defined as follow: $x = (x_1, x_2, x_3, x_4)^T$ with x_1 and x_3 are the angles of rotations of the motor and the link respectively, x_2 and x_4 are their angular velocities.
Matrices $A(x)$, B and C are given as follow:

$$A(x) = \begin{pmatrix} 0 & 1 & 0 & 0 \\ -\frac{K+\beta}{J_m} & 0 & \frac{K}{J_m} & 0 \\ 0 & 0 & 0 & 1 \\ \frac{K}{J_l} & 0 & \xi & 0 \end{pmatrix} B = \begin{pmatrix} 0 \\ 0 \\ \frac{k_T}{J_m} \\ 0 \end{pmatrix}, \; C = \begin{pmatrix} 1\,0\,0\,0 \\ 0\,0\,1\,0 \end{pmatrix} \tag{33}$$

with

$$\xi = -\frac{K}{J_l} - \frac{mgb}{J_l}\frac{sin(x_3)}{x_3} \tag{34}$$

The physical parameters definition and their numerical values are given in the following table: next, we proceed to construct the T-S model from the non-linear system (32). To do we consider the sector of non-linearities of the term $\xi \in [\xi_{min}, \xi_{max}]$ of the matrix $A(x)$. Then, the non-linear term $\xi(t)$ can be transformed under the following form:

$$\xi = F_1\xi_{max} + F_2\xi_{min} \tag{35}$$

where

$$\begin{cases} F_1 = \frac{\xi - \xi_{min}}{\xi_{max} - \xi_{min}} \\ F_2 = \frac{\xi_{max} - \xi}{\xi_{max} - \xi_{min}} \end{cases} \tag{36}$$

Table 1. Model parameters of one-link flexible joint robot

Parameter	Meaning	Value
J_m	Motor inertia	$0.0037\ kg.m^2$
J_l	Link inertia	$0.0093\ kg.m^2$
m	Link mass	$0.21\ kg$
b	Center of mass	$0.15\ m$
K	Elastic constant	$0.18\ N.m/rad$
β	Viscous friction coefficient	$0.046\ kg.m2$
K_T	Amplifier gain	$0.08\ N.m/V$
g	Acceleration due to Gravity	$9.81\ m/s^2$

Then, the global T-S fuzzy model is inferred as:

$$\begin{cases} \dot{x} = \sum\limits_{i=1}^{2} \mu_i(\xi)(A_i x + Bu) \\ y = Cx \end{cases} \tag{37}$$

With

$$A_1 = \begin{pmatrix} 0 & 1 & 0 & 0 \\ -\frac{K+b}{J_m} & 0 & \frac{K}{J_m} & 0 \\ 0 & 0 & 0 & 1 \\ \frac{K}{J_l} & 0 & \xi_{max} & 0 \end{pmatrix}, \quad A_2 = \begin{pmatrix} 0 & 1 & 0 & 0 \\ -\frac{K+b}{J_m} & 0 & \frac{K}{J_m} & 0 \\ 0 & 0 & 0 & 1 \\ \frac{K}{J_l} & 0 & \xi_{min} & 0 \end{pmatrix} \tag{38}$$

The weighting functions are given by:

$$\begin{cases} \mu_1 = F_1 \\ \mu_2 = F_2 \end{cases} \tag{39}$$

The resolution of the LMIs defined in Theorem 3 with $\varepsilon_1 = 8.5$ lead to the following observer and feedback gains:

$$\begin{cases} K_1 = \begin{pmatrix} 25.8648 & 2.1097 & -3.9280 & 7.7194 \end{pmatrix} \\ K_2 = \begin{pmatrix} 17.5796 & 1.4575 & -4.0783 & 4.8719 \end{pmatrix} \\ L_1 = \begin{pmatrix} 3.0890 & 0.0847 \\ -0.7929 & 14.6929 \\ -0.1699 & 3.4540 \\ -1.7301 & 13.6301 \end{pmatrix}, \quad L_2 = \begin{pmatrix} 3.3112 & 0.7589 \\ -0.3938 & 13.2637 \\ 0.6982 & 4.1210 \\ -1.3417 & -19.0193 \end{pmatrix} \end{cases} \tag{40}$$

Simulation results are presented in Figs. 1, 2, 3 and 4 with initial conditions:

$$x_0 = [0\ 0\ \frac{\pi}{2}\ 0]^T\ ;\ \hat{x}_0 = [0\ 0.1\ \frac{\pi}{2}\ 0.1]^T$$

These simulation results show the performance of the proposed observer-based controller designed with the parameters K_1, K_2, L_1, L_2 given by (40). They show that the global asymptotic stability of the closed loop T-S system with the control law is satisfied and the observer gives a good estimation of state variables of the one-link flexible joint robot.

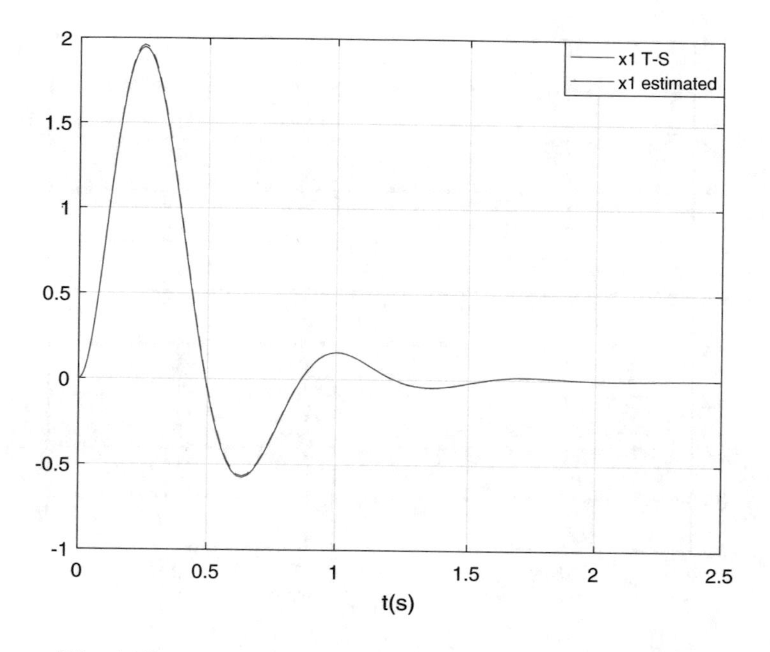

Fig. 1. State x_1 and $\hat{x}_1$ with fuzzy observer-based controller

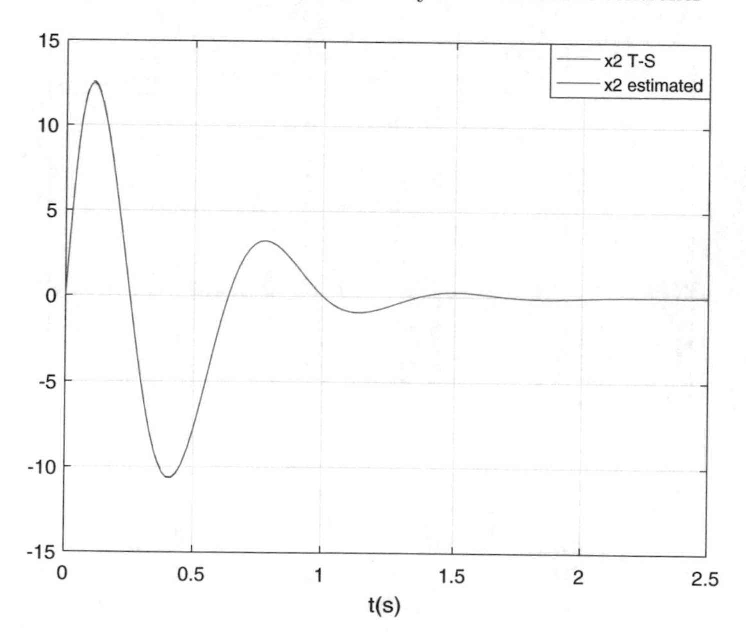

Fig. 2. State x_2 and $\hat{x}_2$ with fuzzy observer-based controller

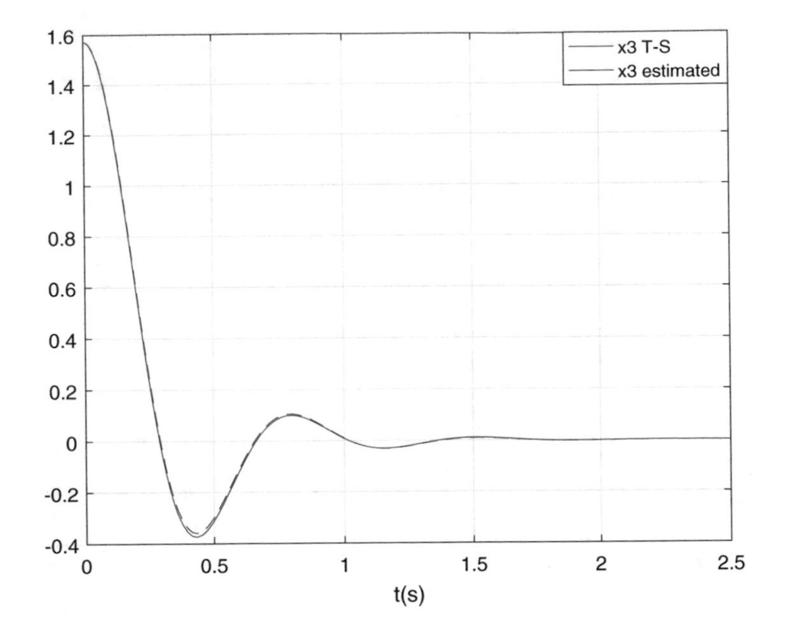

Fig. 3. State x_3 and $\hat{x}_3$ with fuzzy observer-based controller

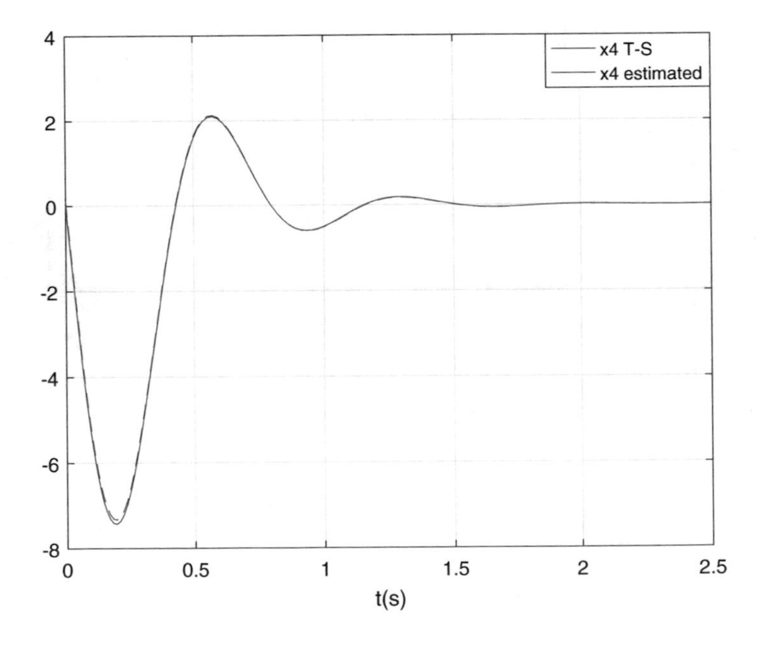

Fig. 4. State x_4 and $\hat{x}_4$ with fuzzy observer-based controller

7 Conclusion

A dynamic output feedback controller design approach for a class of continuous time T-S models with measurable premise variables is proposed in this paper. The convergence of the closed-loop system is studied using the Lyapunov theory and the stability conditions are formulated in LMI terms. Considering a one link flexible joint robot to illustrate the theory effectiveness. The simulation results demonstrate the good performance of the proposed controller design.

References

1. Franklin, P., Powell, J. D.: Feedback control of dynamic systems. Emami-naeini's Feedback Control of Dynamic Systems. 2nd edn. Emami-Naeini (2006)
2. Jonathan, H.: Dynamic output-feedback control architectures matter [Focus on Education]. IEEE Control. Syst. **2**(36), 88–117 (2016). https://doi.org/10.1109/MCS.2016.2602641
3. Takagi, T., Sugeno, M.: Fuzzy identification of systems and its application to modelling and control. IEEE Trans. Syst., Man and Cyber **2**(1115), 116–132 (1985). https://doi.org/10.1109/tsmc.1985.6313399
4. Johansen, T., Shorten, R., Murray-Smith, R.: On the interpretation and identification of dynamic Takagi-Sugeno fuzzy models. IEEE Trans. Fuzzy Syst. **2**(8), 297–313 (2000). https://doi.org/10.1109/91.855918
5. Chadli, M., Maquin, D., Ragot, J.: Observer-based controller for Takagi-Sugeno models. IEEE International Conference on Systems, (SMC2002At) Man and Cybernetics, vol. 5, pp. Hammamet, Tunisia (2002). https://doi.org/10.1109/ICSMC.2002.1176347
6. Taieb, N.H., Hammami, M., Delmotte, F.: A separation principle for Takagi-Sugeno control fuzzy systems. Arch. Control Sci. **29**(2), 227–245 (2019). https://doi.org/10.24425/acs.2019.129379
7. Zoltán, N., Lendek, Z.: Observer-based controller design for Takagi-Sugeno fuzzy systems with local nonlinearities. IEEE International Conference on Fuzzy Systems (FUZZ-IEEE), pp. 1–6 (2019). https://doi.org/10.1109/FUZZ-IEEE.2019.8858937
8. Hu, G., Zhang, J., Yan, Z.: An improved approach to fuzzy dynamic output feedback $H\infty$ control of continuous-time Takagi-Sugeno fuzzy systems. Int. J. Syst. Sci. **53**(14), 1–16 (2021). https://doi.org/10.1080/00207721.2021.2013976
9. Tanaka, K., Wang, H.O.: fuzzy control systems design and analysis: a linear matrix inequality approach. John Wiley & Sons (2001). https://doi.org/10.1002/0471224596
10. Zhang, H., Xie, X.-P.: Relaxed stability conditions for continuous-time T-S Fuzzy-control systems via augmented multi-indexed matrix approach. IEEE Trans. Fuzzy Syst. **2**(19), 478–492 (2011). https://doi.org/10.1109/TFUZZ.2011.2114887
11. Wang, L., Liu, J., Lam, H.-K.: Further study on stabilization for continuous-time Takagi-Sugeno fuzzy systems with time delay. IEEE Trans. Cybern. **51**(11), 5637–5643 (2021). https://doi.org/10.1109/TCYB.2020.2973276
12. Wang, L., Lam, H.-K.: $H\infty$ control for continuous-time Takagi-Sugeno fuzzy model by applying generalized Lyapunov function and introducing outer variables. Automatica **125**(3001), 109–409 (2021). https://doi.org/10.1016/j.automatica.2020.109409

13. Tanaka, K., Sano, M.: On the concept of fuzzy regulators and fuzzy observers. Proceed. Third IEEE Int. Conf. Fuzzy Syst. **2**, 767–772 (1994). https://doi.org/10.1109/fuzzy.1994.343832
14. Tanaka, K., Wang, H. O.:Fuzzy regulators and fuzzy observers: a linear matrix inequality approach. In: 36th IEEE, Conference on Decision and Control 1997, vol. 2, pp. 1315–1320. San Diego (1997). https://doi.org/10.1109/cdc.1997.657640
15. Bergsten, P., Palm, R., Driankov, D.: Observers for Takagi-Sugeno fuzzy systems. IEEE Trans. Syst. Man Cybern.- Part B: Cybern. **1**(32), 114–121 (2002). https://doi.org/10.1109/3477.979966
16. Jamel, W., Khedher, A., Bouguila, N., Othman, K.: Observers design for Takagi-Sugeno models. In: Systems, Automation, and Control, pp. 61–82 De Gruyter, Berlin, Boston (2020). https://doi.org/10.1515/9783110591729-004
17. Boyd, S., El Ghaoui, L., Feron, E., Balakrishnan, V.: Linear matrix inequalities in systems and control theory. PA: SIAM, Philadelphia (1994). https://doi.org/10.1137/1.9781611970777
18. Mohamed, K., Chadli, M., Chaabane, M.: Unknown inputs observer for a class of nonlinear uncertain systems: an LMI approach. Int. J. Autom. Comput. **9**(3), 331–336 (2012). https://doi.org/10.1007/s11633-012-0652-2

Solution Based on Mobile Web Application to Detect and Treat Patients with Mental Disorders

Chaimae Taoussi(✉), Imad Hafidi, and Abdelmoutalib Metrane

Laboratory of Process Engineering, Computer Science and Mathematics, Sultan Moulay Slimane University, Beni Mellal, Morocco
chaimae.taoussi@usms.ac.ma

Abstract. Mental health is an integral part of health and wellness, it covers both an individual and a collective dimension, so it is essential to promote it. Moreover, nowadays the digital era has seen an exponential development of various tools using the Internet. A new solution has been proposed in this article that takes advantage of this technological era to improve the mental health of people in mental distress suffering in silence. The proposed solution is based on a mobile web application available in three different languages, on which the patient will create his account, will then pass internationally recognized psychological tests such as Hospital Anxiety and Depression Scale and Perceived Stress Scale-10, he will instantly receive psychological recommendations designed by professional psychologists based on his score, in addition, to support and psychological help according to the evolution of his psychological state. The main objective of the solution is to promote mental health by diagnosing the maximum of people suffering from mental disorders, to prepare electronic medical records for each of the patients by collecting the maximum of data in order to apply the Big data process afterward and guarantee more efficient and intelligent service.

Keywords: Mobile web application · Mental health · Telemedicine · Mental disorders · Big data

1 Introduction

The multiple technologies available in the mental health service are now being used. This is referred to as e-mental health, which the World Health Organization (WHO) defines as "digital services for the well-being of the individual" and includes health and wellness applications for patients and citizens in general, as well as new digital tools for medical and preventive practices such as consultations and tele assistance [1].

C. Taoussi, I. Hafidi and A. Metrane—Contributed equally to this work. All authors have read and approved the final manuscript.

N. Aboutabit et al. (Eds.): ICMICSA 2022, LNNS 656, pp. 223–231, 2023.
https://doi.org/10.1007/978-3-031-29313-9_20

In this context, the WHO recommends investing in greater integration of mental health into health and social protection systems, thus integrating health coverage into mental health care, so that all people have the greatest possible access to high levels of well-being and mental health, as outlined in the 2030 Agenda for Sustainable Development. So, the definition of mental health in the era of the Sustainable Development Goals (SDGs) seems to offer opportunities for digital health and support a holistic and coherent approach to health by expanding digital mental health and should guide the deployment of digital solutions that are cross-cutting, integrated and focused on the person concerned and their needs [2].

Despite the fact that this technological revolution has allowed the development of digital solutions that have made healthcare more accessible and effective, in the field of mental health there are still blockages and complications. In the majority of countries, especially in the arab world, mental illness is a taboo subject, to this effect psychological disorders such as anxiety, depression, and stress are not detected in advance which causes huge problems that affect both the person as well as his entourage. Furthermore, when it comes to mental disorders, there is no specific type or age range of people who suffer in silence, so it is difficult to diagnose thousands of people at once without the use of technology. Furthermore, even if we are able to diagnose patients and detect the psychological problems they have suffered, it is difficult to provide individualized treatment and psychological support for each person. Finally, the major problem that blocks the evaluation of the progress of the psychological state of each patient in a personalized way is the unavailability of Electronic Medical Records (EMR) well filled and updated after each consultation that contains all the detailed medical information of the patient. That's why this solution based on a mobile web application has been proposed, to respond to the issues mentioned above in order to promote mental health through the detection and treatment of mental disorders.

2 Description of the Proposed Solution

2.1 Technical Presentation of Solution

Medico-call is a company that provides psychological support and monitoring for each person suffering from mental disorders with the help of a team of psychologists of different specialties with a multitude of skills that accompany customers in need to overcome their problems, in complete anonymity, confidentiality, and total availability (24 h/7 days). Therefore, in the context of promoting mental health and community well-being, the main goal is to offer an intelligent solution that exploits the most up-to-date detection technologies and treats patients with mental disorders.

The solution consists in developing a mobile web application available in three different languages (english, french, arabic), which will allow each patient to take internationally recognized psychological tests such as the Hospital Anxiety and Depression Scale (HADS) and the Perceived Stress Scale-10 (PSS-10), depending on the score they receive automatically personalized psychological recommendations written by experienced psychologists in addition to psychological support and follow-up according to the progress of his psychological state. The application aims to promote mental health by diagnosing the maximum number of people suffering from psychological disorders such as anxiety, depression, and stress, in addition, allows the preparation of electronic medical records for each patient by collecting the maximum data to apply the big data process thereafter and guarantee an intelligent service.

In order to guarantee the performance and security of the mobile web application with which the three actors (administrator, psychologist, and patient) interact, each from their portal, the following technologies were used:

- The platform **Node.js** for developing the backend of both the mobile and web versions of the application.
- The open-source JavaScript library **React JS** for developing the front-end of the web version of the application.
- The open-source UI software framework **React Native** for the development of the Front-end of the mobile version of the application.
- A **RESTful API** developed in Python and managed by the Flask framework [3], which aims to request the user's information as well as their test scores and returns personalized psychological recommendations in response.

2.2 Psychological Tests Used in the Solution

Psychology is a field of study that focuses on the assessment and diagnosis of mental, emotional, and behavioral problems. It is also about helping people understand, overcome and manage their problems. As diagnosis is one of the pillars of this field, it is necessary to double the efforts to understand the mental functioning, determine the direction of choice, formulate prognosis and ensure the right diagnosis to many people at the same time with the same efficiency, for this purpose psychological test remains one of the best solutions in this sense, their function is to allow the collection of valid and reliable psychological information while a person is examined.

In the first version of the solution based on a mobile web application for the detection and treatment of patients suffering from mental disorders, the focus was on the detection and treatment of the following three psychological pathologies : anxiety, depression, and stress. The integration of the following two psychological tests known and validated at the international level guarantees a good diagnosis of the pathologies mentioned above: the HADS and the PSS-10.

The Hospital Anxiety and Depression Scale (HADS): The HADS is a questionnaire created by Zigmond and Snaith (1983) [4] to assess an individual's anxiety and depression levels. It is currently a widely used instrument in clinical practice and scientific research. The HADS is a fourteen items self-report scale (seven items for the depression subscale and seven items for the anxiety subscale).

The total score is the sum of the fourteen items, and the score of each subscale is the sum of the seven items of that subscale (ranging from 0 to 21 view that the range for each item is 0 to 3). In addition, subsequent research revealed that the anxiety and depression subscales were separate measures, subsequently, each mood state was divided into four categories: normal, mild, moderate, and severe according to the score range of each patient [5]. An examination of the HADS questionnaire's validity verified the notion that it is a good questionnaire for detecting different dimensions of anxiety and depression, as well as cases of anxiety disorders and depression in non-psychiatric hospital patients [6].

The success of the HADS psychological test, as well as the positive findings it has produced after application to many studies in the medical and psychological fields, is strong evidence of its efficiency. In this context, a first study reveals that anxiety and depression risk scores above the specified thresholds of the HADS are beneficial for detecting anxiety and depression in cancer patients, and therefore could be used in therapeutic practice [7]. Another study found that the HADS helps clinicians provide early treatment and guides future efforts to improve the mental health of people with cardiovascular disease [8]. In addition, based on the HADS test, a first-of-its-kind study was conducted to determine whether some young athletes are at risk for developing anxiety and depression symptoms by examining the distribution and prevalence rates of anxiety and depression symptoms in these young athletes with a focus on detecting clinically relevant consequences [9].

The Perceived Stress Scale (PSS): The PSS is the most extensively used psychological instrument for measuring perceived stress. In terms of current stress levels, the questions in this questionnaire are straightforward, easy to understand, and the answer options are straightforward. They were made to see how uncontrollable, unmanageable, and overburdened respondents' lives are [10]. The PSS is a 14-item self-report measure that takes only a few minutes to complete. Individuals are asked to rate from 0- never to 4 - very often, how much they have experienced each of the feelings and thoughts listed in the past month. Its main advantage is that it is used internationally, with a range of samples such as workers and students, and in a variety of situations. In addition, the PSS was originally created in English by Cohen, Kamarck, and Mermelstein, and has subsequently been translated from English into several other languages such as French [11], Arabic [12], Japanese [13] and Spanish [14]. This allowed us to confirm the test's validity at an international level with the same efficiency and effectiveness.

The Perceived Stress Scale, has been applied in different studies to measure the global level of stress in a series of clinical and research contexts, and it has left a positive imprint and given good results. In this context, the PSS was used to assess stress in cancer patients, yielding clinical data that revealed the importance of perceived stress in major cancer effects such as patient quality of life and treatment adherence [15]. The reliability and validity of the psychometric features of the three versions of the PSS in adult survivors of the suicide death of a family member or loved one were evaluated and confirmed as effective in a final study [16].

2.3 How the Application Works

In order to guarantee the efficiency of detection and treatment of patients with mental disorders, an intelligent mobile web application has been developed with a simple and efficient process of use, which will allow to collect more data later useful for several health care improvement practices:

- First of all, the patient must create his profile in the application to connect afterward and take advantage of the features offered (Fig. 1).
- After the creation of his profile, the patient connects and chooses the test or psychological tests he wishes to take, in the first version of the application the priority was to detect anxiety, depression, and stress by integrating the two international psychological tests: the HADS and the PSS-10 (Fig. 2).
- Depending on the score of the psychological test passed, the patient will automatically receive personalized psychological recommendations every day through his client profile, email, and text message for 40 continuous days. The psychological recommendations sent are simple, short, professional, varied, written by psychologists who have professional experience in psychology and more than seven years of experience in this field (Fig. 3).
- Every day, in addition to the notifications of the recommendations received, the patient is notified of his appointments for scheduled psychological

Fig. 1. Login/Sign up page

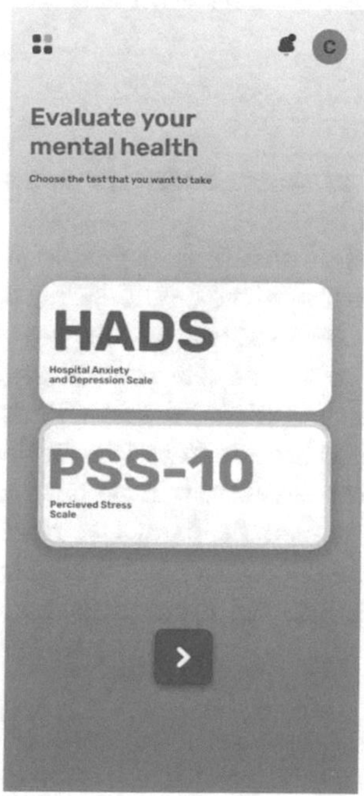

Fig. 2. Psychological tests

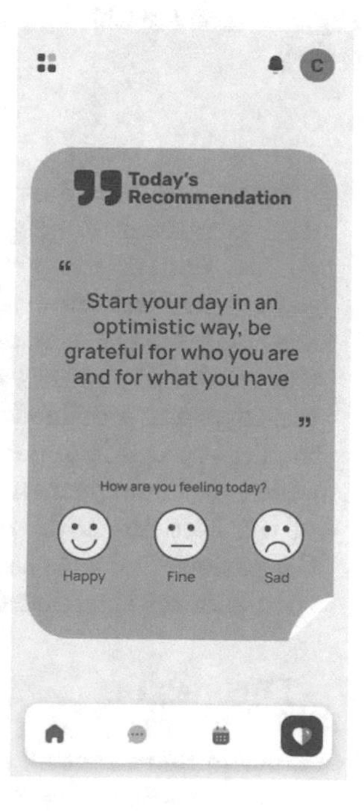

Fig. 3. Psychological recommendations

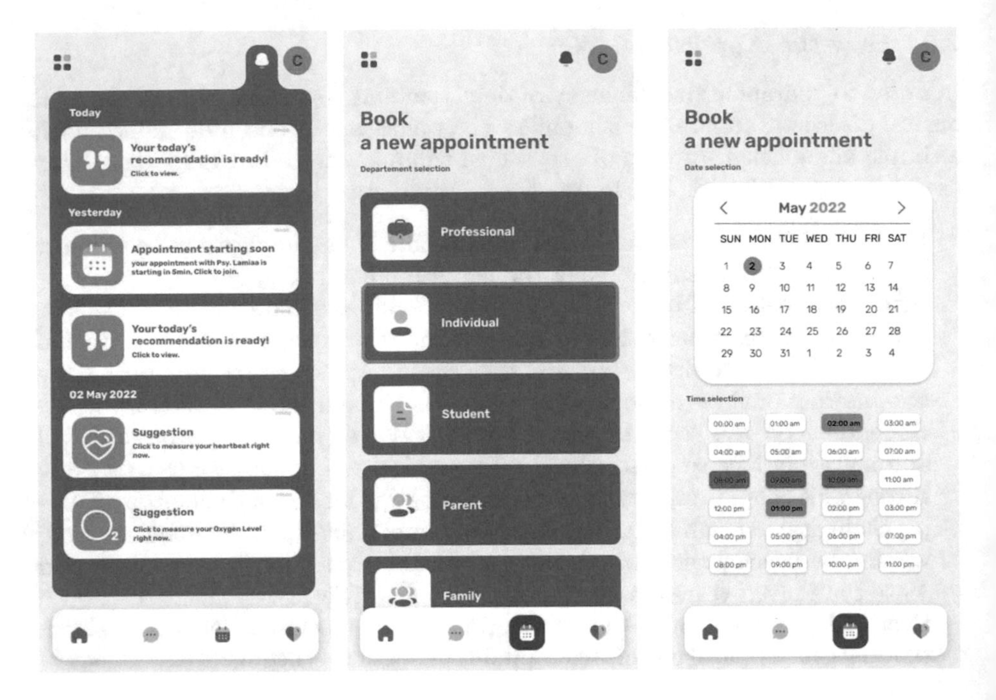

Fig. 4. Patient notifications

Fig. 5. Departments list

Fig. 6. Appointment scheduling

support sessions, as well as suggestions for measuring heartbeats and the oxygen saturation (Fig. 4).

- At the end of the period of sending psychological recommendations, the patient receives an evaluation survey (the same psychological test taken at the beginning) to assess the progress of his psychological state. Based on the results of the survey to assess the mental state of patients, the following steps were defined: If the patient's score improves then their electronic medical record is updated to keep it in the main database, otherwise, personalized psychological support sessions will be carried out by the psychologists of the Medico-call team, to benefit from the psychological support session, the patient first chooses the department in which he is interested (Fig. 5) and then chooses the specialist and the niche that suits him (Fig. 6).

3 Discussion

No one can deny that all areas and practices of our daily lives have experienced real development thanks to technology and digitalization. The proof is that nowadays the use of web and mobile applications has become essential, since the latter has facilitated several tasks by offering efficient services, remotely, for the benefit of everyone, each according to their needs. Although the availability and use of mobile health and mental health web applications have grown exponentially in recent years. During the COVID- 19 pandemic, a "Text4Hope" SMS support program was designed to report on the evolution of stress, anxiety, and depressive symptoms of subscribers, after three months of using this program support, reductions in psychological distress have been observed in subscribers to the "Text4 Hope" program [17]. The proposed innovative solution provides both psychological support and personalized quality monitoring using the latest technologies, which makes it accessible to everyone, especially since it is available in english, french, and arabic. The mobile web application developed is extensible, which makes it possible to work in the future on other pathologies such as addiction and suicide according to the needs of different targets, regardless of individuals, private schools which contain hundreds of students in need of personalized psychological support, as well as companies that need psychological support for their employees who go through frequent periods of stress.

Today, big data technologies and artificial intelligence offer new possibilities for screening and predicting mental problems [18]. On the one hand, the accessibility of powerful computers and the availability of massive data from diverse sources are causing data science approaches in mental health research to become ubiquitous [19]. On one hand, machine learning currently offers the possibility of analyzing disease patterns in complex disease states and could eventually facilitate treatment choices with existing therapies and provide bases for new therapies [20].

The mobile web application developed allows both to provide patients with personalized psychological support through psychological recommendations and clinical sessions, in addition, allows the collection of several medical data, either through the fields filled in by patients (age*, gender*, family situation*, function*, country*, city*, etc.), or through "smartwatches" linked to "smartphones" since the application is connected, which gives an idea of the patient's oxygen saturation and heart rate. In addition, each patient's psychological test scores can be used to categorize patients and facilitate the prediction of subsequent psychological pathologies with better accuracy and in a noticeable delay [21]. Finally, by collecting all this important data, the application will allow healthcare professionals to prepare electronic medical records for each patient to facilitate the follow-up of each case and collect the maximum amount of data, which can then be used to apply the big data process and lead to better improvements in health care in general and mental health in particular.

4 Conclusion

Technology plays one of the most important roles in the future of health care delivery in general and mental health in particular, especially as mental health issues are a huge challenge, requiring both professionalism, and efficiency. To promote mental health by harnessing technology, a solution based on a mobile web application has been proposed, the application is available in three different languages, developed with the most recent technologies, managed by psychologists from the Medico-call company who provides quality personalized psychological support through the psychological recommendations drafted in advance as well as the clinical sessions reserved according to the appropriate time slot for each patient. The application was created on the one hand to improve the mental health of people in mental distress who suffer in silence, on the other hand, to use all the data obtained through the results of psychological tests, connected smartwatches, etc. To prepare electronic medical records for each patient. The future plan is to implement the application and market it to individuals and organizations where there is a large workforce such as private schools and businesses to promote mental health, detect new psychological problems that can be solved, collect more data to apply big data technologies and guarantee more efficient and intelligent service.

5 Declarations

Funding
No funding was received to assist with the preparation of this manuscript.
Conflicts of Interest
The authors declare that they have no conflicts of interest.
Ethical Approval
This article does not contain any studies with human participants or animals performed by any of the authors.

References

1. Santiago, A., Arnaud, M.: Health in the digital age- contributions of big data and new technologies in the prevention and treatment of online gambling addiction. Med. Sci. **35**, 787–91 (2019)
2. World Health Organization: Global Diffusion of EHealth: Making Universal Health Coverage Achievable, Report of the Third Global Survey on EHealth. World Health Organization, Genève, Switzerland (2016)
3. Grinberg, M.: Flask Web Development: Developing Web Applications with Python. Reilly Media, Inc., Sebastopol (2018)
4. Zigmond, A.S., Snaith, R.: The hospital anxiety and depression scale. Acta Psychiatr. Scand. **67**, 361–70 (1983)
5. Snaith, R.: The hospital anxiety and depression scale. Health Qual. Life Outcomes **1**, 1–4 (2003)
6. Dahl, A.A., Haug, T.: The validity of the hospital anxiety and depression scale: an updated literature review. J. Psychosom. Res. **52**(2), 69–77 (2002)
7. Annunziata, M.: Hospital anxiety and depression scale (HADS) accuracy in cancer patients. Support. Care Cancer **28**, 3921–26 (2020)
8. Lemay, K.R., Tulloch, H.E., Pipe, A.L.: Establishing the minimal clinically important difference for the hospital anxiety and depression scale in patients with cardiovascular disease. J. Cardiopulm. Rehabil. Prev. **39**, E6-11 (2019)
9. Weber, S., Lesinski, M.: Symptoms of anxiety and depression in young athletes using the hospital anxiety and depression scale. Front. Physiol. **9**, 182 (2018)

10. Cohen, S., Mermelstein, R.: Perceived stress scale. Measuring Stress: Guide Health Soc. Sci. **10**, 1–2 (1994)
11. Lesage, F.-X., Berjot, S., Deschamps, F.: Psychometric properties of the French versions of the perceived stress scale. Int. J. Occup. Med. Environ. Health **25**(2), 178–84 (2012)
12. Almadi, T., Mansour, A.M.: An Arabic version of the perceived stress scale: translation and validation study. Int. J. Nurs. Stud. **49**, 84–89 (2012)
13. Mimura, C., Griffiths, P.: A Japanese version of the perceived stress scale: translation and preliminary test. Int. J. Nurs. Stud. **41**, 379–85 (2004)
14. Remor, E.: Psychometric properties of a European Spanish version of the perceived stress scale (PSS). Span. J. Psychol. **9**(1), 86–93 (2006). https://doi.org/10.1017/s1138741600006004
15. Golden-Kreutz, D.M., Browne, M.W., Frierson, G.M., Andersen, B.L.: Assessing stress in cancer patients: a second-order factor analysis model for the perceived stress scale. Assessment **11**(3), 216–23 (2004). https://doi.org/10.1177/1073191104267398
16. Mitchell, A.M., Crane, P.A., Kim, Y.: Perceived stress in survivors of suicide: psychometric properties of the perceived stress scale. Rese. Nurs. Health **31**(6), 576–85 (2008). https://doi.org/10.1002/nur.20284
17. Agyapong, V.I., et al.: Text4Hope: receiving daily supportive text messages for 3 months during the COVID-19 pandemic reduces stress, anxiety, and depression. Disaster Med. Publ. Health Preparedness **16**, 1–5 (2021). https://doi.org/10.1017/dmp.2021.27
18. Liang, Y., Daniel, D.: A survey on big data-driven digital phenotyping of mental health. Inf. Fusion **52**, 290–307 (2019)
19. Russ, T.C., et al.: How data science can advance mental health research. Nature Hum. Beh. **3**(1), 24–32 (2019). https://doi.org/10.1038/s41562-018-0470-9
20. Tai, A.M.Y., et al.: Machine learning and big data: implications for disease modeling and therapeutic discovery in psychiatry. Artif. Intell. Med. **99**, 101704 (2019). https://doi.org/10.1016/j.artmed.2019.101704
21. Taoussi, C., Hafidi, I., Metrane, A., Lasbahani, A.: **Predicting psychological pathologies from electronic medical r**ecords. In: Ahram, T., Taiar, R., Groff, F. (eds.) IHIET-AI 2021. AISC, vol. 1378, pp. 493–500. Springer, Cham (2021). https://doi.org/10.1007/978-3-030-74009-2_63

An Unsupervised Voice Activity Detection Using Time-Frequency Features

Hind Ait Mait(✉) and Noureddine Aboutabit

Laboratory of Process Engineering, Computer Science and Mathematics, National School of Applied Sciences Khouribga, University Sultan Moulay Slimane, Beni Mellal, Morocco

hind.ait-mait@usms.ac.ma, n.aboutabit@usms.ma

Abstract. Voice Activity Detection (VAD) is a binary classification problem for separating speech segments from background silence or noise. Over time, many features for the VAD have been proposed. In our study, we applied two features: Short Time Energy (STE) and Spectral Centroid, which belong to temporal and frequency domains, respectively. The goal of applying this VAD method is to use the speech segment extracted as an input of a phonetic segmentation method of Arabic and Moroccan dialect speech signals. We evaluated the method on 400 sentences from the Arabphone corpus recorded in noisy and noiseless environments. The results showed promising accuracy of 85%, which is comparable to some other previously proposed methods.

Keywords: Arabic speech · voiced segment · unvoiced segment · STE · Spectral Centroid · Accuracy

1 Introduction

Voice Activity Detection(VAD) identifies speech and non-speech parts of an input signal. It served as the front end of various speech-processing systems, such as speech enhancement, speech coding, speech synthesis, speech density estimation, speech transmission, and Automatic Speech Recognition (ASR). The performance of the system that uses the VAD is more critical than its output. For example, the quality of speech and, the transmission bandwidth of speech transmission, the word recognition rate of an ASR system. The essential properties of every VAD are accuracy, robustness, latency, computational load, and memory requirements [1]. A typical VAD system is made up of an important part called feature extraction. It concerned quantifying signal characteristics to separate speech from background noise or silence. Early research focused on the acoustic features of the signal. Without prior knowledge to separate speech and non-speech regions, this approach is an unsupervised method. Among the techniques used in this category are the short-time average energy method and zero

Supported by LIPIM.

N. Aboutabit et al. (Eds.): ICMICSA 2022, LNNS 656, pp. 232–240, 2023.
https://doi.org/10.1007/978-3-031-29313-9_21

crossing rate (ZCR) [2], which depended on appropriately selected thresholds and performed efficiently on less noisy speech signals. A VAD method presented in [3] involved a combination of features: ZCR, autocorrelation, Long Term Signal Variability, and energy. The experimental assessment validated the performance of the method in noisy environments. The paper [4] outlined a technique for VAD, where speech was distinguished from stationary and non-stationary noise using a combination of the sub-band temporal envelope characteristic and the sub-band long-term signal variability. The experiments were carried out on datasets from the TIMIT corpus mixed with noisy data from the NOISEX-92 database at different SNRs scenarios. The findings revealed the robustness of the suggested algorithm, showing that it outperformed several baseline methods. The research [5] presented a VAD method based on the k-means clustering technique. Compared with other time-domain voice activity detectors, the outcomes demonstrated its efficiency in terms of non-speech detection rate (HR0) and speech activity detection rate (HR1). The work [6] used information from short-term smoothed log energy and statistical modeling of MFCCs coefficients to detect speech segments. The experiments were applied on the QUT-NOISE-TIMIT corpus under low SNR conditions and in non-stationary noise. The results showed that the suggested method outperformed the systems that presented the outcomes on the same database. In [7], the authors proposed another approach to constructing a robust VAD based on Single Frequency Filtering (SFF). By extracting the temporal variation at specific regularly spaced frequencies across the spectrum, the SFF technique took advantage of this property. Since speech signals include correlations between their samples but noise samples would not, the characteristics of human speech differ significantly from noise at each frequency. Compared to AMR2 [8] and formant-based approaches, the proposed method had better accuracy in detecting speech activity. Another work in developing robust VAD is presented in [9], combining the Gammatone filtering and entropy measurements. Gammatone filtering has been utilized to replicate the human auditory system. Entropy was used as an information-theoretic measure to determine how much information was conveyed by the input signal. Compared to previous VAD methods, particularly in low levels of SNRs, the suggested method showed a noticeable improvement. The work [10] established a robust VAD method in noisy reverberant conditions to reduce the effects of additive noise and room reverberation. It was based on power envelope restoration utilizing the Modulation Transfer Function (MTF) concept. Experiments were carried out in both simulated and real-world noisy reverberant environments to evaluate the performance of the proposed technique. The outcomes showed that it performed considerably better than the conventional techniques. A VAD algorithm based on long-term pitch information was presented in [11]. The Long-Term Pitch Divergence (LTPD) in the introduced system was calculated using features obtained by dividing the audio signal into 88 frequency bands using multi-rate filter banks. According to the experimental results, the suggested method was more effective in non-stationary noisy environments and was more resilient than other VAD algorithms, even at low SNRs levels. In [12], a VAD approach was introduced based on the total spectrum energy in the overlapping speech window frames. The noise energy reduced the noisy speech spectrum in the lower frequency band from the higher frequency band. The speech spectrum energy waveform was also smoothed using a moving average filter. The suggested method was reliable and effective at a range of SNRs levels. Cochleagram features from the Gammatone filter bank-multiple window observation approach were used in [13] to segregate speech from background noise. The finding confirmed the efficiency of the proposed method: Half Total Error Rate (HTER) was much lower than other baseline techniques when speech and non-speech clusters were identified in various noise situations. The research [14] presented an efficient method for hearing aids to detect voice activity based on spectral entropy. It was tested using noisy speech (white, factory, 32-talker babble noise) at SNRs levels between 15 dB and -10 dB. The results demonstrated that the suggested method was more robust under noisy conditions. In addition, it reduced the computational complexity compared to the conventional fast Fourier transform-based spectral entropy approaches. The paper [15] suggested a novel technique for detecting voice activity based on spectro-temporal domain clustering. The auditory model was used to extract these features to distinguish speech from silence. The Gaussian mixture model and WK-means clustering techniques were employed to reduce the dimensions of the spectrotemporal space. According to the results, the suggested method enhanced the segmentation rate for both speech and non-speech when compared to temporal and spectral features in low SNRs. In [16], a novel hybrid architecture for VAD was introduced; it combined layers of end-to-end trained Convolutional Neural Network (CNN) and Bidirectional Long Short-Term Memory (BiLSTM). The results showed that BiLSTM layers improve accuracy over unidirectional layers by 2% and an Area Under the Curve (AUC) of 0.951. A VAD algorithm based on local and global attention was suggested in [17]. Building local attention was the first step in obtaining Long short-term memory to understand more representative links between contextual frames so that the model could concentrate on the most pertinent area of contextual frames. The global attention was then constructed by concatenating multi-head attention layers. The experimental results indicated that the method outperformed existing attention-based VAD methods while using less processing power. It also presented remarkable generalization properties even at very low SNRs like 15dB and in unknown noisy environments. The study [18] suggested a Neural Architecture Search (NAS), which could be used to build a network architecture that increased detection accuracy automatically. The outcomes showed that the proposed NAS framework surpassed the manually state-of-the-art VAD models in various noise-added real-world datasets. In our work, we based on an unsupervised VAD method [19] used two features: Short Time Energy (STE) and Spectral Centroid (SC), which belong to the time and the frequency domains, respectively. Double thresholds were applied to detect the speech areas. It was evaluated using the Arabphone corpus [20], which was recorded in clean and noisy environments. The rest of the paper is organized as follows: Sect. 2 describes the speech corpus used and the proposed

VAD algorithm, in addition to the accuracy as an evaluation measure. Section 3 discusses the experiment results. Section 4 summarizes the paper and discusses future work.

2 Methodology

This section defines the speech corpus used, the VAD algorithm based on STE, Spectral Centroid, and the accuracy as an evaluation metric.

2.1 Speech Corpus

The experiment used 400 Arabic sentences taken from the Arabphone database [20]. It is a Modern Standard Arabic language corpus containing 28 sentences spoken by 30 Algerian adults of both genders from the Wilayas of Annaba, Jiel, and Tarif. Their age is between 20 and 40. This dataset includes 2520 words and 12000 phonemes recorded in various environments. The Table 1 presents some Arabic sequences used in this corpus.

Table 1. Sequences of Arabphone corpus.

Consonant	Sequence
ا	الجنة تحت أقدام الأمهات
ب	يا مقلب القلوب ثبت قلوبنا
ث	ورث ثابت ثلاثة ثياب
ح	جرح الحجر حافر الحيوا ن
خ	أخذ خالد خاتم خديجة
ذ	ذرى الفلاح القمح بالمذراة
ز	زار عزام جزيرة الكرز
س	السمع والبصر من الحواس
ش	لا أشرب الشاي بعد العشاء
ص	سرق اللصوص صندوق الصياد
ض	ضرب الضابط الضربة القاضية
ط	الطائر الوطواط نشيط
ظ	أظافر محفوظ نظيفة
غ	غاب بليغ عند الغروب
ق	إقتربت القافلة من السوق
ل	لا إلاه إلا الله

2.2 Method Description

The suggested method [19] is based on the Short-Term Energy and Spectral Centroid. They were obtained from temporal and frequency domains, respectively. The basic idea is to extract the two feature sequences from the signal, and two thresholds (T1, T2) are dynamically calculated for each feature. Then, the voiced segments were generated of frames with feature values that exceeded the computed T1 and T2. The threshold is obtained from the following formula:

$$T = \frac{W.M_1 + M_2}{W + 1} \tag{1}$$

where:
W is a user-defined parameter.
M1 and M2 are the positions of the first and second histogram local maxima.

Spectral Centroid. The spectral centroid [21] is linked to the brightness measurement of a sound. The "center of gravity" was determined using the magnitude and frequency values from the Fourier transform. The individual centroid of a spectral frame is determined by dividing the average frequency weighted on amplitudes by the sum of the amplitudes as follows:

$$SC = \frac{\sum_{k=1}^{N} KF[K]}{\sum_{k=1}^{N} F[K]}. \tag{2}$$

where F[k] is the amplitude in the Discrete Fourier Transform range corresponding to bin k, and N is the frame length.

Short Time Energy. The energy of the short speech segment is represented by the short-time energy function [20]. It is a fundamental and valuable classification parameter for speech and non-speech regions. It is defined as follows:

$$E(i) = \frac{1}{N} \sum_{n=1}^{N} |x_i(n)|^2. \tag{3}$$

where i is the short-term frame, and N is the frame length.

2.3 Evaluation Measurement

In our study, four parameters concerning the accuracy rate (ACC) are described as follows:
True Positive (TP): Speech segment detected as a correct class.
True Negative (TN): Silence or noise area classified as a non-speech segment.
False Positive (FP): Speech considered as non-speech.
False Negative (FN): Silence or noise rated like speech.
The ACC is computed by the following formula:

$$ACC(\%) = \frac{TP + TN}{TP + TN + FP + FN}. \tag{4}$$

3 Results and Discussion

In this section, we examine the performance of the proposed method using several simulations on 400 sentences recorded in a noiseless and noisy environment (Table 2).

Table 2. Comparison of the accuracy.

Speakers	Accuracy
1	**97.32**
2	**90.1**
3	**89.82**
4	74.10
5	82.09
6	83.25
7	86.84
8	80.86
9	66.94
10	81.90
11	72.52
12	83.48
13	**96.02**
14	**93.83**
Total Accuracy	85%

The experiments have shown that the energy of the speech segments was greater than the energy of the non-speech segments as well as the sequence of spectral centroid for speech segments was highly variable, this is a sufficient reason to explain the miss-classification of some sounds like ح [_H] which considered as silence because of its low energy. The Fig. 1 presents an illustrative example of a sentence composed of three voiced segments that were correctly detected.

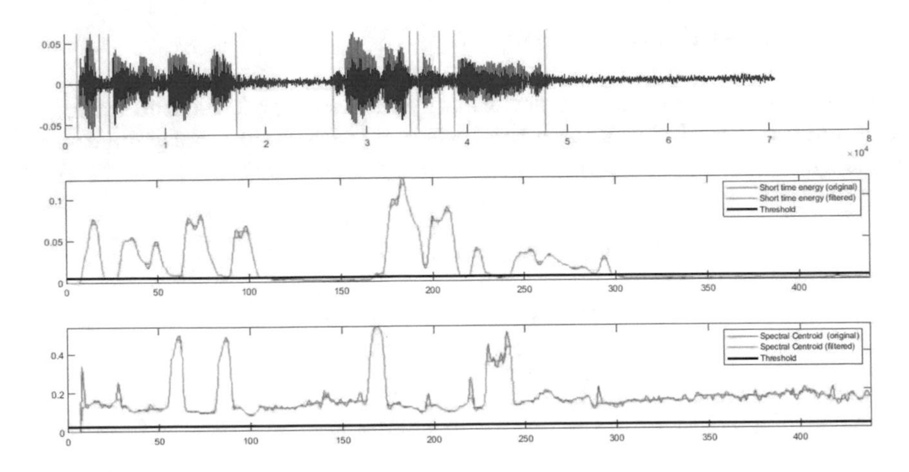

Fig. 1. Detection voiced segments using the proposed algorithm.

Two problems have been identified according to the experimental findings:

Problem 1: Masking effect: As mentioned earlier, the first and second local maximums of the features histogram are used to formulate the threshold for separating speech from noise or silence. Figure 2 depicts that the initial peak was dominated by the second, third, and fourth maximum peaks, especially in the first histogram showing the STE feature.

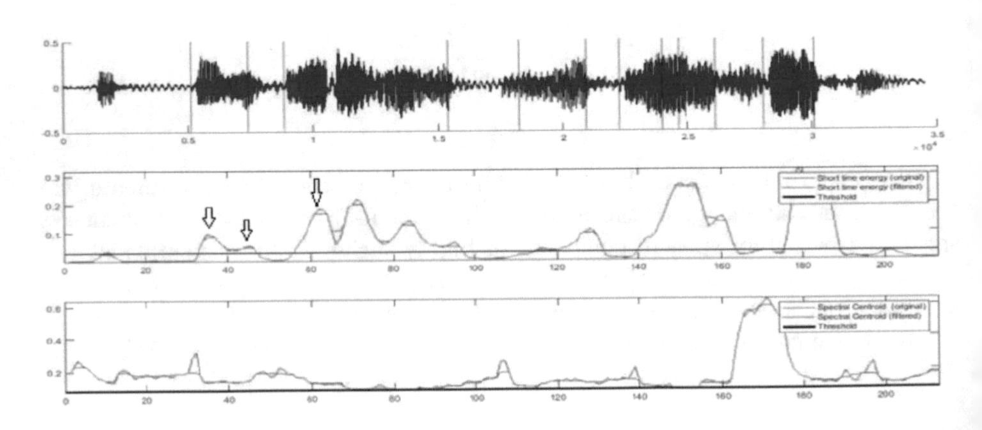

Fig. 2. An illustrative example of the masking problem.

Problem 2: threshold selection: In order to detect speech segments, the values of the two feature vectors must be greater than the threshold T1 and T2. In Fig. 3, we observe that the STE values are higher than the threshold T1 in some intervals, for example, the range [100–150], whereas Spectral Centroid values are lower than T2 in the same interval. Therefore the speech segment was incorrectly classified.

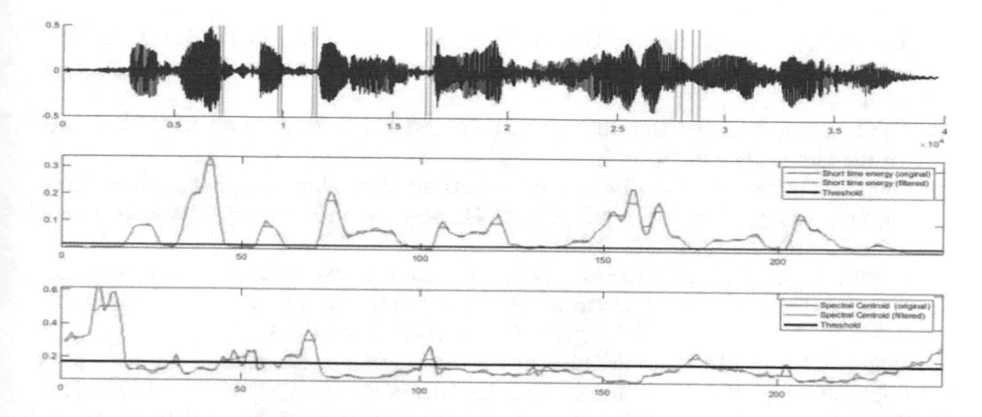

Fig. 3. Illustration of the threshold issue.

4 Conclusion and Future Work

In this article, we have proposed a VAD method based on the STE and the Spectral Centroid. Our experiments were carried out on the Arabphone corpus, which has been recorded in different environments. The suggested algorithm performed well in classifying speech into voiced and unvoiced segments, with an accuracy of 85%. In our study, we found two problems related to the chosen threshold. As a perspective, we will find the appropriate formula of the threshold by applying other modifications, such as including the third local maxima of the features histogram. We intend to combine the two features and add the others like ZCR or the Spectral Flatness Measure to improve the performance of the method to use it as an input of a phonetic segmentation method of Arabic and Moroccan dialect speech signals.

References

1. Bäckström, T.: Speech Coding: with Code-excited Linear Prediction. Springer, Berlin (2017)
2. Lamel, L., et al.: An improved endpoint detector for isolated word recognition. IEEE Trans. Acoust. Speech, Sig. Process. **29**(4), 777–785 (1981)
3. Haghani, S.K., Ahadi, S.M.: Robust voice activity detection using feature combination. In: 2013 21st Iranian Conference on Electrical Engineering (ICEE). IEEE (2013)
4. Liu, B., et al.: Efficient voice activity detection algorithm based on sub-band temporal envelope and sub-band long-term signal variability. In: The 9th International Symposium on Chinese Spoken Language Processing. IEEE (2014)
5. Elton, R.J., Mohanalin, J., Vasuki, P.: A novel voice activity detection algorithm using modified global thresholding. Int. J. Speech Technol. **24**(1), 127–142 (2021). https://doi.org/10.1007/s10772-020-09777-w
6. Sriskandaraja, K., et al.: A model based voice activity detector for noisy environments. In: Sixteenth Annual Conference of the International Speech Communication Association (2015)

7. Aneeja, G., Yegnanarayana, B.: Single frequency filtering approach for discriminating speech and nonspeech. IEEE/ACM Trans. Audio, Speech, Lang. Process. **23**(4), 705–717 (2015)
8. ETSI. Voice Activity Detector (VAD) for Adaptive Multi-Rate (AMR) Speech Traffic Channels. (1999)
9. Ong, W.Q., Tan, A.W.C.: Robust voice activity detection using gammatone filtering and entropy. In: 2016 International Conference on Robotics, Automation and Sciences (ICORAS). IEEE (2016)
10. Morita, S., et al.: Robust voice activity detection based on concept of modulation transfer function in noisy reverberant environments. J. Sig. Process. Syst. **82**(2), 163–173 (2016). https://doi.org/10.1007/s11265-015-1014-4
11. Yang, X.K., et al.: Voice activity detection algorithm based on long-term pitch information. EURASIP J. Audio, Speech, Music Process. **2016**(1), 1–9 (2016)
12. Pang, J.: Spectrum energy based voice activity detection. In: 2017 IEEE 7th Annual Computing and Communication Workshop and Conference (CCWC). IEEE (2017)
13. Mondal, S., Barman, A.D.: Speech activity detection using time-frequency auditory spectral pattern. Appl. Acoust. **167**, 107403 (2020)
14. Liu, F., Demosthenous, A.: A computation efficient voice activity detector for low signal-to-noise ratio in hearing aids. In: 2021 IEEE International Midwest Symposium on Circuits and Systems (MWSCAS). IEEE (2021)
15. Esfandian, N., Jahani Bahnamiri, F., Mavaddati, S.: Voice activity detection using clustering-based method in Spectro-Temporal features space. J. AI and Data Min. **10**, 401–409 (2022)
16. Wilkinson, N., Niesler, T.: A hybrid CNN-BiLSTM voice activity detector. In: ICASSP 2021-2021 IEEE International Conference on Acoustics, Speech and Signal Processing (ICASSP). IEEE (2021)
17. Li, S., et al.: Voice activity detection using a local-global attention model. Appl. Acoust. **195**, 108802 (2022)
18. Rho, D., Park, J., Ko, J.H.: NAS-VAD: neural architecture search for voice activity detection. arXiv preprint: arXiv:2201.09032 (2022)
19. Giannakopoulos, T.: A method for silence removal and segmentation of speech signals, implemented in Matlab, vol. 2. University of Athens, Athens (2009)
20. Frihia, H., Bahi, H.: Embedded learning segmentation approach for Arabic speech recognition. In: Sojka, P., Horák, A., Kopeček, I., Pala, K. (eds.) TSD 2016. LNCS (LNAI), vol. 9924, pp. 383–390. Springer, Cham (2016). https://doi.org/10.1007/978-3-319-45510-5_44
21. Unjung, N.: Spectral Centroid (2001)

Dimensionality Reduction for Predicting Students Dropout in MOOC

Zakaria Alj(✉), Anas Bouayad, Cherkaoui Malki, and Mohammed Mohamed Ouçamah

University Sidi Mohamed Ben Abdellah FSDM, Fez, Morocco
{zakari.alj,anas.bouayad,oucamah.cherkaoui}@usmba.ac.ma

Abstract. This paper aims to find a representation of an initial dataset for predicting MOOCs student drop-out in a more reduced space using dimensionality reduction methods. We used two dimensionality reduction techniques, features selection and data transformation. Our technique uses the contribution of many features selection methods in order to find the pertinent subset of features to keep. The obtained results comparing the models accuracy for the initial dataset and the subset of selected features, encourage the use of these methods in order to reduce the complexity and the computation time.

Keywords: MOOC · Student Drop-out · Dimensionality Reduction · Features Selection

1 Introduction

MOOCs are known by the low completion rate, statistics show that less than 10% of enrolled participants complete the courses. In order to tackle this problem, several researches use machine learning algorithms to predict students in high risk of drop-out. The prediction is based on processing of extracted data from the log trace. The quality of the processing system depends highly on the content of the data. When the dimension is high, solving the problem becomes difficult. Consequently, it is often useful, and sometimes necessary, to reduce the dimension of data. The aim of this work is deal with the curse of dimensionality in KDD cup 2015 dataset using a hybrid features selections and data transformation algorithms.

2 Related Works

Many researchers have recently focused on MOOCs student drop-out. Time consideration is very significant when tackling this problem. Early detection plays a masterful role in reducing the attrition rate [1]. Several studies used machine learning techniques and analysed the log trace the predict students dropout. Gitinabard et al. [2] applied Logistic Regression and Support Vector Machine

N. Aboutabit et al. (Eds.): ICMICSA 2022, LNNS 656, pp. 241–253, 2023.
https://doi.org/10.1007/978-3-031-29313-9_22

(SVM) to detect student at risk. Berens et al. [3] developed predicting system using demographic data and a boosting algorithm combining several ML algorithms: Linear Regression, Neural Network, Decision Tree, AdaBoost. In This paper we used three ensemble ML algorithmes: Random Forest (RF) and Gradient Boosting (GB) since they are outperforming the baseline algorithms used in the literature. The effectiveness of the learning algorithm highly depends on the used data-set. ML algorithms are unable to provide an effective parameter setting method. Therefore, feature selection of parameters is another research content of this paper.

Many other researchers have recently dealt with the curse of dimensionality [4] using features selection and machine learning. These techniques have shown their success in many different concrete applications such as intrusion detection [5], text categorization [6] and information retrieval [7]. Many papers and books are proving the benefits of the feature methods. Such methods are often divided into three categories: filters, wrappers and embedded methods. Works and research are using different strategies such as combining several feature selection methods, which could be done by using algorithms from the same approach, such as two filters [8] or coordinating algorithms from two different approaches, usually filters and wrappers [9,10], combining features selection approaches with other techniques, such as feature extraction [11] or tree ensembles [12]. In the paper we use a hybrid features selection combining algorithms from filters, wrappers and embedded approaches to elect a pertinent subset of features.

3 Data

3.1 Data Presentation

The dataset used in this work comes from KDD cup 2015 10. This dataset contains information extracted from 39 courses of XuetangX which is one of the biggest Chinese MOOC learning platforms. The detailed description of the five parts of the dataset is as the following:

The Table 1 contains informations about the start and the end date of each course the different parts of the dataset.

Table 1. Course Informations

Fields	Type	Description
Course-ID	Nominal	Course Identifier
Course-S	Date	Course Starting Date
Course-E	Date	Course Ending Date

The Table 2 contains informations about the modules of each course, submodules and also their category and their start date.

Table 2. Module Informations

Fields	Type	Description
Course-ID	Nominal	Course Identifier
Module-ID	Nominal	Module Identifier
Module-cat	Nominal	Module Category
Module-child	Nominal	Sub-Module
Module-S	Date	Module Starting Date

The Table 3 contains information about enrollments: an enrollment is a (Student, Course) entry.

Table 3. Enrolment Information

Fields	Type	Description
Enrolment-ID	Nominal	Enrolment Identifier
Student-ID	Nominal	Student Identifier
Course-ID	Nominal	Course Identifier

The Table 4 contains the log trace of every enrollment, log timestamps, and also the source and the event type.

Table 4. Log Informations

Fields	Type	Description	
Enrolment-ID	Nominal	Enrolment Identifier	
Student-ID	Nominal	Student Identifier	
Course-ID	Nominal	Course Identifier	
Event-Time	Timestamps	Time when the event occurs	
Source	Nominal	Source of Event (Server / Browser)	
Event	Nominal	Event Type	Problem Access Video Wiki Discussion Navigate Page Close

The Table 5 contains informations about the real value of the enrollment results (success or drop-out).

Table 5. Enrolment Result

Fields	Type	Description	
Enrolment-ID	Nominal	Enrolment Identifier	
Result	Boolean	Student Result	0 Success 1 Drop-out

3.2 Feature Engineering

Our feature extraction method is based on counting the log of every enrolment, an enrolment is a (student, Course) entry. The Extracted features can be divided into tree parts:

Enrolment History Features: It contains features about the history of interaction with the MOOC, such as the number of successful courses, the number of failed courses, the cumulative number of days spent on the MOOC during old registrations as well as the cumulative number of logs of each event present on the catalogue in Table 4.

Current Enrolment Features: It contains features about the number of days spent on the MOOC during the current enrolment as well as the count of logs of each event. We also extract the count of minutes spent for every enrolment, after examining the log trace we notice that all sessions often start with the **Navigate** event and sometimes with the **Access**. We calculate the difference of time expressed in minutes between one of the two events and the end of session expressed with the **Page_close**. The cumulation of minutes is recorded for each enrolment in the variable m using the following algorithm:

Algorithm 1. Algorithm of connected minutes

Input:
$@ConnectedMin \Leftarrow 0;$
$@BeginDay \Leftarrow$ "00 : 00 : 00";
$@EndDay \Leftarrow$ "23 : 59 : 59";
$@TBegin \Leftarrow 0;$
$@TEnd \Leftarrow 0;$
Output :Connected minutes per enrollement
1: **while** not at the end of enrolment log rows **do** read current
2: **if** Evente='nagivate' or (Evente = 'access' and @TBegin ="") **then** @TBegin= Time-Event;
3: **end if**
4: **if** (@Evente='Page_close') **then** @TEnd= Time-Event;
5: **end if**
6: $Ddiff \Leftarrow DateDiff(MIN, @TBegin, @TEnd)$
7: **if** (@Ddiff < 0) **then**
8: $@x \Leftarrow DateDiff(MIN, @hdebut, @EndDay)$
9: $@y \Leftarrow DateDiff(MIN, @BeginDay, @hfin)$
10: $@Ddiff \Leftarrow @x + @y$
11: **end if**
12: $@ConnectedMin = \Leftarrow @ConnectedMin + @Ddiff$
13: **end while**

Finally we end up with features presented in the Table 6 bellow:

Table 6. Extracted Features

N	Features
1	Enrolment Identifier
2	Count of student previous enrolments
3	Count of student previous succeeded enrolments
4	Count of student previous drop-out enrolments
5	Count of log for the current enrolment
6	Count of log for all previous enrolments
7	Count of days between first and last log
8	Count of days between first and last log for all previous enrolments
9	Count of log for the event : Problem
10	Count of log for the event : Problem for all previous enrolments
11	Count of log for the event : Video
12	Count of log for the event : Video for all previous enrolments
13	Count of log for the event : Navigate
14	Count of log for the event : Navigate for all previous enrolments
15	Count of log for the event : Page-close
16	Count of log for the event : Page-close for all previous enrolments
17	Count of log for the event : Access
18	Count of log for the event : Access for all previous enrolments
19	Count of log for the event : Discussion
20	Count of log for the event : Discussion for all previous enrolments
21	Count of log for the event : Wiki
22	Count of log for the event : Wiki for all previous enrolments
23	Count of logs in the first 10 d of course
24	Count of logs in the second 10 d of course
25	Count of logs in the last 10 d of course
26	Count of active minutes
27	Count of active days
	Enrolment result : Success 0 /Drop-out 1

4 Methodology

Most of classification problems are based on the processing of extracted data structured as vectors. The quality of the classifier depends directly on the correct choice of the content of these vectors. But in many cases solving the problem becomes difficult and very expensive in terms of time and resources owing to the large dimension of these vectors. Consequently, it is often useful, and sometimes necessary to reduce the dimensionality of these vectors to be compatible with resolution methods, even if this reduction may lead to a slight loss of information.

Variable selection and dimensionality reduction techniques provide a natural answer to this problem by eliminating features that do not provide enough predictive information to the model. Reducing the number of explanatory variables has a double advantage. On the one hand, the model will be easily interpretable due to the few number of variables. On the other hand, the prediction error will be reduced by removing the non-informative variables. Methods of dimensionality reduction are generally classified into two categories:

- Dimensionality reduction based on **Features Selection**: Consists on selecting the most relevant features describing the phenomenon studied among all dataset features.
- Dimensionality reduction based on **Data Transformation**: It consists on replacing the initial set of data by a new reduced set built from the initial set of characteristics.

4.1 Features Selection

Feature selection is a critical part of any machine learning pipeline. It aims to isolate the subset of predictors that allow to efficiently explain the target variable. Thus, the performance of a learning algorithm is strongly depending on characteristics used in the learning task. The presence of redundant characteristics can reduce the performance. The methods used to evaluate a subset of features in the selection algorithms can be classified into three main categories: **Filter**, **Wrapper** and **Embedded**.

4.1.1 Filter

This method is considered as a pre-processing step where the evaluation of the relevance of selected features is calculated according to measures that rely on the training data and before the learning phase. One of the advantages of this method is that the features selected are independent from the chosen classifier. Furthermore Filter algorithms are generally coast less computation time since they avoid repetitive executions of learning algorithms on different subsets of variables. However, their major drawback is that they ignore the impact of selecting subsets of features on the performance of the model [13].

Examples: **Chi-square Test, Fisher Score, Correlation Coefficient**

4.1.2 Wrappers

Wrappers methods were introduced by John et al. in 1994 [14]. They generally considered to be better than those filtering methods. They attempt to select a subset of features and evaluate them using a classification algorithm. The evaluation is made by calculating a score compromising between the number of variables eliminated and the success rate of the classifier on the test set. The classification algorithm is called several times during each time a variable is selected, then we calculate the classification rate to judge the relevance of a this variable using a cross validation mechanism. The principle of wrappers is to generate a subset well suited to the classification algorithm. However the major drawbacks of these methods are that they do not provide theoretical justification for the selection and they do not allow to understand the conditional dependencies between variables. Using a classifier to evaluate a subsets of features with cross validation mechanism is very expensive in terms of computation time. The selected subset of features depends on the classifier used and are not necessarily valid if we change the method.

Examples: **Recursive Feature Elimination, Sequential Feature Selection, Genetic Algorithms**

4.1.3 Embedded

Unlike the Wrapper and Filter methods, Embedded methods incorporate the selection of variables during the learning process.

In the Wrapper methods, the dataset is divided into two parts: a learning and a validation set. Embedded methods can use the full set for training, which is an advantage that can improve the results. Another advantage of these methods is their low computation time compared to the other approaches.

Examples: **Lasso Regularization, Random Forest, Decision Tree**

4.2 Data Transformation

The reduction of dimensionality by a data transformation is not made by selecting certain features. The reduction is made by construction of new features. These features are obtained by combining the initial ones. Data transformation risks losing the semantics of the initial set of features and therefore the use of this family of methods is not applicable only in the case where the semantics no longer occurs in the steps following the dimensionality reduction. They are generally grouped into two categories: linear methods and non-linear methods.

Examples: **Principal Component Analysis (PCA), Multi Dimensional Scaling (MDS), T-distributed Stochastic Neighbour Embedding (T-SNE)**

4.3 Used Techniques

Chi-Squared Karl PEARSON (1857–1936). is a statistical test for independence between categorical variables. Independent variables that from the target can be removed from the dataset. The principle is to calculate the difference between two distribution: a calculated distribution and a theoretical distribution obtained if the two variables were completely independent. the difference between the two distribution allows us to accept or to reject the hypothesis of independence H0. Feature selection with Chi-squared can be implemented with Select-K-Best which is a Scikit-learn machine library providing the K most relevant features.

Recursive feature elimination (RFE) the goal of recursive feature elimination (RFE) is to select features by recursively considering smaller and smaller sets of features [15]. First, the estimator is trained on the initial set of features and the importance of each feature is obtained. Then, the least important features are pruned from current set of features. That procedure is recursively repeated on the pruned set until the desired number of features to select is eventually reached. Often used with a validation mechanism, the performance of RFE for feature selection depends highly on the used model and the scoring metric,Since we are facing a class imbalance in our dataset, AUC ROC will be used as a metric to evaluate selection.

In previous work [1] RFE was implemented with Random Forest **(RFE-RF)** and ADABoost Classifier **(RFE-ADA)**. In this paper we will implement RFE with Decision Tree classifier **RFE-CART** and Gradient Boosting **RFE-GB**. The obtained results presented in the next section will justify this choice.

Regularization in machine learning adds a penalty term to the different coefficient multiplying features of the model to reduce their amplitude. This penalty is applied to the coefficient which multiplies each of the characteristics of the linear model. Regularization aims to avoid over-fitting by improving the generalisation of the model. the main types of regularization for linear models are Lasso (L1) and Ridge (L2). It is therefore a method of variable selection and dimensionality reduction: L1 set variables which are not necessary for target prediction to 0 are eliminated. setting coefficient in L1 and approaching them to zero in L2.

Principal Component Analysis (PCA) belongs to a set of unsupervised machine learning that do not require label for the target variable. It is a technique for finding smaller dimensions of variables in where it is possible to observe individuals with maximum of variance. It is essentially an approach transforming the initial set of variables probably correlated with each other, into new variables not correlated and called principal components created from linear combinations of the initial set of variables.

We find Bellow the summary of the used methods in this paper. We will use the following methods since they are producing the best results comparing to the others.

Variable Selection:
Filter: Chi-squared.
Wrappers: RFE-CART, RFE-GB.
Embedded: Lasso, Ridge.

Data transformation:
PCA, T-SNE.

5 Results

We will implement each of the methods presented in the previous section, the obtained scores (ranks, coefficients) will be aggregated and normalized so that they are between 0 for the lowest rank and 1 for the highest. The selection of features to eliminate will be made according to the mean of the obtained score of all the methods [15]. Results are presented in the table bellow.

According to the results found on the Table 7 we will take 0.60 for the Mean value as the statistical threshold for variable selection. The dimensionality of the dataset will be reduced to the following six features:

26 (0.81), 27 (0.80), 25 (0.80), 5 (0.69),11 (0.60), 3 (0.82).

We notice that after changing the selection algorithms used in our former work, the features selected are different and the selection threshold has changed.

Table 7. Features Ranking

N	Lasso	REF-CART	REF-GB	Ridge	Chi-2	Mean
1	0.00	1.00	0.46	0.00	0.00	0.29
2	0.00	0.82	0.87	0.42	1.00	0.62
3	0.00	0.51	0.88	1.00	1.00	0.82
4	0.00	0.71	0.08	0.58	1.00	0.47
5	0.66	1.00	0.90	0.01	0.97	0.69
6	0.05	1.00	0.12	0.01	1.00	0.44
7	0.00	0.50	0.67	0.35	1.00	0.50
8	0.00	1.00	0.71	0.00	1.00	0.54
9	0.00	1.00	0.96	0.01	0.99	0.59
10	0.00	1.00	0.50	0.01	1.00	0.50
11	0.00	1.00	1.00	0.02	1.00	0.60
12	0.00	1.00	0.37	0.01	1.00	0.48
13	0.00	1.00	0.58	0.05	1.00	0.53
14	0.00	0.00	0.00	0.00	1.00	0.20
15	0.00	1.00	0.62	0.09	1.00	0.54
16	0.00	1.00	0.17	0.01	1.00	0.44
17	0.00	1.00	0.33	0.02	0.99	0.47
18	0.00	1.00	0.29	0.01	1.00	0.46
19	0.00	0.14	0.25	0.00	1.00	0.28
20	0.00	0.86	0.67	0.01	1.00	0.51
21	0.00	1.00	0.21	0.06	1.00	0.45
22	0.00	0.43	0.75	0.00	1.00	0.44
23	0.00	1.00	0.87	0.03	0.99	0.58
24	0.00	1.00	0.83	0.01	0.99	0.57
25	1.00	1.00	1.00	0.03	0.99	0.80
26	0.60	1.00	0.87	0.58	1.00	0.81
27	0.00	1.00	1.00	0.98	1.00	0.80

During the feature selection process, validation is the most important step. The validation of the result is often made according to two criteria: the improvement of the precision of the classifier when adding or removing features, as well as the prior knowledge of the important features in the data.

The obtained results comparing the previous work seems more relevant, in the one hand they introduced two important new variables: the number of prior successful registrations and the number of logs. On the other hand the accuracy and the ROC AUC score of the three classifiers used in this paper have been improved.

In order to evaluate the pertinence of our selection we will compare between scores of Accuracy and AUC ROC obtained using all extracted features with and the selected variables with tree commonly used machine learning algorithms: Random Forest, Logistic Regression, Gradient Boosting.

According to the results obtained in the Table 8, we notice that the results are almost the same despite the elimination of twenty features. We notice also that for Logistic Regression, AUC ROC score has been improved. We conclude that the features eliminated are not important to predict the target variable. We can only keep the six selected variables.

Table 8. Models Scores

ML Algorithm	Features	Accuracy	AUC ROC
Logistic Regression	All	0.882	0.873
	Selected	0.870	0.863
Gradient Boosting	All	0.886	0.886
	Selected	0.874	0.877
Random Forest	All	0.885	0.882
	Selected	0.881	0.882

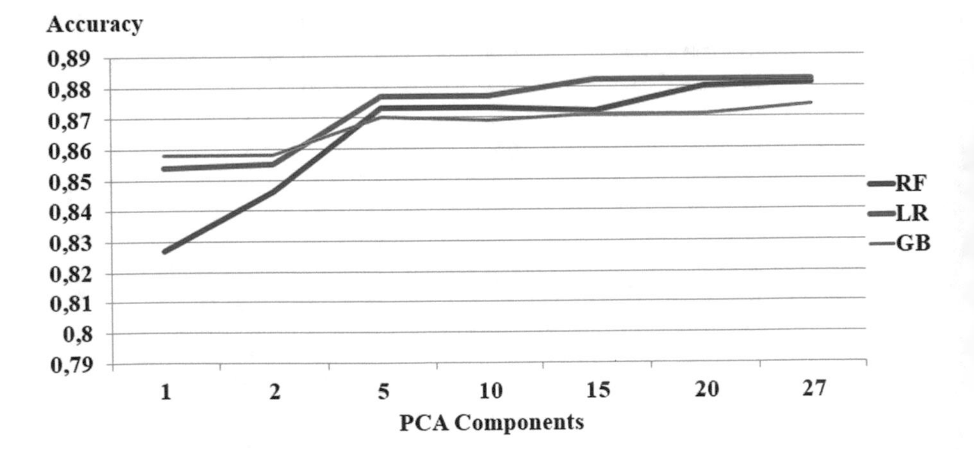

Fig. 1. Accuracy per Components

Figure 1 explains the variation of accuracy as a function of the number of component in the PCA analysis. We notice that for all the models the reduction of dimensionality to 5 neglects only 0.1 of the overall precision if we use all the dimension the use of 10 component gives better precision. For GB and LR from component 10 the score stabilizes respectively at 0.882 and 0.874 which is almost the same score if we use all the features.

The T-Distributed Stochastic Neighbor Embedding (T-SNE) is a non-linear dimensionality reduction methods, it aims to ensure that close points in the starting space have close positions in (2D) two dimensions space. In other words, the measurement of distance between the points in two dimensions space must reflect the measurement of the distance in the initial space. We will implement the T-SNE method on the data extracted features in order to visualise the points in two dimensions.

Visualizing the data allows to develop an intuition about our dataset. One of the reasons to use T-SNE is to validate the extracted features. Indeed, if the

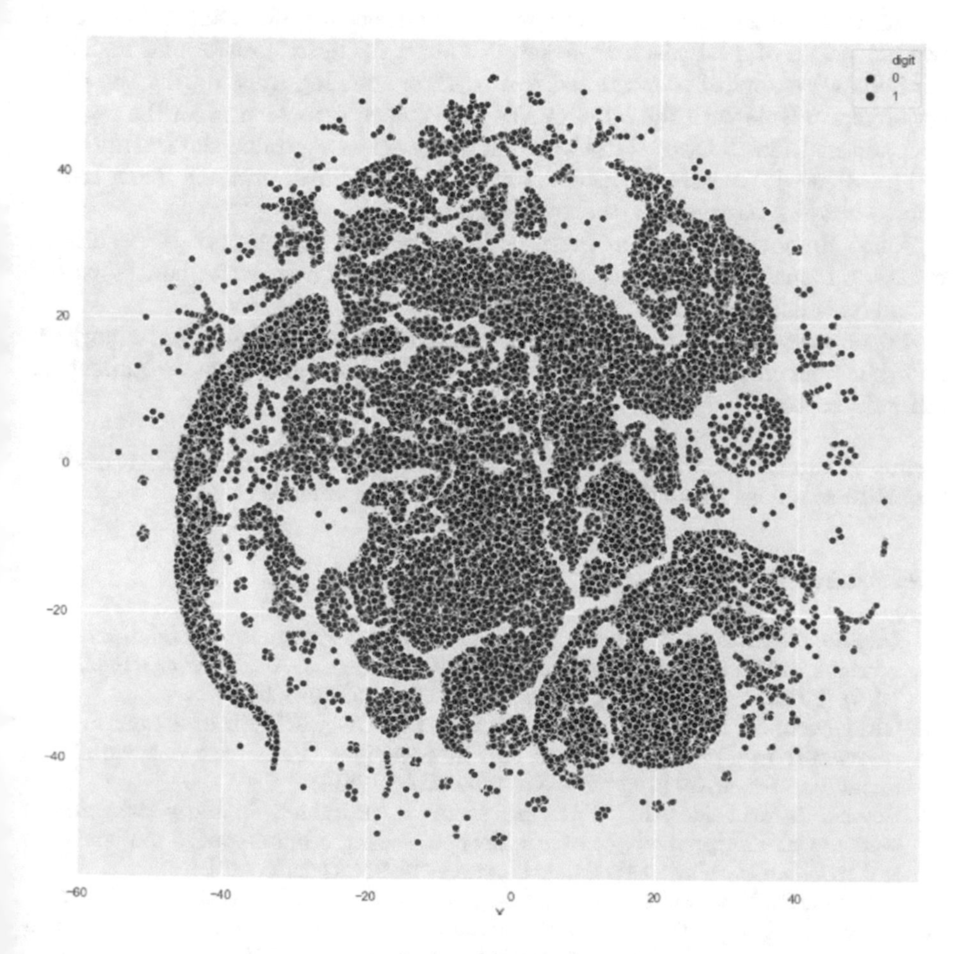

Fig. 2. Space plot Points in 2D

used features are pertinent and representative for the distinctions between data classes, this distinction must be observed when plotting points with T-SNE. If features do not seem to be grouped in a significant way, it may be that they are not so representative of the similarities and distinctions between the data that will allow you to model them.

As we can see in Fig. 2, T-SNE makes it possible to distinguish two different clusters, associated with the student behaviour collected with extracted features from the log trace, red points correspond to enrolment with drop-out result, the purple points correspond to succeed ones. Drop-out points are stained with the same color, grouped together forming a cluster and separated from the purple points. Which shows the similarity and distinction in data. This founding validates the relevance of used features.

6 Conclusion

In this work, we have dealt with the problem of dimensionality reduction through two categories of methods: feature selection and the data transformation. These methods were applied to extracted features from the log trace during the interaction of students with the MOOC. We used the contribution of all the features selection methods presented in this work in order to determine the features that will be selected. We have detailed in the second part linear and non-linear reduction techniques based on data transforming.

Since Removing irrelevant features helps producing competitive prediction results, our main contribution was obtaining encouraging performances with a minimum number of variables.

So as a perspective, we intend to continue to experiment with our algorithm on other benchmark datasets in order to see the behaviour of our approach on other datasets.

Acknowledgement. We would like to gratefully acknowledge the organizers of KDD Cup 2015 as well as XuetangX for making the datasets available.

References

1. Zakaria, A., Anas, B., Oucamah, C.M.M.: Intelligent system for personalised interventions and early drop-out prediction in MOOCs. Int. J. Adv. Comput. Sci. Appl. **13**(9) (2022). https://doi.org/10.14569/IJACSA.2022.0130983
2. Gitinabard, N., Khoshnevisan, F., Lynch, C.F., Wang, E.Y.: Your actions or your associates? predicting certification and dropout in MOOCs with behavioral and social features. arXiv preprint arXiv:1809.00052 (2018)
3. Berens, J., Schneider, K., Görtz, S., Oster, S., Burghoff, J.: Early detection of students at risk–predicting student dropouts using administrative student data and machine learning methods. Available at SSRN 3275433 (2018)
4. Bellman, R.: Dynamic programming. Princeton University Press, NJ 95 (1957)
5. Vasan, K.K., Surendiran, B.: Dimensionality reduction using principal component analysis for network intrusion detection. Perspect. Sci. **8**, 510–512 (2016)

6. Lam, S.L., Lee, D.L.: Feature reduction for neural network based text categorization. In: Proceedings 6th International Conference on Advanced Systems for Advanced Applications, pp. 195–202. IEEE (1999)
7. Ding, C.H.: A probabilistic model for dimensionality reduction in information retrieval and filtering. In: Proceedings of the 1st SIAM Computational Information Retrieval Workshop. Citeseer (2001)
8. Zhang, Y., Ding, C., Li, T.: Gene selection algorithm by combining reliefF and mRmR. BMC Genomics **9**(S2), 27 (2008)
9. Abraham, R.: Dimensionality reduction through bagged feature selector for medical datamining
10. El Akadi, A., Amine, A., El Ouardighi, A., Aboutajdine, D.: A two-stage gene selection scheme utilizing mRmR filter and GA wrapper. Knowl. Inf. Syst. **26**(3), 487–500 (2011)
11. Vainer, I., Kraus, S., Kaminka, G.A., Slovin, H.: Obtaining scalable and accurate classification in large-scale spatio-temporal domains. Knowl. Inf. Syst. **29**(3), 527–564 (2011)
12. Rai, K., Devi, M.S., Guleria, A.: Decision tree based algorithm for intrusion detection. Int. J. Adv. Netw. Appl. **7**(4), 2828 (2016)
13. Kohavi, R., John, G.H., et al.: Wrappers for feature subset selection. Artif. Intell. **97**(1–2), 273–324 (1997)
14. John, G.H., Kohavi, R., Pfleger, K.: Irrelevant features and the subset selection problem. In: Machine Learning Proceedings 1994, pp. 121–129. Elsevier (1994)
15. LLC, M.: MS Windows NT Kernel Description. https://scikit-learn.org/stable/modules/generated/sklearn.feature_selection.RFE.html. Accessed 30 Sept 2010

Remote Heart Rate Measurement Using Plethysmographic Wave Analysis

Zakaria El khadiri(✉), Rachid Latif, and Amine Saddik

Laboratory of Systems Engineering and Information Technology LiSTi, National School of Applied Sciences, Ibn Zohr University Agadir, Agadir, Morocco
zakaria.elkhadiri@edu.uiz.ac.ma

Abstract. Estimating Heart Rate (HR) value is one of the great important things to determine a person's physiological data and monitor the physiological signs. Nowadays, many kinds of research have demonstrated that the most physiological data "e.g., heart rate and respiration rate" can be accurately measured remotely and collected from photoplethysmographic (PPG) signals via digital cameras or webcams. The PPG signal will be subjected to several signals processing algorithms such as EMD, PCA/ICA, FFT, and others, in order to evaluate and extract the appropriate signs that help to extract the vital parameters. Furthermore, many related works have been made to improve the detection of this non-contact technique with the same result measured using contact sensors. In this paper, we present a general approach with familiar processing signal algorithms and their uses to extract heart rate values. Indeed, the current paper is mainly based on three commons signal processing algorithms widely used in the related field, which are the Empirical Mode Decomposition for a signal decomposition, Principal Component Analysis for the signal's dimensionality reduction, moving average algorithm for signal denoising, and Fourier Transform for converting the PPG signal from the time domain into to frequency domain. Our proposed heart rate estimation systems show a good correlation relationship and closeness with a mean of error around of 1.59.

Keywords: Heart rate · PPG signal · Moving average · EMD · PCA

1 Introduction

Heart rate (HR) is one of the vital parameters of a person's health and the most important physiological parameter which is widely used in determining the patient's health condition, these parameters are particularly used for affective computing and in the telemedicine domain. Recently, most of these parameters can be monitored and retrieved at a distance using webcams or RGB cameras.

Supported by organization Laboratory of Systems Engineering and Information Technology LiSTi.

N. Aboutabit et al. (Eds.): ICMICSA 2022, LNNS 656, pp. 254–267, 2023.
https://doi.org/10.1007/978-3-031-29313-9_23

The latter extraction is through photoplethysmographic signals that can be analyzed through different processing signal algorithms such as Continuous Wavelet Transforms (CWT), Independent/Principal Component Analysis (ICA/PCA), Empirical Mode Decomposition (EMD), and among others in order to assess the instantaneous heart rate. Minutely, in the medicine definition, the heart rate is the number of heartbeats per minute (expressed as BPM) which it's established on the number of contractions of the ventricles. The normal heart rate value is between 60 and 100 beats per minute, which can be varied by many factors such as body position, emotions, air temperature, and other factors that can affect the rate and regularity of pulse activity.

In 1995, Costa et al. [1] introduced the first health monitoring system, which they are devoted to extracting physiological data from an RGB camera. As things progress, the field is known for a great evolution and kept undergoing a set of robust algorithms to detect most of these parameters from photoplethysmography (PPG). The accuracy also knows a great increment, as it achieves about 90% in 2015 by Rahman et al. [2] that proposed a system able to detect three physiological parameters heart rate (HR), interbeat (RR), and interbeat interval (IBI). The experiment of this study was founded on two phases, (1): real-time HR estimation simultaneously with the cStress system, (2): HR extraction in offline via recorded film images sequence. The latter proposed experiments have provided a good level of agreement as per the reference measurement with the RSQ and CORREL values were more than 90% of closeness with the reference evaluation. As thing progress, the current field has become the interest of recent research which it's targeted different applications such as newborns and others. Moreover, in 2021, A. Bella et al. [3,4] introduces an overview of the heart rate estimating assessment based on the vitals indicators, also they have provided a brief state-of-the-art of the algorithms employed for this extraction, especially in critical cases such as the Covid-19 pandemic. They reached a significant result that was closer than the sensor's frequency extraction with an error of +/−2 beats/minute.

The current paper proposed three approaches for heart rate estimation, the set of used algorithms are mostly used by processing signals community, which are Empirical Mode Decomposition (EMD), Principal Component Analysis (PCA) [5–7], moving average filtered method, and Fourier Transform, with their uses to determine the physiological signs. In the first stage, our proposed system is firstly based on the famous Viola and Jones method [8], in order to detect and track the part of the human face, the Region of Interest, and to extract the PPG signal from the average of sequence images on the other side. In our case, we will establish our focus on the 3 RGB bands in the estimate of the HR values, then only assess the findings on the green channel as it contains enough PPG data necessary for this extraction. After that, an image pre-processing is performed such as the equalization method [9] in order to adjust the contrast of the digital sequence images and to better distribute the intensities on the whole range of possible values. After gathering and extracting the first noisy PPG signal from the image processing block, we need then to apply the aforementioned signal processing algorithms, in fact, the latter part is considered the main level of our entire heart rate estimating system.

However, the green band of the PPG signal will be directly subjected first to the EMD method in order to decompose the signal into several Intrinsic Mode Functions, which can help to distinct and select the relevant candidate that has more pulse rate data, and to the moving average method in order to remove and eliminate as much as possible the unexpected noises generated by user's environment and illumination, then in order to extract the heart rate value, the filtered PPG signal will be converted from time domain into to frequency domain using the Fourier Transform (FT). In parallel, the three RGB bands of PPG signals will be subjected as well to the Principal Component Analysis (PCA) which served to reduce the signal's dimensionality and to produce the independent Component. Rather, the heart rate value can be easier extracted through the first produced principal component. Our result shows the linear correlation between the three approaches with a standard deviation of 1.59.

The remainder of this paper is organized as follows. In Sect. 2, we will present a global overview of the most recent related works of extracting and monitoring heart rate (HR) values from a webcam. In Sect. 3, we will briefly describe and compare the three processing signal approaches that are used to extract most physiological parameters. In Sect. 4, we will present the results obtained by these approaches. Finally, a conclusion is drawn in Sect. 5.

2 Related Works: Overview

The traditional technique used to measure the heart rate values is by a contact sensor, the latter is not useful at all in the actual cases today for people contaminated by Covid-19. Therefore, many recent types of research show that remote video recording of subjects under ambient light contains rich enough data to extract vital functions such as respiration rate and heart rate. In 2018, P.V Rouast et al. [10] proposed an algorithm dedicate to heart rate monitoring on non-contacting techniques for real-time, this research is based on dimensionality reduction techniques such as Principal Component Analysis (PCA) which is applied to the multidimensional signal in order to produce multiple components, and then to identify the component which contains the high percentage of spectral power. In 2014, Xiaoib et al. [11] presents a proposed method that served to reduce the environmental illumination artifacts from RGB color space signals. The latter study was basically based on the normalized least square (NLMS) adaptive filter, the authors achieved a good agreement and correlation between the reference sensor and the proposed methods.

Again, in 2014, Guo et al. [12] applied the independent vector analysis (IVA) technique to extract HR from a driver's facial video without taking into account the illumination artifacts. Additionally, in [13], the author extracts the same physiological parameters such as HR and HRV via drivers' facial videos both using a multiset canonical correlation analysis (MCCA) approach taking into consideration the motion artifacts issues, the latter work achieved high performance and better than the ICA. Most recently, in 2020, H. Rahman et al. [14] proposed a non-contact approach for physiological parameters extraction using facial video taking into account the illumination, motion, movement, and vibration noises. The latter work extracts four physiological parameters, i.e., heart rate (HR), inter-beat-interval (IBI), heart rate variability (HRV), and oxygen saturation (SpO2) from facial video recordings, they are based on FT/ICA/PCA, normalization, filtering, detrending and smoothing as a processing signal's methods. The evaluation of this method was performed by using a traditional sensor "cStress" which is attached to the hands and fingers of test subjects.

Despite the field undergoing different processing signal algorithms, the issue related to motion artifacts caused by motion, movement and vibration still appear and impacts the extraction result. In this case, several techniques are widely used for removing motion artifacts from images, e.g., FFT methods, SVD methods, wavelet denoising methods, and adaptive filtering methods [15], nonlinear interpolation techniques for eliminating motion artifacts [16]. Again, against the motion issues, the author in [17] used an algorithm fully dedicate to detecting the motion artifacts using a Convolution Neural Network (CNN) in MR images. In 2021, L. Maurya [18]. Presents a review article for non-contact breathing rate monitoring in newborns, however, the latter study provides global advantages, illumination, clinical application, and signal processing algorithms involved in these methods.

3 Proposed Methods

As already mentioned, before getting down to decompose and proceed with the PPG signal, image pre-processing must be done as a prior step. However, the approach outlined in this document is based on the famous Viola-jones object detector as a method to obtain the bounding box for the part of the human face and to track the region of interest (RoI) as well, this method is notably based on the cascades of boosted classifiers and its uses for real-time objects detection with a significant performance detection comparing with the recent techniques. Afterward, through the last bounding box, the first estimated noisy PPG signal can be collected using the average value of each extracted area. On the signal processing side, we used the most known methods as detailed below:

3.1 Reviews of the Existing Methods

Empirical Mode Decomposition. EMD, Empirical mode decomposition is a method used for decomposing a signal into several constituent signal components named Intrinsic Mode Functions (IMFs) [19]. The decomposition is based on frequency and amplitude in a certain time of stationary and non-stationary signals. it produces also a residue signal which will be of no use in this study. On a mathematical side the relationship of the original x(t) signal to the IMF and its residue is expressed as follows:

$$x(t) = \sum_{i-1}^{N} IMF_i(t) + r(t) \tag{1}$$

Then the algorithm to get the IMF is as follows [20]:

1. Determining local minima and maxima of the input signal x(t).
2. Getting the lower envelope of the signal by interpolating all local minima and the same to all local maxima to get the upper envelope.
3. Compute the mean of local maxima and minima so will get the mid-value called m(t).
4. Compute candidates of IMF, IMF I(t) = x(t) - m(t).
5. The IMF is obtained using the following criteria:
   - 5.1. The number of zero crossings and extrema points must be the same or different by at least one.
   - 5.2. The average envelope must be zero at all points.
6. Repeat all steps from 1 to 5 in order to determine the remaining IMFs and finally produce a residue with monotonous values.

Principal Component Analysis. The Principal Component Analysis, is a method for decreasing the dimensionality of a multi-dimension signal and datasets. In fact, the PCA method aims to create a new uncorrelated variable that consecutively maximizes variance. The newly extracted variables are termed the principal component, these components are rather estimated by solving the eigenvalue and eigenvector mathematical problems. In 2016, [21] Ian T. Jolliffe introduced a dedicated article explaining the basic ideas of the PCA algorithm and describing its variants and their application.

Fourier Transform. Fourier transform (FT) is the most known mathematical method for transforming time-domain signals into frequency domains to obtain signal power information at certain frequencies. In this current research, the FFT method is used in order to check the decomposition signal result in the frequency domain, which means, with the FFT tool we will be able to choose whether the signal is the relevant candidate to provide the heart rate value and other physiological data.

3.2 Heart Rate Extraction Method

As aforementioned, this document will depict the results achieved by three signal processing approaches dedicated to extracting the heart rate values. Typically, the global algorithm is mainly divided into 3 blocks, i.e., image processing, signal processing/decomposition, and spectral peak tracking/calculation HR values which are depicted below in Fig. 1:

The image pre-processing and extraction of the PPG part is the first block that can deal directly with the user's faces and provide the first noisy version of the PPG signal. Within this block and after extracting the human faces and RoI areas which consist of the forehead's faces. Then, the finer function involved in this block is equalization which is considered one of the fundamental processes in image analysis and enhancement. However, the principal goal of this methodology is to increase image quality and to improve its histogram in order to keep

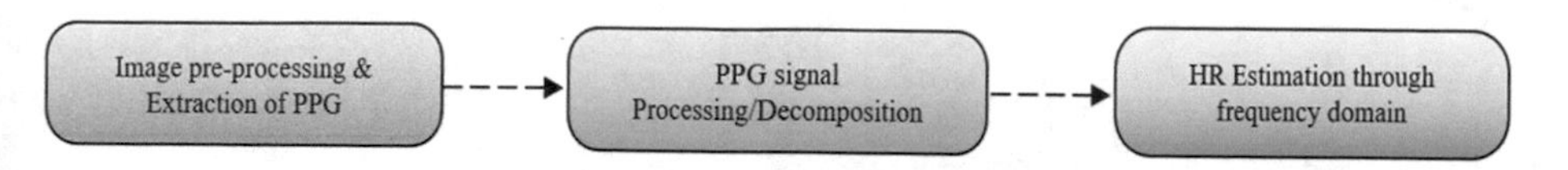

Fig. 1. The three main blocks are used for heart rate estimation.

it in the suitable application. On our side, the approach is based on the same pre-processing image and PPG signal extraction strategy, each one uses its own signal processing method as the following sections depict:

The 1st approach is based on EMD as a method to decompose the PPG signal into a set of signal modes termed IMFs after being performed with a normalization that should be according to Eq. 2. Then, an FFT method is performed on a decomposed signal to determine which model would be the candidate for heart rate information and extraction, the criterion to determine which one is to choose the signal which has a spectral power dominance at frequencies 2 Hz approximatively. Finally, the calculation of the HR values:

$$X_i(t) = \frac{Y_i(t) - \mu_i(t)}{\delta_i(t)} \tag{2}$$

For each $i(= R, G, B)$ signal, μ_i is the mean, δ_i is the standard deviation of Y_i.

While the 2nd approach used the moving average filtered signal as a firsthand method, in order to eliminate the unwanted high and low frequencies noises. Then, a detrending signal was performed as an important step used to remove the long-running trend of signal raw [22]. Finally, similar to the first algorithm, Fourier Transform is applied to PPG filtered signal in order to convert it from a time domain into a frequency domain and to estimate the HR value by the correspondence to the index which is equivalent to the highest spectral power.

The 3rd approach is a little bit different from the second one, indeed, it's inspired by the work proposed by P.V Rouast et al. [10] in 2018 concerning a heart rate monitoring algorithm based on non-contact techniques. This approach is based on three channels "RGB bands" instead of one, which means, it will generate three PPG signals as an input of the processing signal block. After normalizing these signals, the Principal Component Analysis method is applied in order to separate the signals into linearly uncorrelated components and order them, also to isolate the desired part of the signal. Finally, based on the latter components extracted by PCA, the most periodic one has to be chosen to proceed with the rest of the algorithm which is the same as the second approach. The global flow chart of the three main blocks is detailed and shown in Fig. 2:

4 Results and Discussion

Heart rate value extraction was extracted with three methods using C/C++ programing language and the famous OpenCV library, in this section we will present

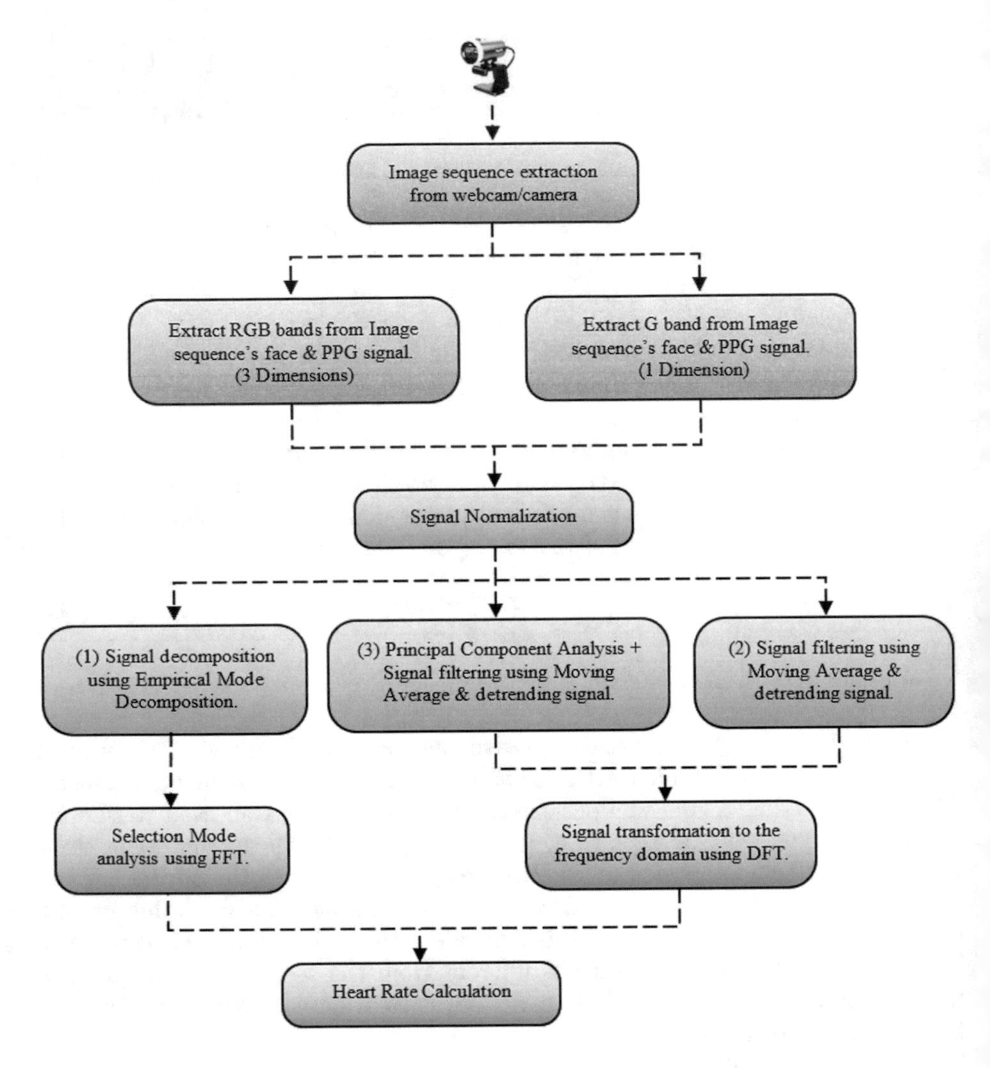

Fig. 2. The stacking layers for Heart Rate values extraction.

the results achieved of these three proposed approaches. On the first hand, the result of the PPG decomposition with the EMD method (1st technique) used is shown in Fig. 3:

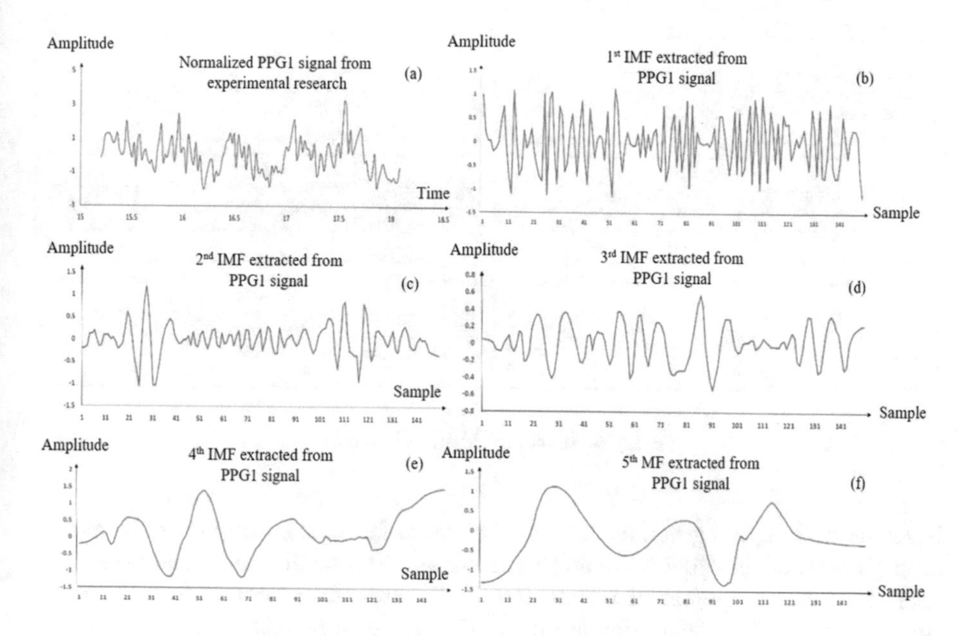

Fig. 3. Result curves extracted from the first approach.

Figure 3 shows the PPG signal with its related-IMFs after applying the EMD. After gathering the noisy PPG signal from the image processing block, the normalized PPG1 signal from experiment research contains some additional unexpected noises which are clearly depicted in Fig. 3-a, those noises are mainly linked to the motion artifacts and illumination environment. The latter signal will be subjected directly to the Empirical Mode decomposition method in order to decompose it into several Intrinsic Mode Functions, the Fig. 3-b until Fig. 3-f show the extracted IMFs from the green band of the PPG signal. Typically, the EMD method separates a time series into a finite number of components, those components are clearly outlined in the following Fig. 4.

These are the IMFs of one of the observed PPG1 signals, each mode was transformed into a frequency domain to pick the appropriate one, which can provide the heart rate value and another physiological parameter. The result of the Fourier Transform is differing from each mode to another, notably in the power spectral which is related to different frequency ranges. In our case, the IMF which has a frequency 1 Hz 2 Hz is picked as the relevant one to provide the heart rate value, which is the IMF4, while the other modes were removed and not considered for HR information. On the other side, the 2nd approach was

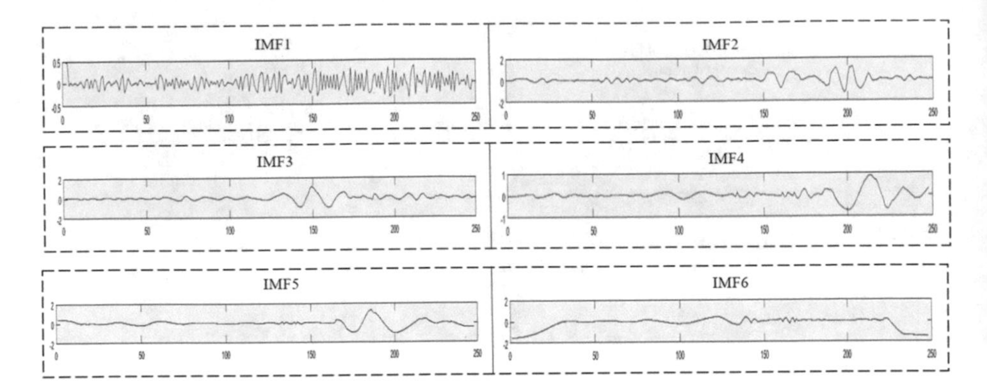

Fig. 4. Intrinsic Mode Function.

based as well on the green channel of the user's faces, then by using the moving average method in order to denoise the signal against the undesired noises, the heart rate values can easier be extracted after transforming the signal from the time domain to frequency domain using the Discrete Fourier Transform (DFT). The results of those approaches can be compared in the Table 1.

Concerning the third method, it's based on the three RGB channels instead of only green space, which means, the image pre-processing and PPG extraction block will provide the PPG signal with three dimensions. The multi-dimension PPG signal will be then subjected to the Principal Component Analysis method in order to reduce its dimension, however, the following Fig. 5 present the result of our third road.

As clearly shown in the figures outlined above, even after performing the pre-processing image to remove the unwanted noises, nonetheless, some noises appeared due to illumination changes or other factors related to the user's environment. Those noises were approximatively removed after detrending the signal and applying the moving average method (2nd approach). After converting the filtered signal from the time domain to the frequency domain using Fourier Transform, the heart rate can be easily estimated using the maximum power response index. Minutely, as shown in Fig. 5 the PCA is indeed applied using 3 dimensions of the input signal resulting from RGB bands, to reduce its dimensionality by producing multiple components, one of these components is considered relevant for heart rate value extracting, and the latter is chosen as the one which has more distinct periodicity, other components are considered as noises due to body movement or lighting changes.

In our experiment research, we have been based on a dataset that contains a very significant PPG signal extracted from experimentation and testing examination and study. For more convenience, we have selected only 20 PPG signal

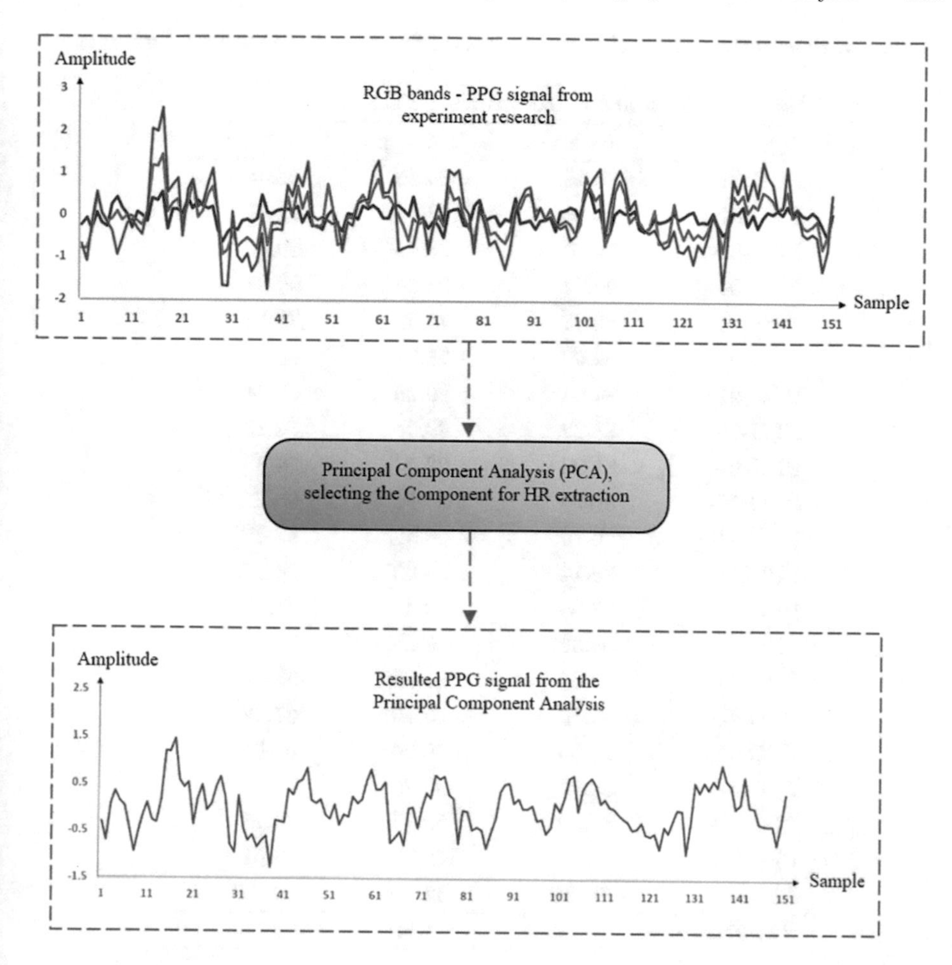

Fig. 5. Principal Component Analysis with selecting the important component.

in order to shed light on the correlation and the agreement between those methods. The following Table 1 present the results achieved using the three approach which are applied rather on the picked-up PPG signals. The Table 1 shows that for each method, the heart rate value is exactly within the normal suitable range for an adult, which is mostly between 60 and 100 beats/min.

Table 1. Heart rate values extracted from 20 PPG signal from experiment research.

Estimated Heart Rate values (HR)			
Signal	1st approach	2nd approach	3rd approach
PPG-01-	83.25	85.15	83.44
PPG-02-	78.83	78.33	78.15
PPG-03-	79.40	83.44	80.79
PPG-04-	93.24	93.66	99.34
PPG-05-	80.96	71.52	78.15
PPG-06-	89.60	71.52	75.50
PPG-07-	90.94	90.26	99.34
PPG-08-	88.22	83.44	74.17
PPG-09-	81.96	83.44	76.82
PPG-10-	66.95	64.90	66.23
PPG-11-	66.79	69.82	62.25
PPG-12-	83.14	84.00	74.17
PPG-13-	83.38	83.44	71.52
PPG-14-	88.32	90.26	79.47
PPG-15-	93.64	91.96	98.01
PPG-16-	87.50	85.15	87.42
PPG-17-	89.55	90.26	83.44
PPG-18-	89.47	90.26	79.47
PPG-19-	87.38	85.15	78.15
PPG-20-	94.52	102.18	98.01
Mean	**84.82**	**83.94**	**81.10**
ST. Deviation		**1.59**	

As additional note, the mean heart rate value extracted using the 1st method is around 84.82 beats/min, and it's around of 83.94 beats/min using the 2nd method, while it's about 81.10 beats/min through 3rd technique. Those heart rate values present a standard of deviation approximately equal 1.59, which proves a good agreement and a significant correlation notably between the 1st method and the 2nd method. Moreover, in order to clearly depict our obtained results, the following Fig. 6 shows the distribution between all proposed systems:

Figure 6 shows the heart rate values distribution based on the selected 20 PPG signals, those signals were thoroughly chosen in order to detail the alignment achieved and to present the slight difference for some PPG signals. The latter difference can be clearly noticed for example on the PPG-12-, PPG-15-, and PPG-20-. Mostly, this variance can be explained due to the motion artifacts and illumination presented in the signal. Even though, our systems show the closeness and the correlation for heart rate extraction with a standard value

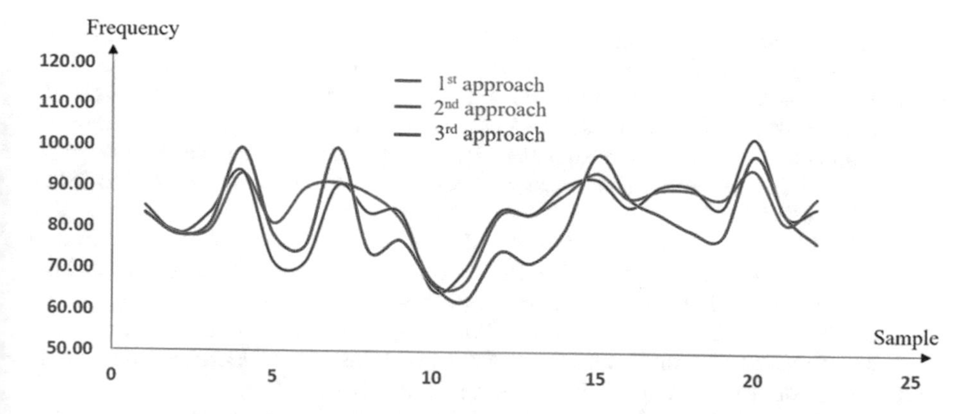

Fig. 6. Heart rate values distribution from the experiment research.

equal to 1.59. Additionally, not only the heart rate values, but rather those approaches are also more suitable to be applied in order to extract other vital signs physiological such as the respiratory rate.

5 Conclusion

This research describes three signal processing approaches used to determine the physiological parameter notably the heart rate value, the simulation process was carried out on MATLAB. The validation test has been carried out by comparing the HR value obtained by the several approaches depicted in the above sections. Additionally, these methods are also useable to collect and extract the remaining vital parameters such as RR, HRV, and arterial blood oxygen saturation. In the end, this research can be the basis for determining which is the appropriate method that should be used in real-time to extract the HR value. Also, it sheds light on another algorithm that will be capable to deal with PPG signal noises the firsthand and optimizing the execution time on the other hand. In future work, we're aiming to choose one of these approaches and decompose each main block into several functional blocks, then try to implement it using a heterogeneous architecture.

References

1. DA COSTA, German. Optical remote sensing of heartbeats. Optics Communications, vol. 117, no 5-6, pp. 395–398 (1995)
2. Rahman, H., Ahmed, M.U., Begum, S.: Noncontact physiological parameters extraction using camera. In: The 1st Workshop on Embedded Sensor Systems for Health through Internet of Things (ESS-H IoT), Oct., (2015)
3. Bella, A., Latif, R., Saddik, A., Guerrouj, F. Z.: Monitoring of physiological signs and their impact on the Covid-19 pandemic. In: E3S Web of Conferences (Vol. 229, p. 01030) (2021). EDP Sciences

4. Bella, A., Latif, R., Saddik, A., Jamad, L.: Review and evaluation of heart rate monitoring based vital signs, a case study: covid-19 pandemic. In: 2020 6th IEEE Congress on Information Science and Technology (CiSt), pp. 79–83. IEEE (2021)
5. Lewandowska, M., Ruminski, J., Kocejko, T.: Measuring pulse rate with a webcam - a non-contact method for evaluating cardiac activity. In: Proceedings of the 2011 Federated Conference on Computer Science and Information Systems (FedCSIS), pp. 405–410 (2011)
6. Balakrishnan, G., Durand, F., Guttag, J.: Detecting pulse from head motions in video. In: Proceedings of the 2013 IEEE Computer Society Conference on Computer Vision and Pattern Recognition, pp. 3430–3437 (2013)
7. Irani, R., Nasrollahi, K., Moeslund, T.B.: Improved pulse detection from head motions using DCT. In: Proceedings of the 9th International Conference on Computer Vision Theory and Applications, pp. 118–124 (2014)
8. Viola, P., Jones, M.: Robust real-time face detection, in computer vision, ICCV 2001. In: Proceedings of the Eighth IEEE International Conference on, pp. 747–747 (2001)
9. Mustafa, et Kader, W.A., Abdul, M.M.M.: A review of histogram equalization techniques in image enhancement application. J. Phys.: Conf. Ser. IOP Publishing, p. 012026 (2018)
10. Rouast, P.V., Adam, M.T.P., Chiong, R., et al.: Remote heart rate measurement using low-cost RGB face video: a technical literature review. Front. Comput. Sci. **12**, 858–872 (2018)
11. Li, X., Chen, J., Zhao, G., Pietikainen, M.: Remote heart rate measurement from face videos under realistic situations. In: Proceedings of the IEEE Conference on Computer Vision and Pattern Recognition (pp. 4264–4271) (2014)
12. Guo, Z., Wang, Z.J., Shen, Z.: Physiological parameter monitoring of drivers based on video data and independent vector analysis. In: 2014 IEEE International Conference on Acoustics, Speech and Signal Processing (ICASSP) (pp. 4374–4378). IEEE (2014)
13. Qi, H., Wang, Z. J., Miao, C.: Non-contact driver cardiac physiological monitoring using video data. In: 2015 IEEE China Summit and International Conference on Signal and Information Processing (ChinaSIP) (pp. 418–422). IEEE (2015)
14. Rahman, H., Ahmed, M.U., Begum, S.: Non-contact physiological parameters extraction using facial video considering illumination, motion, movement and vibration. IEEE Trans. Biomed. Eng. **67**(1), 88–98 (2019)
15. Yadhuraj, S.R., Sudarshan, B.G., Prasanna Kumar, S.C.: GUI creation for removal of motion artifact in PPG signals. In: 2016 3rd International Conference on Advanced Computing and Communication Systems (ICACCS) (Vol. 1, pp. 1–5). IEEE (2016)
16. Rareş, A., Reinders, M.J., Biemond, J.: Image sequence restoration in the presence of pathological motion and severe artifacts. In: 2002 IEEE International Conference on Acoustics, Speech, and Signal Processing (Vol. 4, pp. IV-3365). IEEE (2002)
17. Meding, K., Loktyushin, A., Hirsch, M.: Automatic detection of motion artifacts in MR images using CNNS. In: 2017 IEEE International Conference on Acoustics, Speech and Signal Processing (ICASSP) (pp. 811–815). IEEE (2017)
18. Maurya, L., Kaur, P., Chawla, D., Mahapatra, P.: Non-contact breathing rate monitoring in newborns: a review. Comput. Biol. Med. **132**, 104321 (2021)

19. Huang, N.E., Shen, Z., Long, S.R., Wu, M.C., Shih, H.H., Zheng, Q., Liu, H.: The empirical mode decomposition and the Hilbert spectrum for nonlinear and non-stationary time series analysis. In: Proceedings of the Royal Society of London. Series A: Mathematical, Physical and Engineering Sciences 454(1971) 903-995 (1998)
20. Chen, X., Shao, J., Long, Y., Que, C., Zhang, J., Fang, J.: Identification of Velcro rales based on Hilbert-Huang transform. Phys. A: Stat. Mech. Appl. **401**, 34–44 (2014)
21. Jolliffe, I.T., et CADIMA, J.: Principal component analysis: a review and recent developments. In: Philosophical Transactions of the Royal Society A: Mathematical, Physical and Engineering Sciences, vol. 374, no 2065, p. 20150202 (2016)
22. Tarvainen, M.P., Ranta-Aho, P.O., Karjalainen, P.A.: An advanced detrending method with application to HRV analysis. IEEE Trans. Biomed. Eng. **49**, 172–175 (2002)

IoT and Networks

An Online Model for Detecting Attacks in Wireless Sensor Networks

Hiba Tabbaa(✉) and Imad Hafidi

Laboratory of Process Engineering, Computer Science and Mathematics, National School of Applied Sciences Khouribga, University Sultan Moulay Slimane, Beni Mellal, Morocco
hiba.tabbaa@usms.ac.ma, i.hafidi@usms.ma

Abstract. The Internet of Things (IoT) has been employed in a variety of critical fields, including healthcare, geriatric surveillance, self-driving vehicles, and energy management. The most prevalent infrastructure for these applications is Wireless Sensor Networks (WSNs). Although WSNs have intriguing characteristics, the security of such networks is a major concern, especially for applications where confidentiality is extremely crucial. In order to set up WSNs securely, any type of intrusion should be identified before attackers potentially harm the network. However, research findings have demonstrated that existing approaches are ineffective, particularly in detecting attacks in real-time, mostly owing to the accumulation of enormous amounts of data through interconnected devices. Within this interpretation, our intention is to construct a robust Intrusion Detection System (IDS) for analyzing real-time network traffic of WSNs using online learning algorithm while taking into account the network's resource constraints. To achieve this goal, we investigated the use of an online classifier, namely the Hoeffding Adaptive Tree (HAT), along with the selection of relevant attributes to distinguish four kinds of DoS attacks among normal network traffic: the attacks considered are Blackhole, Grayhole, Flooding, and Scheduling attacks. Among the experimental findings, we determined that utilizing the HAT classifier along with the Chi-squared method of feature selection, detection rate percentages were 86.75%, 80.02%, 94.92%, 99.12% and 99.03% respectively for Flooding, Scheduling, Grayhole, Blackhole and normal case attacks. With an overall accuracy of 99.03%. Based on these findings, it is indeed possible to infer that the HAT classifier is extremely beneficial for categorizing attacks, as it has managed to secure a high detection rate despite the existence of many threats.

Keywords: Intrusion Detection System · Wireless Sensor Network · Online Learning · Attack Detection

1 Introduction

Wireless Sensor Networks (WSNs) have become increasingly popular in the recent years, attracting the interest of researchers and industry alike, they were

N. Aboutabit et al. (Eds.): ICMICSA 2022, LNNS 656, pp. 271–282, 2023.
https://doi.org/10.1007/978-3-031-29313-9_24

initially designed and deployed for military applications, and yet their usefulness has since expanded to also include the environmental surveillance, health care, home automation, transportation field, and industrial observation applications, and many others. [1,2].

WSN is composed of a requisite number of intelligent sensor nodes, which are inexpensive, energy-restricted, able to carry out a variety of tasks, wirelessly interconnected and scattered over an area of interest labeled a sensor field [3,4]. Most of the WSNs are distributed in a brutal environment where the presence of human beings is impractical, difficult or almost impossible.

These widely dispersed networked sensor nodes gather valuable data within the monitored area and convey the observed data to an authorized sink node, which acts as a gateway between both the sensor network and the outside world either via single-hop or multi-hop communication for necessary operations such as preprocessing and feature extraction.

Due to its ad-hoc nature, a WSN is subject to many internal and external assaults. Attacks like Distributed Denial of Service (DDoS), blackhole, wormhole, and grayhole assault are among the typical attack types employed in the WSN [5]. For a number of reasons that are unique to these networks, guaranteeing the security of WSNs from aggressive attacks is a challenging undertaking. Additionally, there are a number of conceivable vulnerabilities with the sensor nodes, and the medium of communication is also unreliable. Maintaining the state of WSNs free from threats is therefore both a challenging and significant mission [6].

Accordingly, different approaches known as intrusion detection methods have been developed to deal with intrusions and attackers. These techniques are in charge of monitoring the happenings occurring in the network to identify anything abnormal. An Intrusion Detection System (IDS) has been described in [7] as a reliable system that continuously examines activity traffic to detect whether there is an attack or if these events constitute a legitimate use of the system. When preventative strategies are ineffective in stopping a network attack, IDS serve as a second-line defense.

The methodologies broadly employed to construct IDSs and put them to use for attack detection in the historical present are immensely related to machine learning techniques (MLT). Though the majority of methods rely on batch or offline machine learning, which demands the preservation of the entire set or at least a subset of historical data in computer memory. Due to the time-critical and high-speed data streaming nature of some WSN applications, real-time (or near real-time) online IDS is required to cope with this kind of data with high detection accuracy. These types of data analytics are envisaged to be supported by online machine learning (OL) techniques.

In this paper, a new intrusion detection technique based on online machine learning that models various security incursion types and is appropriate for execution on resource-constrained devices is proposed and evaluated. The resulting IDS can be thought of as a solid decision support system which can present essential information about potential security breaches in a WSN.

This work is sorted out when pursues: In Sect. 2 the literature survey is presented. Online learning for attack detection in WSNs is introduced in Sect. 3. In Sect. 4 the methodology is thoroughly discussed. Section 5 provides the results of the experiment as well as performance analyses in comparison to other relevant methodologies. Finally, Sect. 6 concludes our paper.

2 Literature Survey

Increasing research efforts are being made to develop suitable intrusion detectors for WSNs that are tailored to their individual properties, as well as a diverse set of machine learning algorithms that can detect malicious attacks in WSNs with great precision. Considering that small-scale work has been accomplished to develop IDS using the concept of online learning to detect commonly known and novel attacks in WSNs, this section discusses some of the previous works and methodologies.

Almomani et al. [8] created a novel WSN dataset in 2016, which is known as WSN-DS. It comprised typical network traffic as well as many DoS scenarios (flooding, grayholes, blackholes, and scheduling attacks). A data mining toolkit (WEKA) was used to implement an artificial neural network (ANN) to identify and categorize the four assaults. According to the academic research, the technicality of using the ANN algorithm enabled a higher classification of DoS assaults by omitting the grayhole attack since its detection rate is quite low in comparison to the others.

In 2018, Park et al. [9] proposed a new approach to predict DoS attacks in WSNs. They used the Random Forest classifier to recognize categorical DoS attacks in the WSN-DS dataset. From the results of their experiments, they concluded that the random forest classifier outperformed the other classifiers.

Alqahtani et al. [10] proposed the GXGBoost model in 2019 as a novel intrusion attack detection system that is based on the genetics algorithm and a gradient boosting classifier (XGBoot). The latter is a gradient enhancement model aimed at improving the effectiveness of existing models in detecting minority classes of attacks in WSN data flows that are very imbalanced.

In 2020, Chandre et al. [11] designed an effective IDS by exploiting MLT for WSNs. On the WSN-DS dataset, this article conducts comparison research and performance evaluations of several MLTs. The comparison study reveals that the KNN classifier surpasses the other classifiers in terms of intrusion detection.

Dong et al. [12] suggested an anomaly detection model in 2020 using the principle of the ensemble learning algorithm founded on the information gain ratio and the bagging algorithm. First, the traffic data attribute is chosen using the information gain ratio approach. Second, the bagging approach is utilized to construct an ensemble classifier, which is then used to train numerous upgraded C4.5 decision trees.

To recognize various types of DoS attacks, Samir et al. [13] proposed an online learner that uses the information gain ratio to select relevant features of sensor data with an online passive aggressive algorithm. The authors carried out the experiment using the WSN-DS dataset.

3 Online Learning for Attack Detection in WSNs

According to [14], there exist two perspectives for machine learning techniques: batch/offline learning and online learning. Conventional learning-based approaches, as with offline learning, are unable to alter so as to fit their structure and parameters in order to gather insights from continuous traffic instances. Offline models are not ideal candidates for change detection because they are usually updated less frequently, resulting in considerable detection delays. When extensive training sets are provided, a major drawback is that the algorithm must be retrained pretty much every time the network's normal behavior changes. Given the dynamic nature of WSNs, this is likely to occur frequently during the network's lifespan. As a consequence, we seek novel data analytic techniques that can adapt to additional data entering in real-time. This is particularly an OL functionality [15], and it thus would be an effective choice for WSNs.

Online learning is a type of learning-based approach that collects data sequentially in a potentially endless series $\mathrm{S} = (s_1, s_2, \ldots, s_t, \ldots)$ of data $s_i = (x_i, y_i)$ arrives one at a time and uses limited memory resources, and the goal is to build the classifier h, run the model, and frequently update it with latest instances. At the beginning of the learning process, the whole training set is indeed not available in OL. The model is regularly adjusted as more samples are received [16].

OL operates on a sequence of data examples with time stamps. At each step t, the learner receives an incoming example $x_t \in \mathcal{X}$ in a d-dimensional feature space, that is $\mathcal{X} = \mathbb{R}^d$. First it tries to anticipate the target class $\hat{y}_t$ for each incoming instance x_t after receiving it:

$$\hat{y}_t = \operatorname{sgn}\left(f\left(x_t, w_t\right)\right) \in \mathcal{Y} \tag{1}$$

where $\hat{y}_t$ is the anticipated target class, x_t is a sample, w_t is the weight assigned to the sample, $sgn()$ is the sign function returning $\{0, 1\}$, f is a function mapping x_t, w_t into a real number $r \in \mathbb{R}$, and $\mathcal{Y}$ is a class label. After making the prediction, the true label $y_t \in \mathcal{Y}$ is revealed, and the learner then computes the loss $\ell(y_t, \hat{y}_t)$ based on some criteria, the difference between the learner's prediction and the revealed true label y_t is calculated. Based on the results of the loss, the learner finally determines when and how to modify the classification algorithm at the ending of each learning step.

The algorithmic framework presented below gives an overview of the majority of online learning algorithms, which were described in [17]. In general, different OL algorithms are distinguished by their different definitions and implementations of the loss function $\ell(\cdot)$ and their update functions $\Delta(\cdot)$.

The online accuracy for a sequence up to the current time t is given by [18]:

$$E(S) = \frac{1}{t}\sum_{i=1}^{t} 1 - \ell(h_{i-1}(x_i), y_i) \qquad (2)$$

where h_{t-1} is the previously learned model.

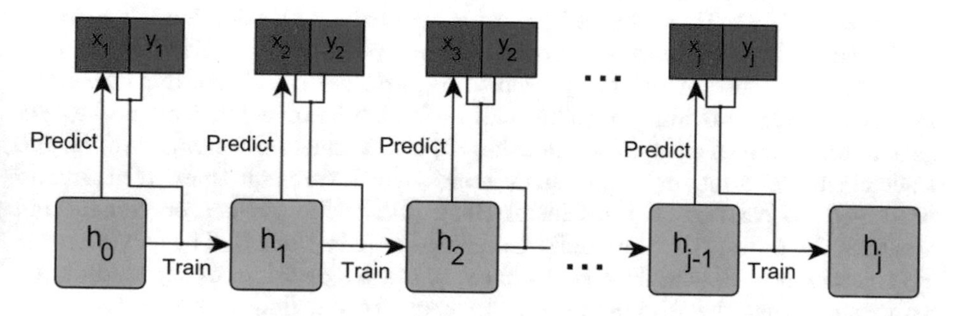

Fig. 1. The online learning scheme [18].

Algorithm 1. Online Learning

Require: $w_1 = 0$
1: **for** $t = 1, 2, \ldots T$ **do**
2: The learner gets an entering example: $x_t \in \mathcal{X}$;
3: The class label is predicted by the learner: $\hat{y}_t = \text{sgn}\left(f\left(x_t, w_t\right)\right)$;
4: The environment reveals the true class label: $y_t \in \mathcal{Y}$;
5: The classification model is updated by the learner: $\ell(w_t; (x_t, y_t))$;
6: **if** $\ell(w_t; (x_t, y_t)) > 0$ **then**
7: The learner updates the classification model:
8: $w_{t+1} \leftarrow w_t + \Delta(w_t; (x_t, y_t))$
9: **end if**
10: **end for**

We chose to work with the **Hoeffding Adaptive Tree (HAT)**, an online learning algorithm, in this article. When standard data analytics approaches encounter difficulties due to the sheer growth of network traffic, HAT can be utilized to process large data flows. This is particularly crucial for detecting newly designed attacks, as malevolent individuals are constantly developing new threats. Because the underlying distribution of the data in streaming applications can change rapidly over time, this Hoeffding Tree uses the adaptive windowing (ADWIN) algorithm to monitor the performance of branches on the tree and replace them with new branches when their accuracy decreases. This is known as concept drift [19].

4 Methodology

4.1 WSN Detection System Design

An important criterion for the realization of IDS mechanisms in the WSN is the location of these agents in this type of network. WSN has two main types of network topologies: plane structure and cluster-based structure [20]. In a flat topology, the sender node relies on multi-hop communication to reach the base station (BS). This process leads to a high communication load. Therefore, integrating IDS agents into this type of topology is not an effective solution.

Clustering sensor nodes has proven to be particularly efficient during sensing and monitoring, providing benefits such as fault tolerance, efficient data aggregation, and reduced energy consumption. In Fig. 2, clusters are linked to the BS. Each cluster is formed by some sensor nodes, which transmit their collected and processed information to the Cluster-Head (CH). The CH sensor gathers and analyzes the data prior to actually transmitting it to the BS. The server on the BS collects these data and processes them to make a decision. In our study, since we presume that the CHs have a greater capacity, our detection occurs in both the cluster heads and the base station.

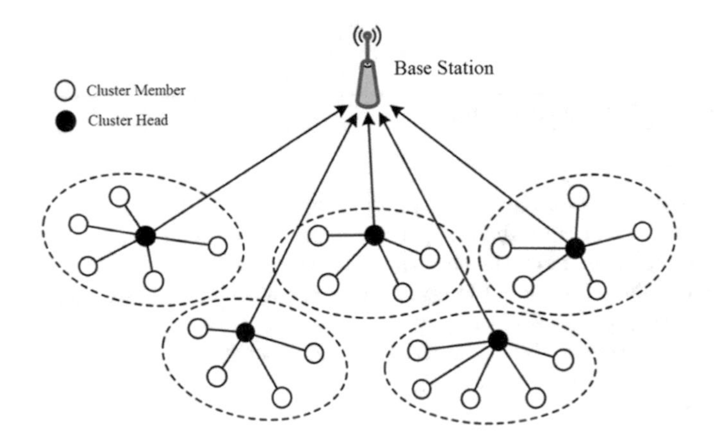

Fig. 2. The WSN network topology [21].

4.2 The Structure of the Proposed Model

The main methodology of the suggested model, outlined in Fig. 3, consists of using the study of online learning methodologies to classify the network traffic flow of WSNs. This methodology is based on using an online algorithm that is resilient to concept drift in tandem with a data pre-processing engine, with the model being modified after each new instance, in which it enhances intrusion detection effectiveness by observing unusual activity and selecting only the

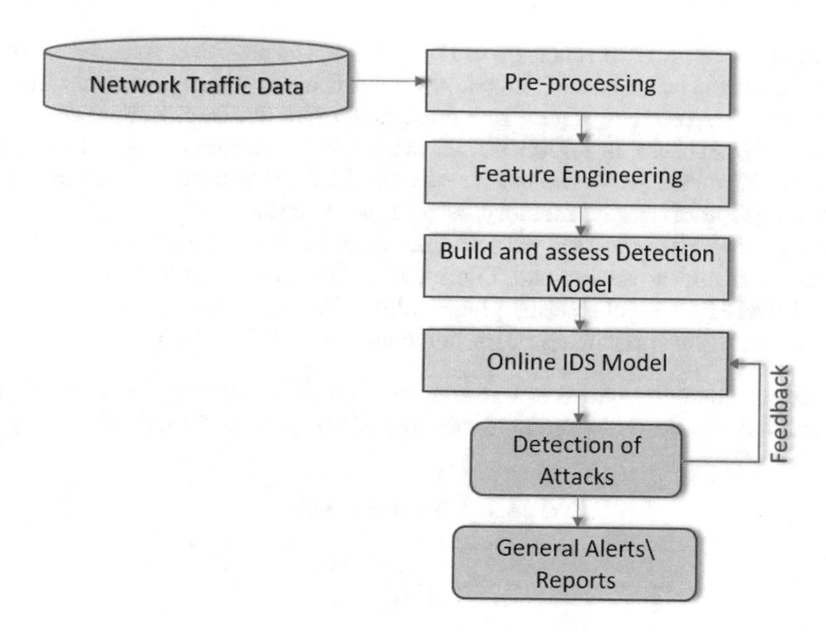

Fig. 3. The structure of the proposed model.

relevant attributes according to the feature selection method, with real-time classification of each packet as normal or as an attack.

5 Experimental Results and Comparisons

5.1 Dataset Descriptions

The WSN-DS dataset was used for evaluating the proposed approach [8]. This is an intrusion detection dataset developed in Network Simulator 2 (NS-2) using the LEACH (Low Energy Adaptive Clustering Hierarchy) protocol. The dataset file contains only 19 features, which are stated as follows: Id, Time, Is_CH, who_CH, Dist_To_CH, ADV_S, ADV_R, JOIN_S, JOIN_R, ADV_SCH_S, ADV_SCH_R, Rank, DATA_S, DATA_R, Data_Sent_BS, Dist_CH_BS, Send_code, Consumed_Energy, including the class label Attack_Type. WSN-DS contains four types of denial-of-service attacks: blackhole, grayhole, flooding, and scheduling besides the normal behavior. The following is a brief description of the attacks :

- ***Blackhole attack:*** DoS attack in which the attacker impersonates the CH. Then the compromised node refuses to forward data packets and starts dropping them, so that no packet reaches the sink node.

- ***Grayhole attack***: In this type of DoS attack, the assaulter masquerades as a CH for all the other units. The fraudulent CH electively or randomly discards packets after receiving them, preventing real packets from being sent.
- ***Flooding attack:*** Is a DoS attack where the attacker sends a large number of CH messages to the sensor via the LEACH protocol, resulting in the consumption of energy, memory, and network traffic.
- ***Scheduling attack:*** DoS attack that involves the attacker taking the role of the CH and adjusting the TDMA schedule from broadcast to unicast at the LEACH protocol's setup phase, allocating all nodes the same time slot to transmit data, resulting in packet collision and data loss.

Table 1 shows the entire statistical characteristics of the WSN-DS dataset that was used to evaluate the performance of the proposed method.

Table 1. WSN-DS dataset.

Type of attack	Quantity	Percentage(%)
Normal	340066	90.77
Grayhole attack	14596	3.90
Blackhole attack	10049	2.68
Scheduling attack	6638	1.77
Flooding attack	3312	0.88
Total traces	**374661**	**100**

The technological features of the computer used throughout the implementation process are as follows:

- Central Processing Unit: Intel(R) Core(TM) i7-4610M CPU @ 3.00GHz 3.00GHz
- Random Access Memory: 8 GB.
- Operating System: Windows 10 Pro 64-bit.

5.2 Performance Metrics and Evaluation

Performance metrics such as Accuracy, Precision, Recall and F1-score were used to evaluate the proposed methodology. The mathematical equations for the performance metrics are shown in Table 2, where:

- True Positive Values : TP.
- True Negative Values : TN.
- False Positive Values : FP.
- False Negative Value : FN.

Table 2. Calculation formulas for performance metrics.

Performance metrics	Mathematical expression
Accuracy	$\frac{TP+TN}{TP+FP+FN+TN}$
Recall or Detection rate	$\frac{TP}{TP+FN}$
Precision	$\frac{TP}{TP+FP}$
F1-Score	$\frac{2\times Precision\times Recall}{Precision+Recall}$

5.3 Results and Findings

To obtain better computational efficiency of the detection method and improve the detection rate in WSN, the existing approaches employ dimensionality reduction and methods such as feature selection (FS). The following are the main methods for FS and dimensionality reduction: Correlation, Linear discriminant analysis, Information gain, Information gain ratio, Chi-squared, and other methods.

For WSN-DS data, FS is used in this experiment. First, features with little to no impact were excluded using the FS methods of information gain ratio and chi-squared. Then, using the features selected with the online HAT algorithm to verify the selection results and especially the impact of the use of FS on the performance of the algorithm, the performances are presented in Table 3.

Table 3. Comparison of methods for selecting characteristics.

Feature selection method	Accuracy with HAT (%)
Chi-squared	99.03%
Information gain ratio	98.74%
Without feature selection	98.67%

Through experience, as shown in Table 3, using accuracy as a performance index with the HAT algorithm, when using the information gain ratio method, the Accuracy was 98.74%. When the chi-squared FS method was used, the Accuracy has increased to 99.03%. Intrusion detection using the HAT algorithm has higher classification accuracy when choosing chi-squared as the FS method along with the feature set: S= {Id, Time, Is_CH, who_CH, Dist_To_CH, ADV_S, ADV_R, JOIN_R, ADV_SCH_S, Rank, DATA_S, DATA_R, Data_Sent_BS, Dist_CH_BS}.

The Accuracy was 98.67% when all of the data was used without using any of the FS methods along with the online HAT algorithm. This demonstrates how using feature selection as a pre-processing step aids in the classification of attacks by attaining improved performance and better results.

Table 4. HAT algorithm confusion matrix.

	Normal	Grayhole	Blackhole	Scheduling	Flooding
Normal	339026	461	25	310	243
Grayhole	737	13855	4	0	0
Blackhole	63	25	9961	0	0
Scheduling	1256	1	67	5312	2
Flooding	413	0	4	22	2873

Table 4 shows the confusion matrix for the online HAT algorithm. For example, there are 10049 records for the blackhole attack, as shown in Table 1. 9961 were correctly identified as a blackhole attack, 63 samples were classified as no attack, 25 records were identified as a grayhole attack, and no packets were identified as scheduling or flooding assaults.

Table 5. A summary of the HAT algorithm's results using the Chi-squared selection method.

	Precision	Recall	F1-Score
Normal	99.28	99.69	99.49
Grayhole	96.60	94.92	95.76
Blackhole	99.01	99.12	99.07
Scheduling	94.12	80.02	86.50
Flooding	92.14	86.75	89.36
Weighted average	99.01	99.03	99.01
Overall accuracy		99.03%	

Table 5 illustrates the classification report of the proposed WSN intrusion detection model for both an attack and a normal scenario. The total Accuracy is 99.03% and by analyzing each class label, we observe that the detection performance for the normal case is higher than the abnormal case. The model achieves a score of 99.69% in detecting normal traffic, and 99.12%, 94.92%, and 86.75% for Blackhole, Grayhole, and Flooding attacks, respectively. With the model scoring the lowest in detecting Scheduling attacks 80.02%. These experimental findings reveal that the algorithm accurately determines if the WSN is still in normal mode or is vulnerable to any type of attack.

5.4 Comparative Analysis

To compare our work with related recent work on the same data set, the true positive rate (TPR), or the recall, is used as a uniform metric to assess this. Figure 4 compares the proposed approach's TPR values to results from related

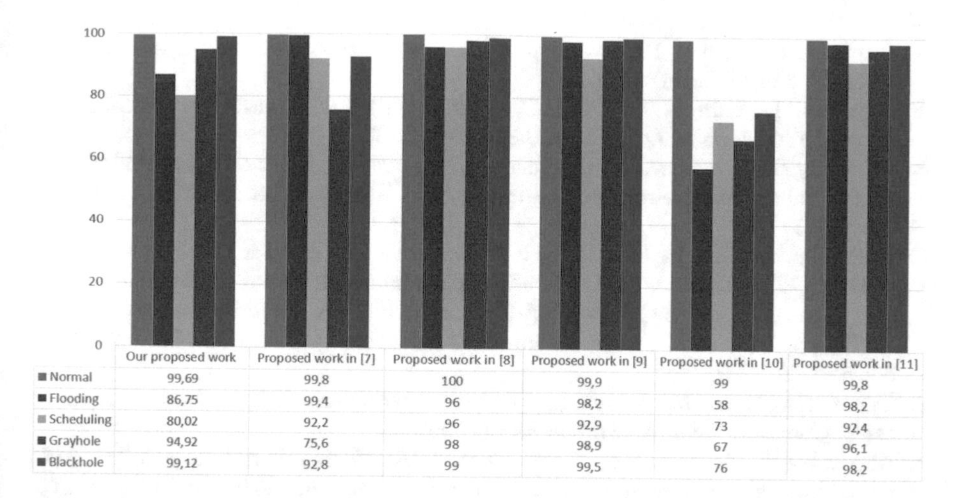

	Our proposed work	Proposed work in [7]	Proposed work in [8]	Proposed work in [9]	Proposed work in [10]	Proposed work in [11]
Normal	99,69	99,8	100	99,9	99	99,8
Flooding	86,75	99,4	96	98,2	58	98,2
Scheduling	80,02	92,2	96	92,9	73	92,4
Grayhole	94,92	75,6	98	98,9	67	96,1
Blackhole	99,12	92,8	99	99,5	76	98,2

Fig. 4. The proposed approach's TPR percentage values compared to the results of related work.

work. However, related works employ offline learning. Individual online learning methods are generally characterized by reduced recall. Although, compared to offline training, the accuracy of HAT improves over time with respect to iterations since, at each iteration, the classifier attempts to better train the model.

6 Conclusion

The main challenge in the development of an IDS for a wireless sensor network is acknowledging the existence of attacks with exalted accuracy while meeting the restrictions and challenges required to extend the system's life. Certainly this objective might be accomplished in a variety of ways in the event that more awareness is paid to the detecting techniques utilized. As a result, in this research, we focus on the development of an online classifier that is well trained using hierarchical clustering models by selecting and pre-processing an appropriate data set. Resulting in the proposal of an appropriate, intelligent, and lightweight intrusion detection system based on online learning capable of processing massive amounts of data streams in real-time and determining the presence of an attack using a cluster-based topology. The proposed robust model detects intrusions efficiently and avoids wasting WSN resources.

References

1. Kumar, D.P., Amgoth, T., Annavarapu, C.S.R.: Machine learning algorithms for wireless sensor networks: a survey. Inf. Fusion **49**, 1–25 (2019)
2. Kurniabudi, K., et al.: Network anomaly detection research: a survey. Indonesian J. Electrical Eng. Informat. (IJEEI) **7**(1), 37–50 (2019)

3. Gupta, S.K., Sinha, P.: Overview of wireless sensor network: a survey. Telos 3.15µW, 38mW (2014)
4. Hasan, S., Hussain, Z., Singh, R.K.: A survey of wireless sensor network. Int. J. of Emerg. Technol. and Adv. Engin 3.3, 487-492 (2013)
5. Chander, B.: Kumaravelan, One class SVMs outlier detection for wireless sensor networks in harsh environments: analysis. Int. J. Recent Technol. Eng. **7**(4), 294–301 (2018)
6. Ifzarne, S., Hafidi, I., Idrissi, N.: Secure data collection for wireless sensor network. In: Ben Ahmed, M., Mellouli, S., Braganca, L., Anouar Abdelhakim, B., Bernadetta, K.A. (eds.) Emerging Trends in ICT for Sustainable Development. ASTI, pp. 241–248. Springer, Cham (2021). https://doi.org/10.1007/978-3-030-53440-0_26
7. Debar, H., Dacier, M., Wespi, A.: Towards a taxonomy of intrusion-detection systems. Comput. Netw. **31**(8), 805–822 (1999)
8. Almomani, I., Al-Kasasbeh, B., Al-Akhras, M.: WSN-DS: a dataset for intrusion detection systems in wireless sensor networks. J. Sensors 2016 (2016)
9. Park, T., Cho, D., Kim, H.: An effective classification for DoS attacks in wireless sensor networks. In: 2018 Tenth International Conference on Ubiquitous and Future Networks (ICUFN). IEEE (2018)
10. Alqahtani, M., et al.: A genetic-based extreme gradient boosting model for detecting intrusions in wireless sensor networks. Sensors 19.20, 4383 (2019)
11. Chandre, P.R., Mahalle, P.N., Shinde, G.R.: Deep learning and machine learning techniques for intrusion detection and prevention in wireless sensor networks: comparative study and performance analysis, pp. 95–120. Singapore, Design frameworks for wireless networks. Springer (2020)
12. Dong, R.-H., Yan, H.-H., Zhang, Q.-Y.: An intrusion detection model for wireless sensor network based on information gain ratio and bagging algorithm. Int. J. Netw. Secur. **22**(2), 218–230 (2020)
13. Ifzarne, S., et al.: Anomaly detection using machine learning techniques in wireless sensor networks. J. Phys.: Conf. Ser. vol. 1743, no. 1. IOP Publishing (2021)
14. Lohrasbinasab, I., et al.: From statistical-to machine learning-based network traffic prediction. Trans. Emerg. Telecommun. Technol. 33.4, e4394 (2022)
15. Lobo, J.L., et al.: Spiking neural networks and online learning: an overview and perspectives. Neural Netw. **121**, 88–100 (2020)
16. Gama, J., et al.: A survey on concept drift adaptation. ACM Computing Surveys (CSUR) 46.4, 1–37 (2014)
17. Hoi, S.C.H., Wang, J., Zhao, P.: Libol: a library for online learning algorithms. J. Mach. Learn. Res. 15.1, 495 (2014)
18. Losing, V., Hammer, B., Wersing, H.: Incremental on-line learning: a review and comparison of state of the art algorithms. Neurocomputing **275**, 1261–1274 (2018)
19. Tsymbal, Alexey: The problem of concept drift: definitions and related work. Computer Science Department, Trinity College Dublin **106**(2), 58 (2004)
20. Rachburee, N., Punlumjeak, W.: A comparison of feature selection approach between greedy, IG-ratio, Chi-square, and mRMR in educational mining. In: 2015 7th international conference on information technology and electrical engineering (ICITEE). IEEE (2015)
21. Sreedevi, P., Venkateswarlu, S.: An efficient intra-cluster data aggregation and finding the best sink location in WSN using EEC-MA-PSOGA approach. Int. J. Commun. Syst. 35.8, e5110 (2022)

An IoT Ecosystem-Based Architecture of a Smart Livestock Farm

Khalid El Moutaouakil(✉), Hamza Jdi, Brahim Jabir, and Noureddine Falih

LIMATI Laboratory, Polydisciplinary Faculty, University of Sultan Moulay Slimane, Mghila, BP 592, Beni Mellal, Morocco
elmoutaouakil.kh@gmail.com

Abstract. Livestock farm managers and farmers must adopt new methods to boost their productivity and meet the growing demand for food. In this regard comes this work to offer an IoT ecosystem-based architecture of a smart livestock farm that aims to help farmers to improve their farm performance and save a lot of waste and prevent economic losses. The architecture that can be applied in cattle, small ruminants and poultry farms focuses on the continuous monitoring of livestock health, performance, reproduction, feed, milking, and behavior alongside identification and location tracking, all through advanced technologies like the Internet of Things, Artificial Intelligence and Big data and using IoT sensors and other smart devices.

Keywords: Agriculture 4.0 · IoT · Smart farm · IoT sensors · Smart livestock farming · Precision livestock farming

1 Introduction

Farmers need to multiply their production and boost their performance to meet the increasing demand for food and keep up with the rising requirements due to the continuously increasing population [1]. To do that, they should rely on new technologies based on the Agriculture 4.0 standard and adopt new techniques to make their livestock farms more smart, efficient and easier to manage through relying on real-time automatic monitoring [2].

The goal of smart livestock farming is to develop a management system based on ongoing, automated real-time monitoring of the health, the welfare and the control of production and reproduction through the use of advanced technologies like the IoT (Internet of things), AI (artificial intelligence) and Big data [3].

Smart farm management methods guarantee a performance boost through continuous monitoring and real-time tracking of vital parameters using IoT sensors, video recordings, acoustic systems, robots, and other smart devices [4].

In this work we propose a smart livestock monitoring farm design that uses the aforementioned smart farm management methods and advanced technologies for the efficient management of livestock farms. It can be applied in cattle, small ruminants (sheep and goats) and poultry farms.

N. Aboutabit et al. (Eds.): ICMICSA 2022, LNNS 656, pp. 283–293, 2023.
https://doi.org/10.1007/978-3-031-29313-9_25

In the following sections, we will take an in-depth look at the existing core IoT technologies, the IoT applications for the smart livestock farm, the proposed smart livestock farm architecture and the livestock system modeling alongside a discussion from several points of view.

2 Smart Livestock Farming Using IoT Technologies

2.1 IoT Technologies

The Internet of Things (IoT) is a system of interconnected computing objects, animals, and other linked things that may exchange data across a network without human intervention. A wide range of sectors including agriculture are increasingly using IoT to run more efficiently, better understand and control their products to provide better performance and services, boost decision-making, and raise the value of their work [5].

Web-enabled smart devices that use embedded systems to gather and send data from their environment make up an IoT ecosystem. The IoT devices connect to an IoT gateway or other edge devices to share the data they collect and make it available for users.

There are four core IoT technologies which are radio frequency identification technology, sensors, sensors network and the internet (Fig. 1). These four pillars serve as the foundation for identification, sensing, processing, and delivery of information, respectively [6].

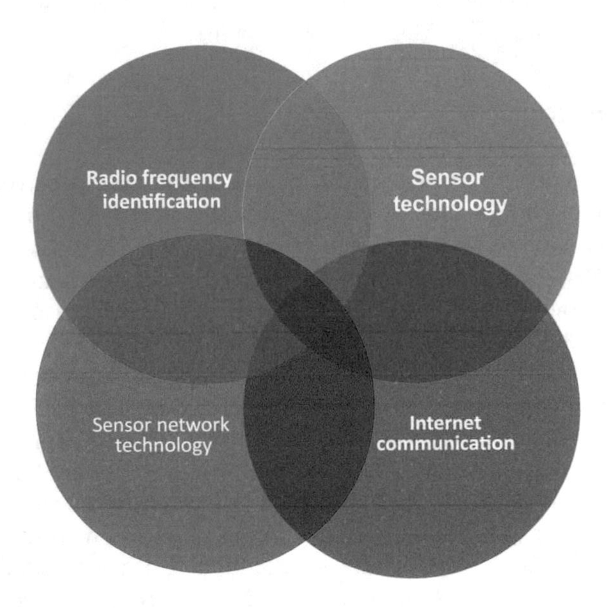

Fig. 1. Core IoT technologies

2.2 IoT Applications for the Smart Livestock Farm

The proposed smart livestock farm consists of eight monitoring units to keep track of the health, reproduction, feed, milking, performance and behavior, in addition to the identification and location tracking of the livestock herd (Fig. 2).

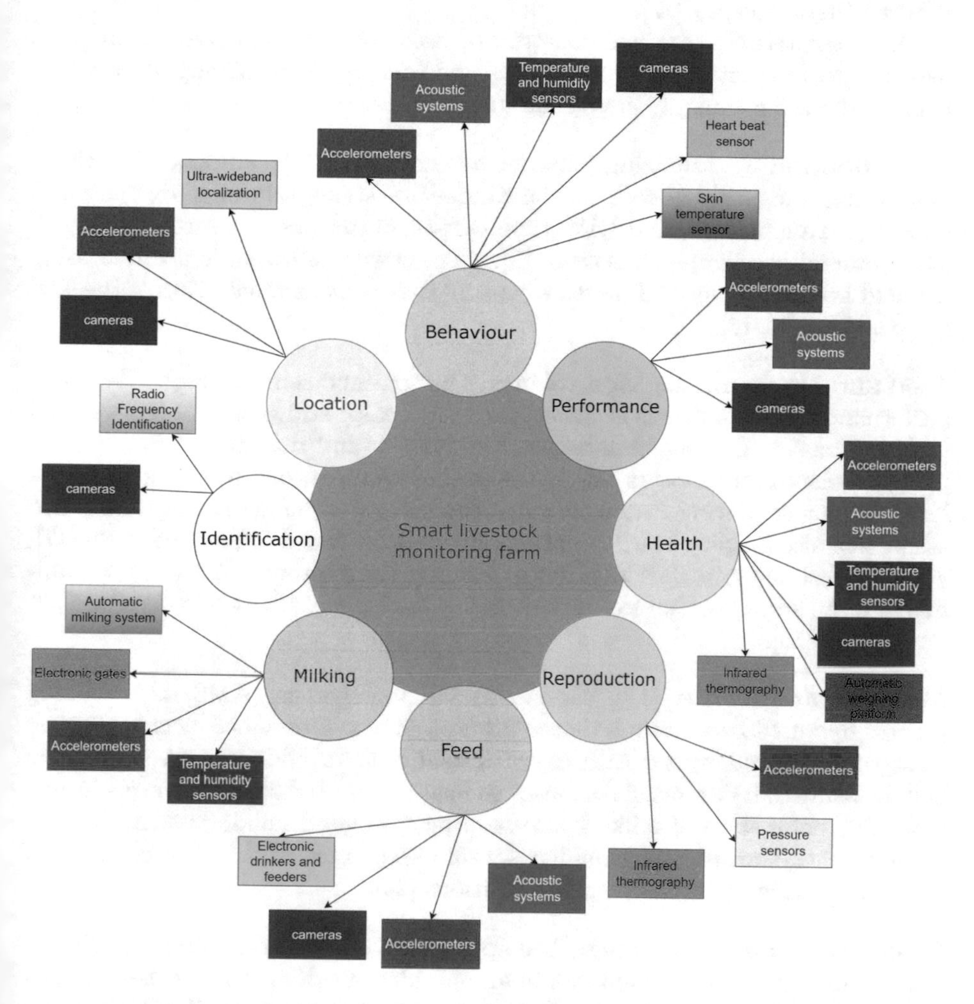

Fig. 2. Overview of the IoT applications for the smart livestock farm

We will highlight the sensors and the smart devices proposed for each monitoring and tracking unit, and the role of each one of them in the following subsections.

Health Monitoring. For health monitoring, the goal is to early detect diseases by using advanced technologies and sensors. Accelerometers, temperature and humidity sensors enable keeping track of the vital parameters of the animal's health [7,8]. Tracking coughing through the acoustic systems make the early detection of instances of respiratory infection possible [9,10] while using AI and computer vision algorithms on videos permits detecting the first signs of diseases using video scenes [7,11], in addition to automatic weighing platforms and infrared thermography [12].

All these technologies and sensors are used for early diagnosis of diseases like lameness for cattle, small ruminants and poultry. Also, calving illnesses are another disease commonly present on farms.

Reproduction Monitoring. Estrus detection through automated activity monitoring and analysis such as the number of steps, rate of movement, frequency of ruminations, and lying time can accurately be measured through a neckmounted accelerometer sensor [13]. Also, estrus detection is possible using infrared thermography and pressure sensors that enables monitoring of the sexual behavior [14,15].

Feed and Performance Monitoring. The continuous monitoring of feed and performance of the animals is critical for their health and welfare. Farmers need to keep track of grazing, feed intake, rumination and drinking of the animals using accelerometer sensors [16] by detecting accelerations resulting from the head movements during grazing, feeding, ruminating, drinking and resting. Video scenes are also keeping track of eating, rumination and drinking activities [17], while Acoustic systems are used for measuring the grazing behavior of the animals [18]. In addition, the electronic drinkers and feeders are used for feed poultry monitoring [19].

Milking Monitoring. The Automatic Milking System (AMS) is a milking process operated by machines that guarantee good performance by choosing the right time for milking for each dairy animal [20]. In addition, accelerometers help measure dairy animals' responses to heat stress [21,22]. Also, the electronic gates for automation of milking also help achieve good milking results [23]. In addition, temperature and humidity sensors keep track of the vital parameters of dairy animals under heat stress conditions [24].

Identification and Tracking. The application of computer vision algorithms allows the development of automatic animal identification systems using image shots and video scenes. Also, Radio frequency identification (RFID) technology is an efficient method for identification through RFID tags [23].

For tracking the location of the animals, accelerometers are in use for geolocation enhancement [22], in addition to Ultra wideband (UWB) and video scenes [25,26] for real-time location tracking.

Behavior Monitoring. The identification of animal behavior like feeding, walking and standing is an important thing for the farmer or the barn manager. Machine learning algorithms can classify multiple behaviors using the accelerometer [27], temperature and humidity sensors [28] alongside the use of videos scenes [29] and acoustic systems [30]. Also, stress can be detected using skin temperature sensors [31].

3 The Smart Livestock Farm Architecture

3.1 Schematic Diagram of the Smart Livestock Farm

Smart livestock monitoring in connected farms ensures that animals are healthy, disease free and performing well. It also aims at preventing or managing outbreaks of serious animal diseases and offering farmers a remote control method to keep track of their farmed animals (Fig. 3).

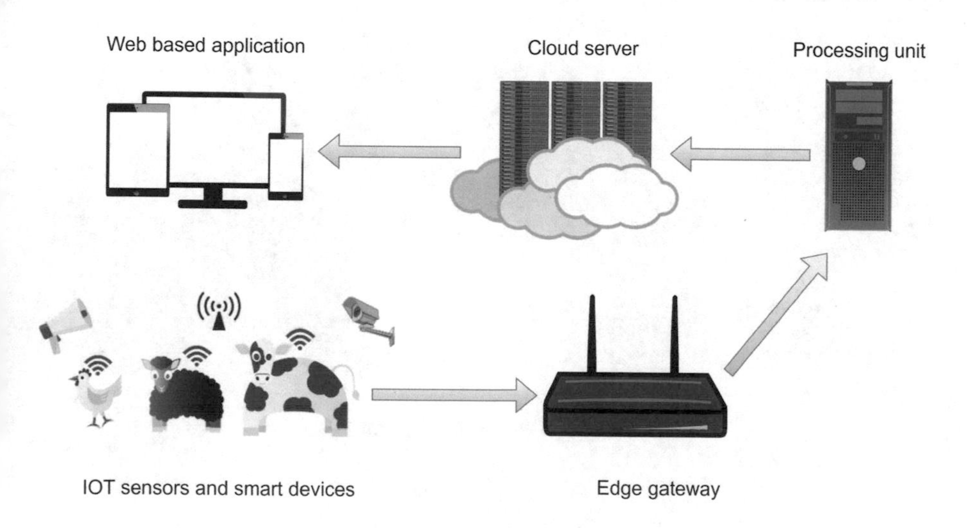

Fig. 3. Schematic diagram of the smart livestock farm

The data generated by the IoT sensors (temperature, humidity, heartbeat, accelerometers and pressure sensors) in addition to the other devices (cameras, acoustic systems, automatic weighing platforms, Infrared thermography, electronic drinkers and feeders and automatic milking systems) are sent to a local edge gateway that aggregate the data for a selected time interval, and transfer it to the processing unit that operates artificial intelligence [32] and big data [33] techniques to process the data, identify anomalies and make the necessary calculations then making the results available in the cloud server that stores the data and put it into service of a web-based application that the farmer can access remotely through a desktop or mobile device. The farmer or the bran manager and the veterinary physician can also get notifications in emergency cases detected by the system through the analysis of the data gathered from the sensors and the other devices. In addition, they have full access to the history of the data of the animals.

3.2 Data Flow Chart

This diagram shows the flow of information in the system starting from the data extracted from the IoT sensors and smart devices then the data collection phase through the edge gateway and the data analysis by the processing unit and finally the visualization via the web-based application of the data stored in the cloud server (Fig. 4).

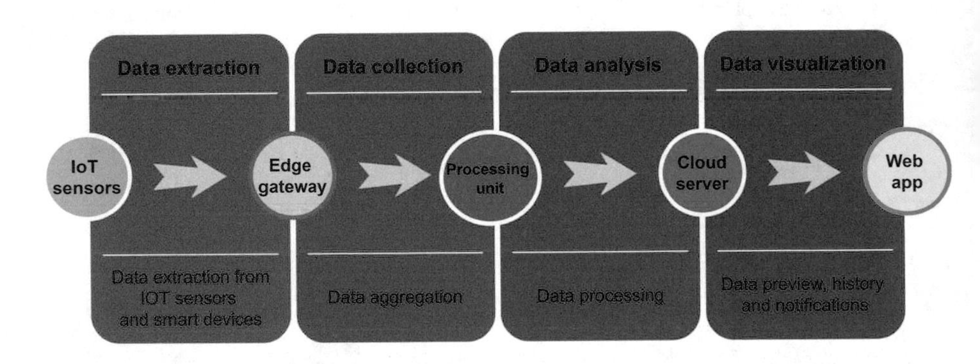

Fig. 4. Data flow diagram

3.3 Illustration of the Multi-layer IoT-based Smart Farm

The smart livestock farm consists of three layers, namely the physical, middle, and application layers. The physical layer is made up of IoT sensors and other devices that can be attached or from a distance monitors the animals in the barns. The middle layer consists of two units, the edge gateway where the generated data is aggregated, and the processing unit where the computational activities take place with the use of artificial intelligence and big data techniques. The Application layer is made of the cloud server for data storage and database management and the web-based application that allow access to the data from a desktop or mobile devices (Fig. 5).

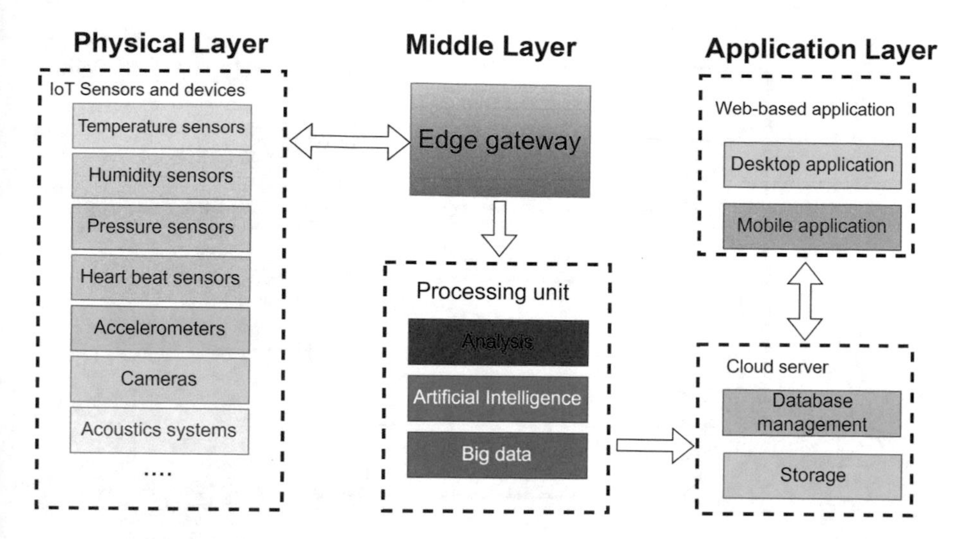

Fig. 5. Illustration of the multi-layer IoT-based smart farm

4 Livestock Monitoring System Modeling

4.1 Use Cases Overview

In the following figure, there is an overview of the interactions that the farmer will have with the software. The main goal of the proposed smart livestock farm software system is to provide the farmer or the bran manager an easy way to insert and update the existing herd, automatically generate the corresponding identification codes, view the monitoring statistic coming from the sensors and track the location of the animals in real-time. The monitoring statistic includes performance, health, reproduction, feed, and milking stats, alongside behavior reports. In addition, the farmer can access the data archive and the history in the software and get notifications in case the system detected an emergency that needs quick intervention (Fig. 6).

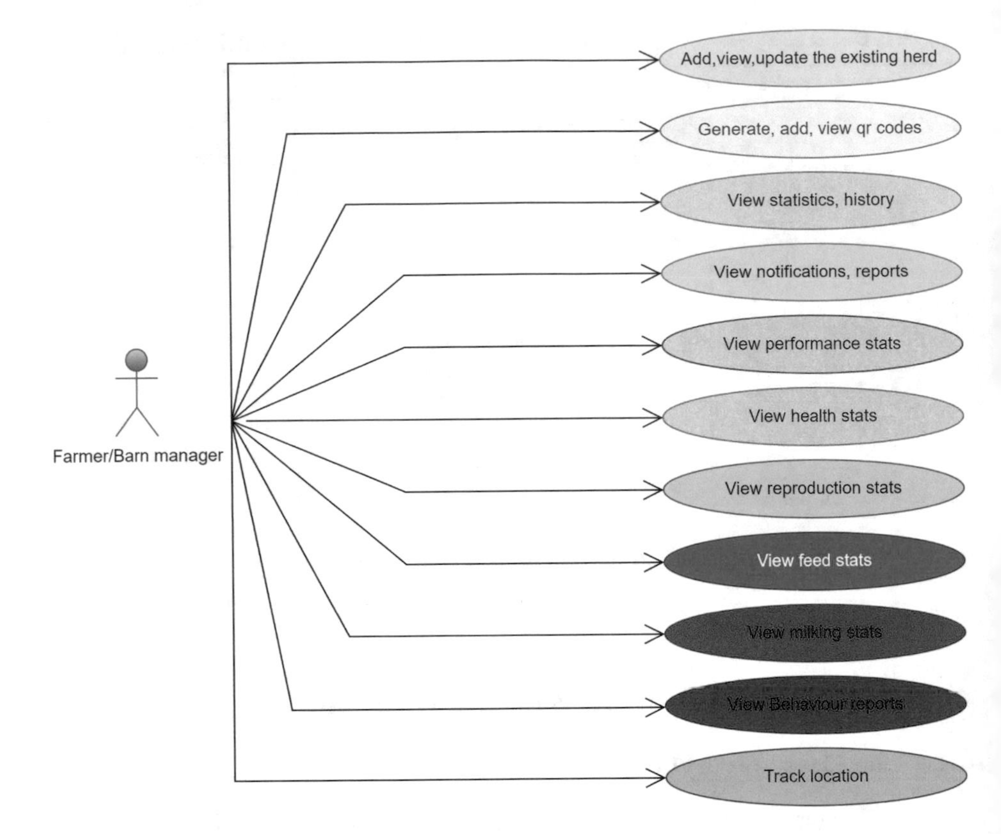

Fig. 6. Use case diagram

4.2 System Process Interactions

The proposed system operates through the operational process shown in Fig. 7. The data is collected at certain intervals through the IoT sensors and the other devices in the livestock barn and get transmitted to the edge gateway. The local edge gateway aggregates the data received and send it to the processing unit to make the necessary calculations and stores it in the database on the cloud server. The web application provides monitoring information and notification service during abnormal situations. In addition, the farmer or the barn manager can monitor the livestock farm remotely through a desktop or mobile device.

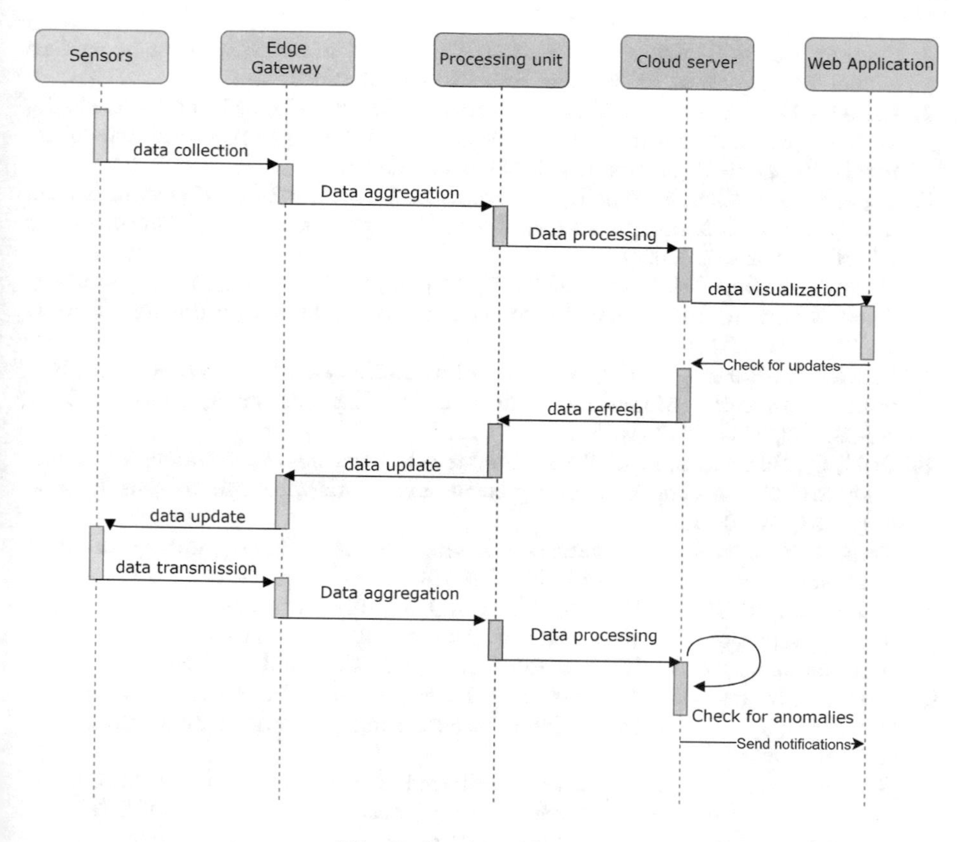

Fig. 7. Sequence diagram

5 Conclusion

The efforts of all actors should be united to confront the challenges that face farmers and livestock managers today. In this context comes this paper to provide farmers, livestock farm managers, and veterinarians with a smart livestock farm architecture based on cutting-edge technologies that will enable them to track and monitor every aspect of their livestock in order to boost their productivity and meet the growing demand. In our upcoming works, we intend to test the modules of this system in a real local livestock farm using the various aforementioned devices and technologies then share the results in a new paper.

References

1. Calicioglu, O., Flammini, A., Bracco, S., Bellù, L., Sims, R.: The future challenges of food and agriculture: an integrated analysis of trends and solutions. Sustainability **11**(1), 222 (2019)
2. Liu, Y., Ma, X., Shu, L., Hancke, G.P., Abu-Mahfouz, A.M.: From industry 4.0 to agriculture 4.0: current status, enabling technologies, and research challenges. IEEE Trans. Ind. Inf. **17**(6), 4322–4334 (2020)

3. Neethirajan, S.: The role of sensors, big data and machine learning in modern animal farming. Sens. Bio-Sensing Res. **29**, 100367 (2020)
4. Akhigbe, B.I., Munir, K., Akinade, O., Akanbi, L., Oyedele, L.O.: IoT technologies for livestock management: a review of present status, opportunities, and future trends. Big Data Cogn. Comput. **2021**(5), 10 (2021)
5. Gope, P., Gheraibia, Y., Kabir, S., Sikdar, B.: A secure IoT-based modern healthcare system with fault-tolerant decision making process. IEEE J. Biomed. Health Inf. **25**(3), 862–873 (2020)
6. Morrone, S., Dimauro, C., Gambella, F., Cappai, M.G.: Industry 4.0 and precision livestock farming (PLF): an up to date overview across animal productions. Sensors **22**(12), 4319 (2022)
7. Mandel, R., Harazy, H., Gygax, L., Nicol, C.J., Ben-David, A., Whay, H.R., Klement, E.: Detection of lameness in dairy cows using a grooming device. J. Dairy Sci. **101**(2), 1511–1517 (2018)
8. Grilli, G., Borgonovo, F., Tullo, E., Fontana, I., Guarino, M., Ferrante, V.: A pilot study to detect coccidiosis in poultry farms at early stage from air analysis. Biosyst. Eng. **173**, 64–70 (2018)
9. Carpentier, L., et al.: Automatic cough detection for bovine respiratory disease in a calf house. Biosyst. Eng. **173**, 45–56 (2018)
10. Mahdavian, A., Minaei, S., Yang, C., Almasganj, F., Rahimi, S., Marchetto, P.M.: Ability evaluation of a voice activity detection algorithm in bioacoustics: a case study on poultry calls. Comput. Electron. Agric. **168**, 105100 (2020)
11. Lowe, G., Sutherland, M., Waas, J., Schaefer, A., Cox, N., Stewart, M.: Infrared thermography-a non-invasive method of measuring respiration rate in calves. Animals **9**(8), 535 (2019)
12. Gelasakis, A.I., et al.: Evaluation of infrared thermography for the detection of footrot and white line disease lesions in dairy sheep. Vet. Sci. **8**(10), 219 (2021)
13. LeRoy, C.: Estrus detection intensity and accuracy, and optimal timing of insemination with automated activity monitors for dairy cows (Doctoral dissertation, University of Guelph) (2016)
14. Marquez, H.P., Ambrose, D.J., Schaefer, A.L., Cook, N.J., Bench, C.J.: Infrared thermography and behavioral biometrics associated with estrus indicators and ovulation in estrus-synchronized dairy cows housed in tiestalls. J. Dairy Sci. **102**(5), 4427–4440 (2019)
15. De Freitas, A.C.B., et al.: Surface temperature of ewes during estrous cycle measured by infrared thermography. Theriogenology **119**, 245–251 (2018)
16. Ikurior, S.J., Marquetoux, N., Leu, S.T., Corner-Thomas, R.A., Scott, I., Pomroy, W.E.: What are sheep doing? Tri-axial accelerometer sensor data identify the diel activity pattern of ewe lambs on pasture. Sensors **21**(20), 6816 (2021)
17. Ruuska, S., Kajava, S., Mughal, M., Zehner, N., Mononen, J.: Validation of a pressure sensor-based system for measuring eating, rumination and drinking behaviour of dairy cattle. Appl. Animal Behav. Sci. **174**, 19–23 (2016)
18. Werner, J., et al.: Evaluation of the RumiWatchSystem for measuring grazing behaviour of cows. J. Neurosci. Methods **300**, 138–146 (2018)
19. Batuto, A., Dejeron, T.B., Cruz, P.D., Samonte, M.J.C.: e-poultry: an IoT poultry management system for small farms. In: 2020 IEEE 7th International Conference on Industrial Engineering and Applications (ICIEA) (pp. 738–742). IEEE (2020)
20. Bonora, F., Benni, S., Barbaresi, A., Tassinari, P., Torreggiani, D.: A cluster-graph model for herd characterisation in dairy farms equipped with an automatic milking system. Biosyst. Eng. **167**, 1–7 (2018)

21. Benni, S., Pastell, M., Bonora, F., Tassinari, P., Torreggiani, D.: A generalised additive model to characterise dairy cows' responses to heat stress. animal, 14(2), 418-424 (2020)
22. Mozo, R., Alabart, J.L., Rivas, E., Folch, J.: New method to automatically evaluate the sexual activity of the ram based on accelerometer records. Small Ruminant Res. **172**, 16–22 (2019)
23. Alejandro, M.: Automation devices in sheep and goat machine milking. Small Ruminant Res. **142**, 48–50 (2016)
24. Bovo, M., Agrusti, M., Benni, S., Torreggiani, D., Tassinari, P.: Random forest modelling of milk yield of dairy cows under heat stress conditions. Animals **11**(5), 1305 (2021)
25. Meunier, B., Pradel, P., Sloth, K.H., Cirié, C., Delval, E., Mialon, M.M., Veissier, I.: Image analysis to refine measurements of dairy cow behaviour from a real-time location system. Biosyst. Eng. **173**, 32–44 (2018)
26. Qiao, Y., Su, D., Kong, H., Sukkarieh, S., Lomax, S., Clark, C.: BiLSTM-based individual cattle identification for automated precision livestock farming. In: 2020 IEEE 16th International Conference on Automation Science and Engineering (CASE) (pp. 967-972). IEEE (2020)
27. Carslake, C., Vázquez-Diosdado, J.A., Kaler, J.: Machine learning algorithms to classify and quantify multiple behaviours in dairy calves using a sensor: moving beyond classification in precision livestock. Sensors **21**(1), 88 (2020)
28. Unold, O., et al.: IoT-based cow health monitoring system. In: Krzhizhanovskaya, V.V., et al. (eds.) ICCS 2020. LNCS, vol. 12141, pp. 344–356. Springer, Cham (2020). https://doi.org/10.1007/978-3-030-50426-7_26
29. Montalcini, C.M., Voelkl, B., Gómez, Y., Gantner, M., Toscano, M.J.: Evaluation of an active LF tracking system and data processing methods for livestock precision farming in the poultry sector. Sensors **22**(2), 659 (2022)
30. Bishop, J., Falzon, G., Trotter, M., Kwan, P., Meek, P.: Sound analysis and detection, and the potential for precision livestock farming-a sheep vocalization case study. In: Proceedings of the 1st Asian-Australasian Conference on Precision Pastures and Livestock Farming, Hamilton-New Zealand (pp. 1-7) (2017)
31. Cui, Y., Zhang, M., Li, J., Luo, H., Zhang, X., Fu, Z.: WSMS: Wearable stress monitoring system based on IoT multi-sensor platform for living sheep transportation. Electronics **8**(4), 441 (2019)
32. El Moutaouakil, K., Jabir, B., Falih, N.: A convolutional neural networks-based approach for potato disease classification. In: International Conference on Business Intelligence (pp. 29-40). Springer, Cham (2022). https://doi.org/10.1007/978-3-031-06458-6_2
33. Rabhi, L., Falih, N., Afraites, A., Bouikhalene, B.: Big data approach and its applications in various fields. Procedia Comput. Sci. **155**, 599–605 (2019)

Energy-efficient Next Hop Selection for Topology Creation in Wireless Sensor Networks

Said El Hachemy(✉), Abdellah Boulouz, and Yassin Eljakani

Faculty of Sciences, Ibn Zohr University, Agadir, Morocco
said.elhachemy@edu.uiz.ac.ma

Abstract. Topology creation in wireless sensor networks (WSNs) has been one of the main criteria for a better routing experience. Recently, many fields have adopted WSNs for better ease of data collection and monitoring applications. However, most of these applications oblige the long-term deployment of the network and less frequency of sensor maintenance. Hence, it is necessary for the WSN to be sustainable in order for a longer lifetime and failure immunity. One of the main contributors in these goals is topology formation. Faster and efficient topology formation in WSNs allows for faster and reliable packet delivery, as well as reduced energy consumption and faster network deployment. However, as WSN applications have become more diverse, WSNs require efficient and faster topology creation methods. Thus, in this paper, we present the design of a novel method for next hop selection in WSNs for optimal topology creation. The approach aims to improve energy consumption, increase network lifetime, and improve packet delivery ratio. The method offers three modes for hop selection based on distance, energy, and a balanced mode that takes into consideration both of these measures. The main point of this approach is limiting the number of next hop candidates for each node by focusing on neighbors that are in the direction to the Base Station (BS). This is done by narrowing the angle at which the node is allowed to choose its next hop. The algorithm was tested and was able to extend network lifetime as well as increase the number of packets delivered to the BS.

Keywords: Topology Creation · Energy-Efficient · WSNs

1 Introduction

Internet of things (IoT) has become a daily part of our everyday lives, and the number of IoT devices and the amount of generated data have been on an exponential increase every year. Amongst the branches of IoT is the usage of low energy devices named sensors in wireless sensor networks (WSNs) for several types of tasks. These sensors are used for environment monitoring, threat detection, weather and atmosphere metrics measurement, patient monitoring, surveillance and security, industrial automation, military applications and many more [1].

However, sensors have many constraints regarding computation power, energy levels, and memory capacity. Hence, it is essential for WSN protocols to

N. Aboutabit et al. (Eds.): ICMICSA 2022, LNNS 656, pp. 294–305, 2023.
https://doi.org/10.1007/978-3-031-29313-9_26

take these limitations into account for a seamless routing experience. The most popular protocols used for WSN are RPL, AODV, DSR, LEACH, and OLSR. Each of these protocols has a set of mechanisms that allow for low energy consumption and longer network lifetime while maintaining good qualities of service (QoS) such as latency, packet delivery ratio (PDR), and packet loss rate (PLR). Amongst the factors that help increase network performance is the next hop selection algorithm used in these protocols. In most cases, multi-hop routing is essential for the network to function for longer periods of time. Thus, hop selection has to be swift and optimal. For these reasons, our aim is the creation of an algorithm that takes into account the different energy limitations as well as the dynamicity of the network for an optimal next hop selection. The devised method is based on limiting the number of neighbors a node can choose its next hop from, by creating a polygon inside of which the optimal hop would be chosen. The method allows for the usage of three modes of hop selection, each mode with its energy and QoS trade-offs. The algorithm resulted in a tree-like topology with a number of hops that reduces the PLR, and increase network lifetime and PDR. The research highlights of our proposed method are mentioned as:

- A next hop selection algorithm for efficient topology creation in WSNs. The selection allows for efficient energy consumption and increased PDR.
- The creation of three modes for next hop selection:
  - Distance-based selection: Where the node chooses the closest node in the polygon as its next hop.
  - Energy-based selection: The next hop in this case is the node with the highest energy level among the nodes inside of the polygon.
  - The balanced next hop selection: Where a cost metric is computed for each node inside of the polygon and the node with the lowest cost is chosen as the next hop.
- The proposed approach was analysed and the performed simulations illustrated significant improvements in terms of network lifetime and PDR.

This paper is organized as follow, in Sect. 2 we discuss several previous works that treat the next hop selection problem for energy efficiency. Section 3 describes the design of our proposed method and its different mechanisms. We then evaluate our approach using experiments and simulations in Sect. 4. After which we conclude with a discussion in Sect. 5

2 Related Works

Energy-efficiency is a major requirement in WSNs. As devices have low energy levels, it is essential for the routing protocols used to be as optimal as possible. In [2], the author presented an efficient multi-hop routing protocol for efficient data dissemination in hierarchy-based WSNs. The proposed method was based on analysing the residual energy of nodes and the distance to BS, to assign ranks to nodes, and consequently to choose the route for data exchange. The BS then performs periodic checks on the nodes to keep the network communication accessible at all times. The method is said to improve PDR and increase the overall network lifetime. However, looking at the simulation parameters used, the chosen packet size of 256 bytes, and the nodes' initial energy levels of 1000 J J seem to be higher by magnitudes compared to the standard values used in state-of-the-art works. Hence, the method needs further validation.

From another perspective, the authors in [3] proposed a solution for WSNs named CIDF-WSN. This approach was based on developing a neighbor information database that allows every node to choose its optimal next hop in order to forward packets. The selection of relay hops depends associating each node with a cost that represents the node's potential to become the next hop. The cost is computed based on node's workload, number of successful transmissions, computational resources, and residual energy using a multi-criteria decision making procedure. The method has the ability to treat incoming packets using a normal mode used for times when the network is stable. This mode aims for a smooth packet forwarding experience while maintaining nodes and updating selection cost. As well as a recovery mode that is used in cases where the network is not as reliable as before. This mode's priority is to ensure the delivery of packets regardless of the network state, while taking into consideration node needs. As the approach takes into account the resource-constrained nature of WSN devices, The method gave good results compared to recent works in this field in regard to energy consumption, latency, and control overhead.

In the same context, the authors in [4] presented a routing protocol named Enhanced Hybrid Multipath Routing (EHMR) for next hop selection in low level energy networks. The protocol depends on hierarchical clustering for the computation of cost metrics based on maximum residual energy and minimum hop count to allow for swift packet delivery, longer network lifetime, and efficient packet lifetime. The results given by this approach demonstrated low latency rate as well as low end-to-end delay. Energy consumption was also reduced and a high packet delivery ratio was achieved. However, the conducted experiments assumed the availability of higher data rates, a high energy per node of 50 J, and a very high transmission range of 500 m which is high compared to state of the art studies in WSN applications. This implies the need for further experiments.

Another study that aims for energy efficiency in WSNs is the method devised by CHUAN et al. in [5]. In order to maximise the network lifetime, the authors proposed a protocol for region source routing called ER-SR. In this protocol, instead of nodes performing their path calculations, the protocol assigns the task of path computation to the nodes with the highest residual energy level, so called source nodes. This selection is done dynamically considering the nature of the network. Although this alleviates the energy load off of normal nodes, it still impacts source nodes the most during the network communication. The method also uses a distance-based ant colony optimization algorithm to find the global transmission path for each node. This mechanism would indeed reduce the energy consumption, however, it will certainly increase complexity and reduce the protocol's capability to be included in real-time applications.

The authors in [7] proposed a routing protocol that aims to reduce delay and enhance energy efficiency in WSN. The method was based on using two-hop information to allow nodes to make better forwarding decisions that minimize delay and balance load traffic across the network. The approach also adopts the potential relay information (PRI) metric to guarantee that the selected next-hop would ensure an efficient delivery and preserve quality of service. The PRI used takes into consideration the residual energy, distance, delay, and the link quality of neighbor nodes. The approach was able to have good results compared to similar methods. However, the approach increases the computational complexity cost in terms of node-to-node communication which could impact the energy consumption in some cases. The method also suffers massively from the WSN constraints in large scale networks.

Similarly in [8], the authors developed a clustering algorithm named Enhanced Energy Proficient Clustering (EEPC) for application in the field of sensor tracking to improve energy consumption in networks with both stationary and mobile nodes. The method's goal is finding the optimal next hop node among the stationary nodes based on mobile nodes' velocity and location, and a metric is devised based on which the selection is done. EEPC's main contributions revolve around optimal relay node selection in a hierarchical routing protocol, reducing control overhead and complexity by using a track-by-track data aggregation approach, and cluster head selection using balanced clustering. Simulations were conducted and have validated the performance of the approach. However, further experiments are needed for highly dense networks as the approach was tested using only 100 nodes in an area of 1000×1000 m. From another point of view, the fixed nodes in the network were said to be close to the base station, and their distribution was not declared. This could imply that the positions of the fixed nodes were handpicked, which could also lead to biased results.

As the majority of studies regarding WSN aim to increase lifetime and energy efficiency, it is apparent that the field still has a need for further development and research regarding resource management and energy consumption. The common problem in this regard is the energy/QoS trade-off. The performance requirements oblige reduced energy consumption but not at the cost of quality of service such as latency, packet delivery ratio, packet loss rate and others.

3 Proposed Method

In this paper we have proposed a method for topology creation in WSNs. The approach's aim is to find the best next hop for nodes to the base station, and to create a topology that improves the different energy and QoS constraints in the network. The next hop selection is based on distance and energy measures.

For every node N in position P_n, we compute two points in space P_1 and P_2. With the position of the BS as P_{bs}, we choose the next hop among the nodes that are inside the polygon $P_n P_1 P_{bs} P_2$ (Fig. 1).

3.1 Polygon Computation

For every node N in position P_n, we compute two points A and B such that the distances between the points P_{bs}, P_n, A, and, B satisfy the following : $|P_{bs}B| = |P_{bs}A| = \frac{|P_nP_{bs}|}{2}$ and the angles $\angle P_nP_{bs}A = \angle P_nP_{bs}B = \frac{\pi}{2}$. Next, we compute two points in space P_1 and P_2 where $P_1 \in P_nA$ and $P_2 \in P_nB$ such that the angles $\angle P_nP_{bs}P_1 = \frac{\pi}{8}$ and $\angle P_nP_{bs}P_2 = \frac{\pi}{8}$ (Fig. 1).

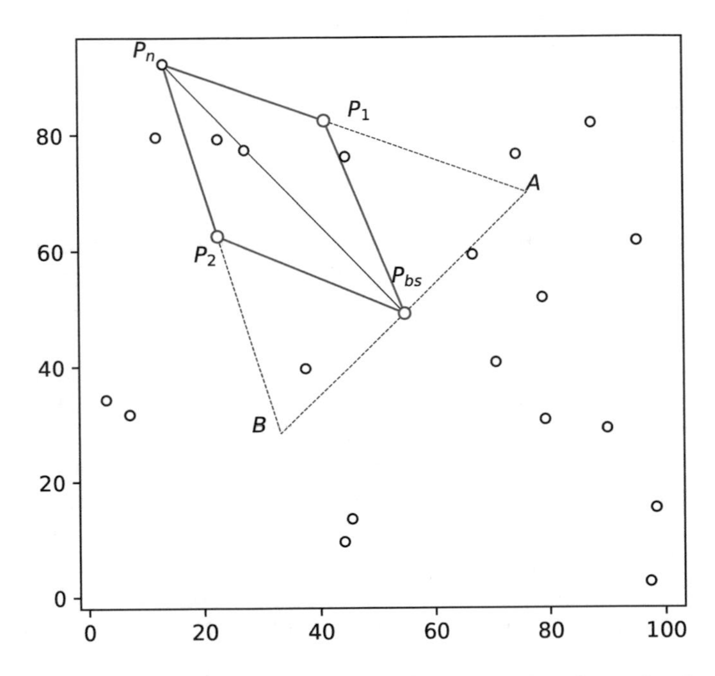

Fig. 1. Creation of the $P_nP_1P_{bs}P_2$ Polygon for Next hop selection

The creation of the polygon aims to limit the number of the next hop candidates for each node, and the narrow angles at which the polygon is created would allow for the selection to be focused towards the BS. Hence, the resulting network topology would be more precise. In this case, node N will have to choose its next hop among the nodes that are inside of the polygon $P_nP_1P_{bs}P_2$ based on distance and energy measures as discussed in the next subsection.

3.2 Next Hop Selection

After finding the set of nodes N_p that exist inside of the polygon, the node N can choose its next hop using three modes: Distance-based selection, Energy-oriented selection, and Balanced selection.

Distance-based Selection. In case of a distance-based next hop selection, the node N chooses the closest node that exists within the polygon (Fig. 2a).

The distance used in this case (and in this paper in general) is the euclidean distance defined in 1.

$$d(N_1, N_2) = \sqrt{(x_{N_1} - x_{N_2})^2 + (y_{N_1} - y_{N_2})^2} \tag{1}$$

Energy-oriented Selection. For an energy-oriented routing experience, the node N chooses the node with the highest energy level as its next hop. At the beginning of the deployment, the energy of nodes is practically similar, hence the node N ends up choosing the closest node. However, in advanced stages of the deployment, the node chooses the next hop with the highest energy (Fig. 2b).

Balanced Selection. For a more balanced experience (Fig. 2c), the node chooses its next hop based on the measure defined in 2:

$$N_{hop} = argmin(Cost(N, N_p)) \tag{2}$$

where

$$Cost(N, N_i) = d(N, N_i)/(E_{N_i} + E_N) \tag{3}$$

where N is the current node, N_p is the set of nodes N_i within the polygon. $d(N, N_i)$ is the distance between N and N_i, and E_{N_i} and E_N are the energy levels of nodes N_i and N respectively. In this case N_{hop} should be the node that minimizes the Cost function (Fig. 2c). The aspects of the cost function is taking the distance between the nodes as well as both nodes' energy levels into consideration to make a next hop selection. The larger the distance between the two nodes, the more energy they require to be able to connect.

In case a polygon does not contain any nodes, the node N chooses its next hop from its neighbors using the corresponding mode. In case no neighbors exist, the node routes its packets to the Base Station directly. Figure 3a, 3b, and 3c visualize the topologies created via the different modes used.

In the energy-based mode there is a trend where a cluster of nodes is targeting a single node as its next hop. These clusters form at nodes that are closest to the BS. The interpretation for this is that the usage of data aggregation is the factor with the highest potential to form efficient topologies. As for the balanced mode, the same trend occurs but in smaller frequencies as the cost function used depends on both the distance and the residual energy of nodes.

4 Experiments and Results

We have generated 100 nodes randomly and we have applied the algorithm in an area of $100 \times 100\,m^2$, while positioning the BS at the center point of the area $(50, 50)$. To have a better understanding of the topology created. We have applied the algorithm on 20, 50, 75, and 100 nodes. Figures 4a, 4b, 4c, 4d illustrate the resulting topologies. From the figures, the creation is performed in a bottom-up manner starting from leaf nodes and reaching the BS. The topology

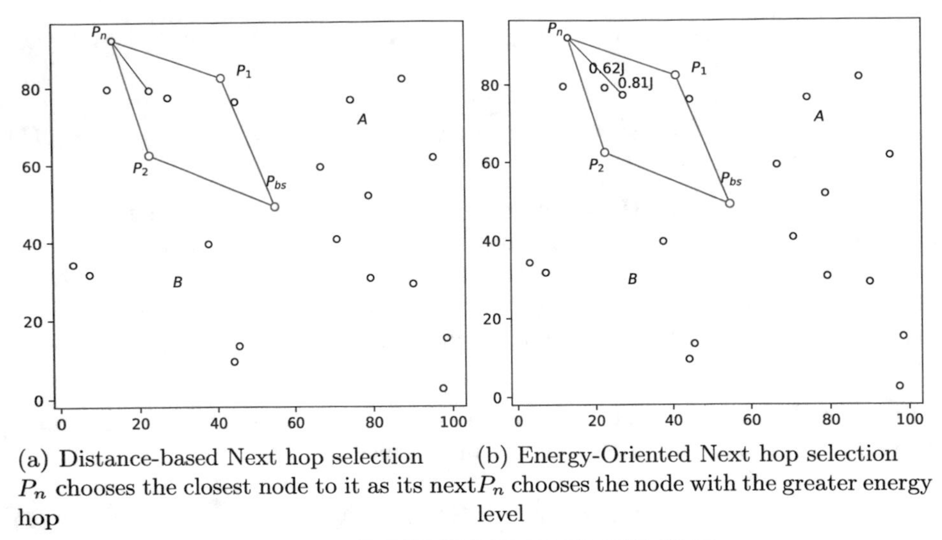

(a) Distance-based Next hop selection P_n chooses the closest node to it as its next hop

(b) Energy-Oriented Next hop selection P_n chooses the node with the greater energy level

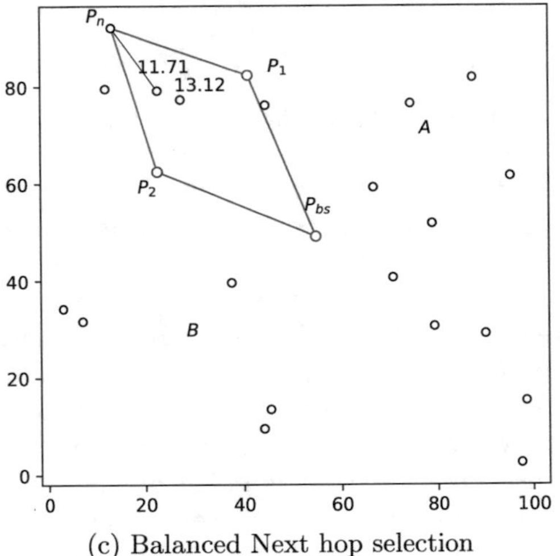

(c) Balanced Next hop selection P_n chooses the node with the lowest cost metric

Fig. 2. Next hop selection modes

automatically accounts for the number of hops without any mechanisms taking place. This would ensure a rise in PDR. The node density is also automatically taken into account as the algorithm's resulting tree-like paths are as optimal as possible.

To evaluate the algorithm, we have run different simulations corresponding to the different topology creation modes. The simulations were conducted on 100

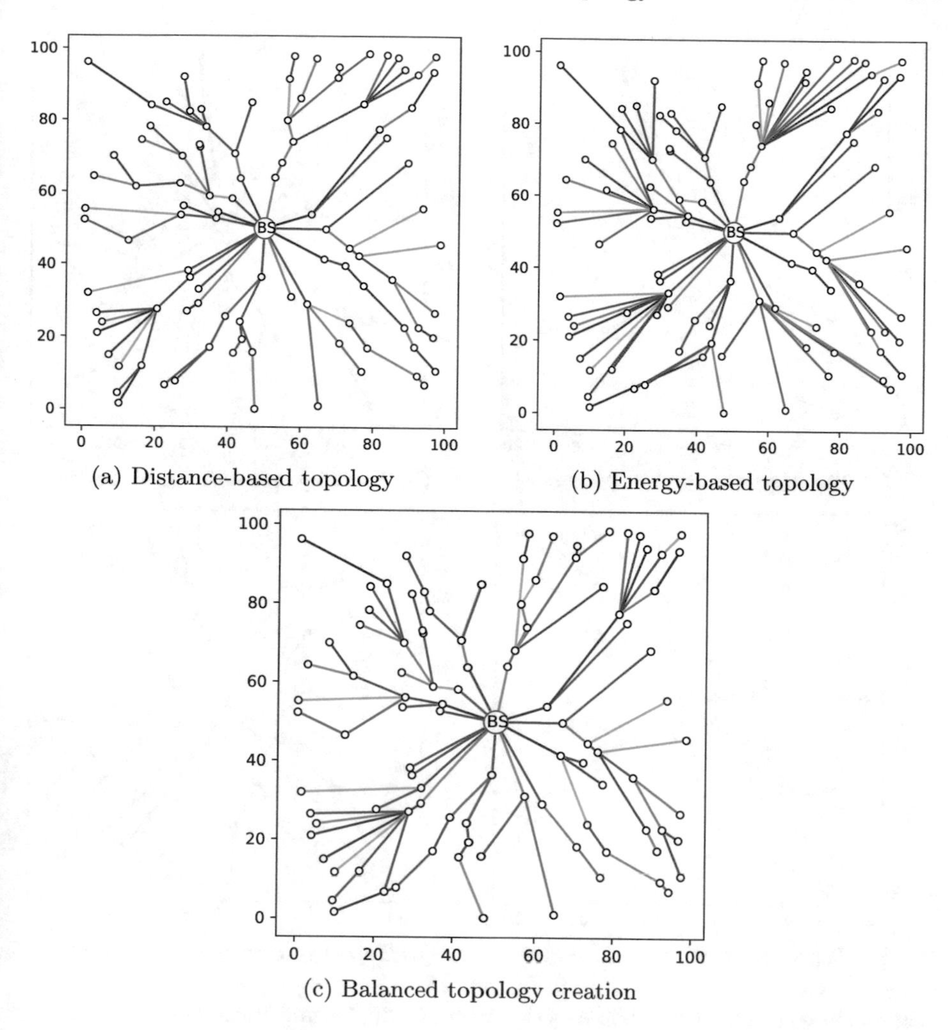

(a) Distance-based topology

(b) Energy-based topology

(c) Balanced topology creation

Fig. 3. Topology creation modes using 100 nodes on an area of 100×100m

nodes in an area of $100 \times 100\,m^2$. The radio dissipation model used to conduct the experiments is the same as the one used in [6–8] and in many other papers. The reason for choosing these specific values is because they are used by the majority of researchers. This would allow for more authentic comparisons and evaluations. In our experiments, we use a transmission range of 75m as it is on the high end of the spectrum, and to accommodate for bigger packets in terms of size, we used a packet size of 4000 bits. We initialized our nodes with 0.5J and 1J to have a broader perspective of the capabilities of the algorithm. The remaining of the simulation parameters are described in Table 1.

To resume the results of the simulations, we have performed a thorough statistics count of the different measures concerning energy and packet delivery in

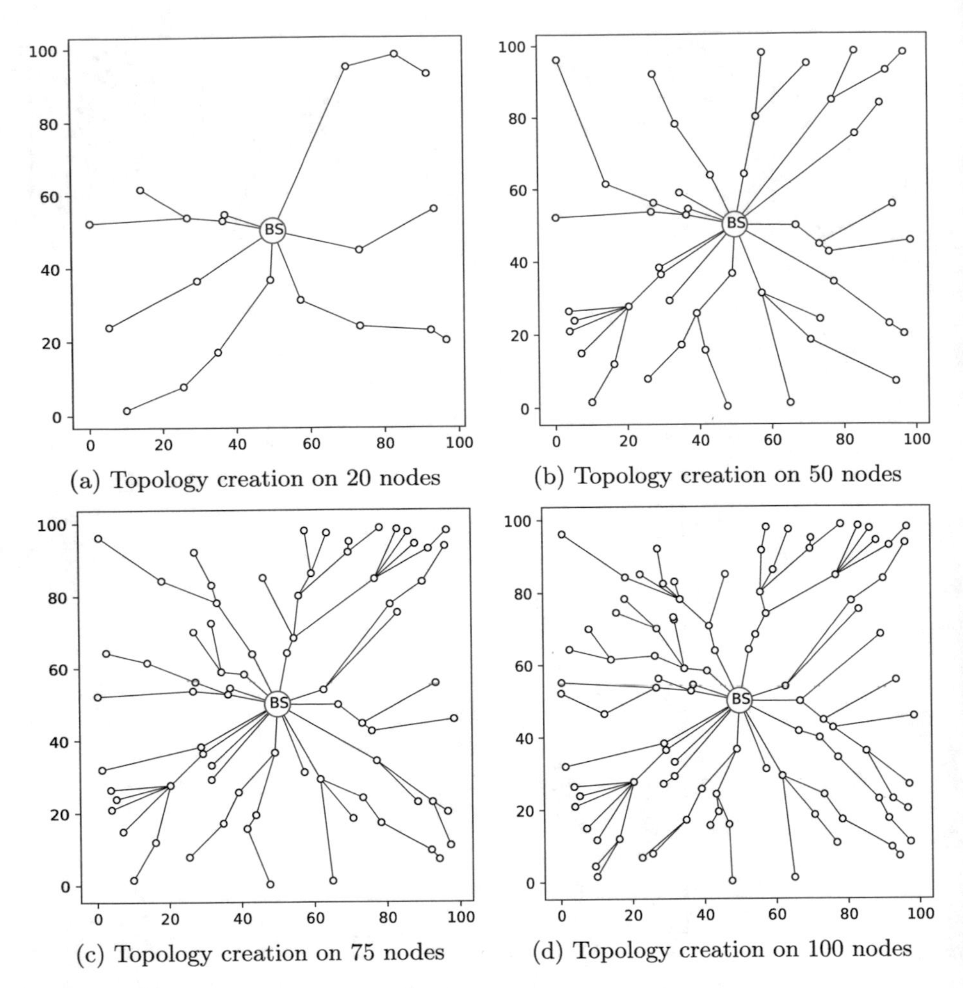

Fig. 4. Distance-based next hop selection using 20, 50, 75, and 100 nodes with a random seed of 80

respect to the different modes used. The results demonstrated that our algorithm highly impacts the energy usage. This is clear from the energy usage per bit of data, as well as the corresponding size of data delivered to the base station after data aggregation. The packet delivery ratio was measured at an average of 85%, and has known stability in both cases of 50 and 100 nodes regardless of the mode used, which is due to realistic nature of the simulation. The detailed statistics are stated in Table 2.

We have compared our method with the LEACH protocol [6] known for its hierarchical nature and energy-oriented routing mechanism, as well as the method proposed in [7] for its two-hop relay mechanism. The parameters used in this experiments are the same as described in Table 1. Figure 5 illustrates the

Table 1. Simulation parameters

Parameter	Value
Network size	$100 \times 100\ m^2$
Number of nodes	50, 100
Transmission range	$75\ m$
Packet size	4000 bits
Initial energy, E0	0.5 and 1 J/node
Transmitter energy, ETX	50 nJ/bit
Receiver energy, ERX	50 nJ/bit
Amplification energy for short distance, Efs	$10\ pJ/bit/m^2$
Amplification energy for long distance, Emp	$0.0013\ pJ/bit/m^4$
Data aggregation energy, Eda	$5\,nJ/bit$
Data collection energy, Edc	$5\,nJ/bit$

Table 2. Result metrics for each selection mode

Parameters	Distance		Energy		Balanced	
	$0.5j$	$1j$	$0.5j$	$1j$	$0.5j$	$1j$
Energy consumption per bit ($\mu J/bit$)	0.1883	0.1884	0.1865	0.1171	0.1859	0.1820
Simulation time	$10.6s$	$13.5s$	$8.9s$	$13.7s$	$11.6s$	$12.1s$
Total packets dropped	22163	50483	20625	49027	20643	44749
Total packets sent	147744	295566	147221	326379	147347	297613
Total packets received	125581	245083	126596	277352	126704	252864
Total data received at BS (MB)	31.64	63.24	31.94	101.70	32.04	65.47
PDR	84.99%	82.92%	85.98%	84.97%	85.98%	84.96%

number of nodes alive during the simulation of our method using the distance based hop selection of our approach, the leach protocol, and the method in [7].

According to Fig. 5, we see that our algorithm does not retain nodes compared to LEACH or to [7] as the first node death occurred in earlier rounds compared to the other two methods. However, our method focuses on the long term packet delivery to the BS. The network lifetime in our case was a third higher than the LEACH protocol, and the number of nodes alive that is capable of communication was 100% more than its counterpart in [8]. The Table 3 resumes the evaluation metrics such as the energy consumption per bit, as well as the PDR for the three methods. We notice that our approach offers better energy consumption than both of the other methods, and a higher PDR than Leach, however, our approach's PDR is 3% lower than [8]. Overall, we can conclude that our approach has better trade-offs compared to both of the other methods.

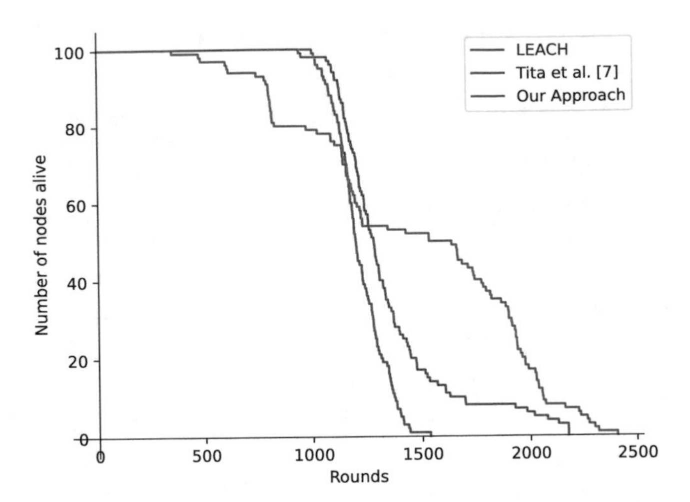

Fig. 5. Network lifetime comparison

Table 3. Comparison performance metrics

Metric	Our approach	Leach	Tita et al. [8]
Energy consumption per bit ($\mu J/bit$)	0.1884	0.2580	0.2129
Total packets dropped	50483	24726	17089
Total packets sent	295566	121908	122092
Total packets received	245083	97182	105002
Total data received at BS (MB)	63.24	46.20	55.97
PDR	82.92%	79.17%	86.00%

5 Conclusion

In this paper we have designed an algorithm for topology formation in wireless sensor networks. The nature of these networks related to insufficient resources and lower energy level, as well as node dynamicity, call for the need for an energy-efficient method that takes into account the trade-off between energy and QoS. The devised approach have performed relatively well in this regard and gave good results concerning network lifetime and PDR. The method also provides three different modes for next hop selection where each mode has its mechanism and advantages. However, additional experiments and comparisons with other state of art approaches are needed to further validate the results of our method, and the next step in this regard is to take into consideration the calculation time of the polygon used for the next hop selection as the calculation are done at the nodes. As well as the control overhead that happens in case of node deaths. This might put a load on the resources, however, this load is minimal compared to the amount of data circulating in the network. Our future work would build on this approach to further improve network lifetime, increase scalability, introduce other qualities of service such as latency, and to discover the performance of the algorithm in other scenarios, such as networks composed of higher numbers of devices, and in heterogeneous networks.

References

1. Kumar, S., Tiwari, P., Zymbler, M.: Internet of Things is a revolutionary approach for future technology enhancement: a review. J. Big Data **6**(1), 1–21 (2019). https://doi.org/10.1186/s40537-019-0268-2
2. Altowaijri, S.: Efficient next-hop selection in multi-hop routing for IoT enabled wireless sensor networks. Future Internet. 14. 35. (2022)
3. Salah, M., Atif, M., Kim, B.: CIDF-WSN: a collaborative interest and data forwarding strategy for named data wireless sensor networks. Sensors **21**(15), 5174 (2021)
4. Naushad, A., Abbas. G., Shah. S., Abbas, H.: Energy efficient clustering with reliable and load-balanced multipath routing for WSNs. In: 3rd International Conference on Advancements in Computational Sciences, ICACS, pp. 1–9. Springer, Heidelberg (2020). https://doi.org/10.1109/ICACS47775.2020.9055957
5. Xu, C., Xiong, Z., Zhao, G., Yu, S.: An energy-efficient region source routing protocol for lifetime maximization in WSN. IEEE Access 7 (2019)
6. Radhika, M., Sivakumar, P.: Energy optimized micro genetic algorithm based LEACH protocol for WSN. Wireless Netw. **27**(1), 27–40 (2020). https://doi.org/10.1007/s11276-020-02435-8
7. Tita, E.D., Nwadiugwu, W., Lee, J., Kim, D.: Real-time optimizations in energy profiles and end-to-end delay in WSN using two-hop information, Comput. Commun. 172 (2021)
8. Guleria, K., Verma, A., Goyal, N., Sharma, A., Benslimane, A., Singh, A.: An enhanced energy proficient clustering (EEPC) algorithm for relay selection in heterogeneous WSNs, Ad Hoc Networks, 116 (2021)

Physical Layer Parameters for Jamming Attack Detection in VANETs: A Long Short Term Memory Approach

Yassin El Jakani(✉), Abdellah Boulouz, and Said El Hachemy

Computer Systems and Vision Laboratory, Faculty of Sciences, Ibn Zohr University, Agadir, Morocco
yassin.eljakani@edu.uiz.ac.ma

Abstract. Nowadays, the reliance on intelligent traffic management is increasing. Vehicular ad hoc networks (VANETs), which belong to mobile ad hoc networks MANETs, are the focus of researchers as a promising approach to building intelligent transportation systems. One of the leading security requirements of VANETs is availability. Thus, adversaries take advantage of the physical layer vulnerabilities of these wireless sensor networks to launch radio frequency (RF) jamming attacks that cause a denial of services which compromises drivers' safety. This paper proposes a recurrent neural network (RNN) architecture based on Long-short Term Memory (LSTM) for jamming attack detection, using a physical layer parameter defined as the Received Signal Strength Indicator (RSSI). We compared our model with multiple machine learning algorithms, including random forest, support vector machine, and K Nearest Neighbor. The proposed approach overcomes other models with a 78% accuracy and a precision of 100% related to malicious RSSI signals.

Keywords: Physical layer · vanets · denial of service · IoT · Deep learning · LSTM

1 Introduction

In recent years, VANETs have gained a particular interest in the intelligence transport area [1]. It represents an important field of research regarding its benefits to improve transport conditions, including travel efficiency, passengers comfort, and road safety [2].

A vehicular Ad-hoc Network (VANET) is a collection of vehicles equipped with sensors that form a wireless sensor network (WSN). VANET uses dynamic vehicle-to-vehicle (V2V) and Vehicle-to-Infrastructure (V2I) communications. The carried data over those communications are mostly safety-related information that helps vehicles make real-time decisions. Therefore, contacts require fast processing with an optimal delay. Thus, any defection on this real-time communication system will directly impact people's safety on the road. Therefore, Denial of Service attacks (DoS) has gained a particular interest in research area [3].

N. Aboutabit et al. (Eds.): ICMICSA 2022, LNNS 656, pp. 306–314, 2023.
https://doi.org/10.1007/978-3-031-29313-9_27

The radio frequency jamming attack is a type of DOS attack that works under the physical layer, affecting delay-sensitive vehicular networks. With this critical vulnerability, attackers take advantage of the radio frequency information received on the physical channel to send jamming signals. This vulnerability requires accurate jamming attack detection. This paper proposes a novel approach based on a deep recurrent neural network that exploits the RSSI signal measured over receiving nodes. We adopt Long-short Term Memory, a well-used model to capture temporal patterns, which are, in our case, the malicious RSSI signals produced by jamming signals. Thus, the main contributions of this paper are:

- Investigating the importance of the physical layer parameter (Received signal strength indicator - RSSI) in detecting radio frequency jamming attacks.
- Exploiting the capability of Long-short Term Memory on extracting valuable patterns from RSSI signal.
- Building a deep learning architecture (LSTM) for jamming attack detection using RSSI time series signals. Moreover, finding the best set of hyperparameters.

This paper is organized as follows: we start with a literature review of jamming attack detectors in Sect. 2. Then we will explain the proposed approach in Sect. 3. The Results will be the topic of Sect. 4, with detailed interpretations and discussions. Lastly, we conclude and present future works in Sect. 5.

2 Related Works

Security became the primary constraint of Wireless sensor networks. VANETs represent a crucial area of WSNs where the communication between vehicles is vulnerable to Denial of services attacks, especially Jamming attacks. Therefore, many papers were proposed in the literature to address VANETs security limitations and reduce their effects on the expected behavior of VANETs.

In [4], the authors proposed a Jamming Attack detector based on the Artificial Bee Colony (ABC) algorithm using four parameters: packet delivery ratio (PDR), energy consumption (EC), distance, and packet loss ratio (PLR). The ABC is a meta-heuristic algorithm that depends on swarm intelligence of the foraging behavior of honey bee colonies based on three entities: employee foragers, onlookers, and scouts. Each node in the sensor network is simulated as a source food with multiple thresholds (fitness function) to be respected, such as high PDR and low EC, where each violation of the given entries means a jamming attack is in place. The employee bees and onlooker bees report and update the status of each sensor node in the WSN. The scout bees communicate the jammed bees to the neighbor's nodes. Their approach showed promising results compared with the ant system and particle swarm optimization (PSO).

The paper [5] targets jamming detection in cluster-based networks based on PDR and PLR parameters. In their approach, the cluster head (CH) computes the PDR, and the individual CMs calculate the PLR; then, a fuzzy-based piggybacking method (JDFPA) is used to detect the jamming in the network. The fuzzy-based algorithm searches for the optimal range of PDR and PLR for accurate jamming detection. Their combination of PDR, PLR, and JDFPA gives higher detection accuracy compared to using each PDR or PLR separately. A PDR threshold is also proposed in [6] to detect jamming occurrences. The threshold is computed as shown in Eq. 1:

$$Down_{pdr} = \frac{previous_{pdr} - current_{pdr}}{currenttime - previous_{time}} \quad (1)$$

where $current_{pdr}$ is the new PDR value at the current time, $previous_{pdr}$ is the PDR at a previous time, $current_{time}$ is the present moment, and $previous_{time}$ is the previous time. Jamming is reported as soon as the change of PDR is greater than the given threshold.

On the other hand, recurrent neural networks have been adopted in multiple papers in recent years to improve WSN security, thanks to their ability to extract temporal attack patterns from time series signals.

In [7], the authors have built a distributed denial of service attack detector on Internet of Vehicles (IoV) based on a hybrid model of two recurrent neural network architectures: LSTM and gated recurrent unit (GRU). The used dataset combines CIC DoS 2016, CICIDS 2017, and CSE-CIC-IDS 2018. The input features are packet-based parameters such as flow ID (To indicate the Packet Flow Context associated with a Logical Link Control) and ACK flag (used to acknowledge the successful receipt of a packet). The model learns the patterns from packet content to address DOS attacks.

The literature review gathers a variant of used approaches, from using optimization methods to advanced deep learning models. However, the parameters of physical layers are less used, which make it challenging to recognize radio frequency jamming attack, which will be the main focus of this paper.

3 Methodology

3.1 Overview

The proposed approach introduces a recurrent neural network for Deny of services attack detection on VANETs. Most DoS attacks are time-related, where the attacker sends a jamming radio signal to intercept and jam communication between vehicles. The Jamming attacks usually affect signal strength, affecting the expected behavior of VANETs. That is why we have built our approach by exploring the time series data gathered from the physical layer of WSN, represented by the received signal strength indicator (RSSI). Thus, our methodology will use LSTM, a state-of-the-art recurrent neural network that can extract temporal-related patterns from the RSSI parameter. We split the signal into timestamps where the target will be a "normal" or an "abnormal" signal. The hyper-parameters of the LSTM network are calculated using the Bayesian optimization [8] method to achieve optimal accuracy. The methodology overview is given in Fig. 1.

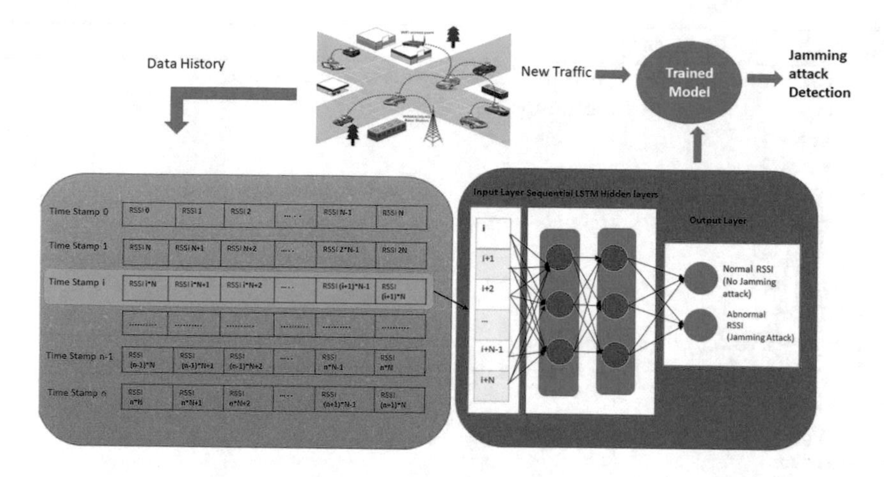

Fig. 1. Model overview

3.2 Data Preparation

The data used in this paper is the received signal strength indicator measured in two scenarios: with and without a jamming attack. Our approach is based on formulating a supervised learning model by splitting the RSSI signal into timestamps and setting the occurrence of a Jamming attack as a target as in the equation :

$$target = f(rssi_i, rssi_{i+1}, rssi_{i+2}, ..., rssi_{i+timestamp}) \tag{2}$$

where

- *target* is the attack status (0: no jamming attack, 1: jamming attack)
- *timestamp* is the number of timestamps
- $rssi_i$ is the value of RSSI at the timestamp i
- the function f is the proposed recurrent neural network model.

The input of the data preparation algorithm in Algorithm 1 is two RSSI signals for both malicious and benign captured during a jamming attack and regular traffic, respectively. The aim is to get the correct input format for the recurrent neural network and a list of timestamps with their respective attack status (malicious or benign). Lines 3 to 7 iterate over malicious RSSI signal, each member of the array X will be a 1D-array with a length of *timestamps* labeled as 1 (malicious). The second loop goes over the RSSI benign signal. The timestamps extracted are added to the arrays X and y. Then, the RSSI values are normalized and transformed to a range between 0 and 1 according to the equation on line 14. In the end, we will have two collections that will serve as input and output for the LSTM learning algorithm we will build.

Algorithm 1. Data preparation

Require: $malicious\_RSSI, benign\_RSSI, number\_timestamps$
Ensure: X : RSSI input timestamps , y : Attack status
1: $X \Leftarrow list()$
2: $y \Leftarrow list()$
3: **for** $i \leftarrow 0$ to $length(malicious\_RSSI) - number\_timestamps$ **do**
4: $\quad x \Leftarrow malicious\_RSSI[i : (i + number\_timestamps)]$
5: $\quad X[i] \Leftarrow x$
6: $\quad y[i] \Leftarrow 1$
7: **end for**
8: **for** $j \leftarrow i$ to $length(benign\_RSSI) - number\_timestamps$ **do**
9: $\quad x \Leftarrow benign\_RSSI[j - i : (j - i + number\_timestamps)]$
10: $\quad X[j] \Leftarrow x$
11: $\quad y[j] \Leftarrow 0$
12: **end for**
13:
14: $X \Leftarrow \frac{X - min(X)}{max(X) - min(X)}$
15:
16: **return** X, y

3.3 Model Architecture

The RNN architecture adapted in this paper is Long Short-Term Memory (LSTM) with four layers. The first layer is an LSTM layer with 14 hidden units, followed by a dense layer of 9 hidden layers. Then a regularization layer is added to reduce overfitting with a probability of 0.5. We compile the model using a Root Mean Squared Propagation (RMSprop) optimizer with a learning rate of 0.00038 based on categorical cross-entropy loss. The mentioned hyperparameters' values were found using the Bayesian Optimization algorithm [8]. A summary of each layer's hyper-parameters and their optimal values are given in Table 1.

Table 1. Model hyper-parameters

hyper-parameter	Options	Optimal value
LSTM hidden units	[1 to 16]	14
Dense hidden layers	[1 to 10]	9
Regularization Layer	[0.0 to 0.9]	0.5
Learning rate	[1e–4 to 1e–2]	0.00038

The model was set to be trained for 100 epochs using an early stopping mechanism based on validation accuracy with a batch size of 32. The used activation functions are Rectified Linear Unit (Relu) at LSTM layers and softmax at fully connected layers.

4 Results and Discussion

This section will present the dataset used, the performance metrics adapted, and our model results compared with other machine learning approaches. In the end, we will make interpretations concerning our initial problem.

4.1 Dataset

This paper used a public dataset from [9]. The data contains traces of 802.11p packets collected in an open space with and without jamming attacks. Multiple patterns were adopted, including constant and reactive. The open space is a filled parking lot with a total length of 500 m. The jammer in the experiment was located 180m from the north end of the parking lot and slightly over 300m from the south end. On a parallel roadway 30 m away, a few passing automobiles occasionally traveled north to south and vice versa throughout the measurements. In the experiment, different jamming profiles and vehicle motion scenarios were employed.

From the data traces, we extracted around 27330 RSSI measures calculated with the jamming signal's presence and 28669 associated with the expected behavior of the VANET network. Figure 2 shows the distribution of the first timestamp ($x_{i,0}$ where i is between 0 and the number of rows) for two RSSI signals (normal and abnormal).

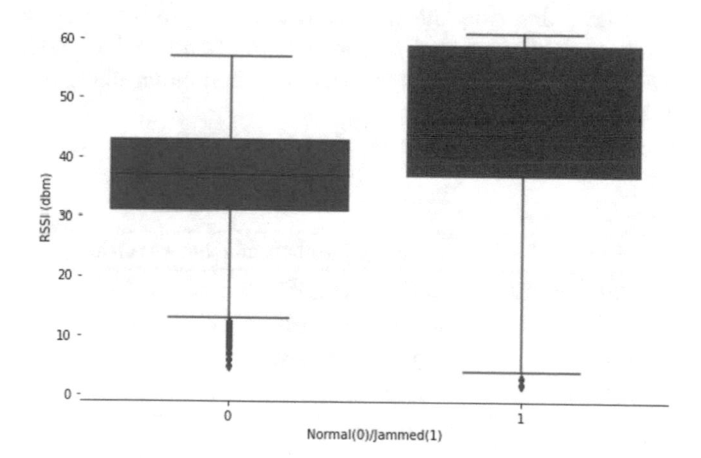

Fig. 2. Box plot of RSSI signals

The typical scenario of VANETs demonstrates the overall RSSI values distributed in a short range between 31 dbm and 43 dbm, where the median is equal to 37 dbm, which indicates a normal distribution. On the other hand, RSSI values are distributed more extensively between 37 dbm and 57 dbm when the jamming attack is in place, where the median is 44 dbm. The expected signal behavior changed, and the larger values of RSSI indicate signal weakness.

4.2 Results

The metrics adapted in this paper are precision and accuracy. The balance in the number of measurements in both malicious and benign RSSI signals makes it possible to consider accuracy as a representative and valuable metric. Accuracy is the total test samples correctly classified over the total number of samples in the test data:

$$Accuracy = \frac{TP + TN}{TP + TN + FP + FN} \tag{3}$$

where,

- TP: True positive
- TN: True negative
- FP: False positive
- FN: False negative

Nevertheless, as we are more concerned about detecting malicious signals (Jamming attacks), we also considered precision as defined in the following equation:

$$Precision = \frac{TP}{TP + FP} \tag{4}$$

Precision gives an advanced interpretation of how the model performs regarding TP samples.

Table 2 and Fig. 3 describe the precision and accuracy values for multiple machine learning models compared to the results of the LSTM model. For precision, we consider the one associated with the malicious class related to the jamming attack.

Table 2. Results

Model	Accuracy	Malicious class precision
Random Forest	71.81%	51.1%
SVM	70.51%	38.0%
KNN	65.02%	52.8%
LSTM	77.90%	100%

The LSTM architecture achieved an excellent accuracy score compared to the other models, followed by the random forest algorithm. The early stopping mechanism interrupts the training when the model starts to overfit based on validation error after four iterations of no improvement. Figure 4 shows the training and validation accuracy and cross-entropy error.

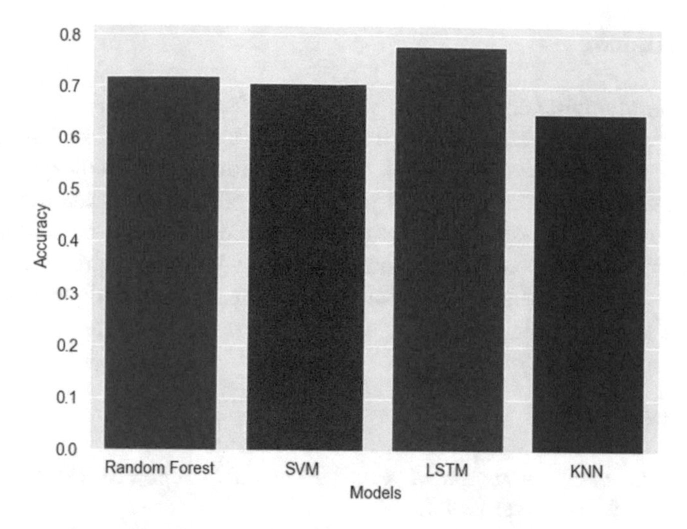

Fig. 3. Accuracy results

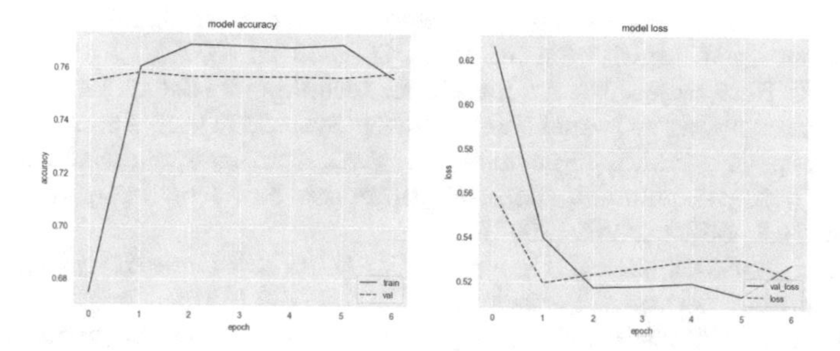

Fig. 4. Train loss/accuracy **vs** Validation loss/accuracy

4.3 Discussion

The outcome of this paper has introduced a new approach to detecting malicious RSSI signals on VANETS. The high precision and accuracy scores project the capability of LSTM to extract time-related patterns from RSSI time series signals because of the mechanism used by LSTM that can maintain the information for long periods. Furthermore, the low complexity of our model will give real-time predictions while reducing the vulnerabilities associated with the physical layer of VANETS. Any jamming signal against the VANET will apply changes to the RSSI in the receiving nodes. These changes will consequently be captured by the LSTM model. Therefore the attack will be detected, and the appropriate actions will be taken.

5 Conclusion

Security is a challenging constraint in VANETS. Especially with the vulnerability of the physical layer against radio frequency jamming attacks. In this paper, we proposed a new approach based on long short-term memory: a recurrent neural network to recognize temporal patterns of RF jamming attacks from RSSI signals. The results showed a higher precision score related to malicious calls. As a result, the security of Vehicle networks will increase against RF jamming attacks. Investigating other parameters from different network layers like MAC and network layers might improve detection accuracy in future work.

References

1. Mahi, M.J.N., et al.: A review on vanet research: perspective of recent emerging technologies. IEEE Access (2022)
2. Nirbhay Kumar Chaubey: Security analysis of vehicular ad hoc networks (vanets): a comprehensive study. Int. J. Secur. Appl. **10**(5), 261–274 (2016)
3. Quyoom, A., Mir, A.A., Sarwar, A.: Security attacks and challenges of vanets: a literature survey. J. Multimedia Inf. Syst. **7**(1), 45–54 (2020)
4. Sasikala, E., Rengarajan, N.: An intelligent technique to detect jamming attack in wireless sensor networks (wsns). Int. J. Fuzzy Syst. **17**(1), 76–83 (2015)
5. Kanagasabapathy, P.M.K., Poornachary, V.K., Murugan, S., Natesan, A., Ponnusamy, V.: Rapid jamming detection approach based on fuzzy in wsn. Int. J. Commun. Syst. **35**(2), e4205 (2022)
6. Mokdad, L., Ben-Othman, J., Nguyen, A.T.: DJAVAN: Detecting jamming attacks in vehicle ad hoc networks. Perform. Eval. **87**, 47–59 (2015)
7. Ullah, S., et al.: Hdl-ids: A hybrid deep learning architecture for intrusion detection in the internet of vehicles. Sensors **22**(4), 1340 (2022)
8. Jia, W., Chen, X.-Y., Zhang, H., Xiong, L.-D., Lei, H., Deng, S.-H.: Hyperparameter optimization for machine learning models based on Bayesian optimization. J. Electr. Sci. Technol. **17**(1), 26–40 (2019)
9. Puñal, O., Pereira, C., Aguiar, A., Gross, J.: CRAWDAD dataset uportorwthaachen/vanetjamming2012 (v. 2014-05-12). https://crawdad.org/uportorwthaachen/vanetjamming2012/20140512 (2014)

Big Data and Business Intelligence

An Improved Active Machine Learning Query Strategy for Entity Matching Problem

Mourad Jabrane(✉), Imad Hafidi, and Yassir Rochd

Laboratory of Process Engineering, Computer Science and Mathematics, National School of Applied Sciences Khouribga, University Sultan Moulay Slimane, Beni Mellal, Morocco
mourad.jabrane@usms.ac.ma

Abstract. Entity matching (EM) is crucial step in data integration. Supervised machine learning (SML) approaches have attained the SOTA performance in EM. In real - world scenarios SML suffers from absence or lack of large labeled data for training. Active machine learning (AML) for EM minimize the number of training data required and tries to reduce the amount of Hand labeling by picking just the helpful pairs. ALL AML approaches use just one of the two criteria - informativeness or representativeness - for query selection, Which limit their efficacy. In this work, we propose a Combined Score Sampling (CSS) that combines informativeness and representativeness selection criteria. We evaluate the CSS using the benchmark e-commerce data-sets pair Abt-Buy and AML with ensemble learning as SML model and we demonstrate that it effectively addresses the issue. Comparing our strategy to SML, we demonstrate that it lead to overall enhanced F1 score and stability of the learnt models.

Keywords: Entity matching · Supervised machine learning · Active machine learning · data integration

1 Introduction

Entity matching is a sophisticated procedure that reduces the semantic gap between duplicate representations of the same real-world item, as well as the amount of storage required and the number of bytes that must be sent between endpoints. This issue is described by machine learning as a binary problem classification. The input is two profiles with their similarity and the output is the result (matching or non-matching). They use a labeled data-sets including both positive and negative record pairs to solve this problem. For building a robust and effective model, ML requires a huge number of labeled data; yet, in real-world issues, unlabeled data are typically available but labeled data are rare since the labeling data process is costly, time-consuming, and exceedingly tedious. Active machine learning (AML) tackles this problem by querying for the label

N. Aboutabit et al. (Eds.): ICMICSA 2022, LNNS 656, pp. 317–327, 2023.
https://doi.org/10.1007/978-3-031-29313-9_28

of the most useful pair. In general, AL use either informativeness or representativeness as selection criterion. The informativeness evaluates its capacity of the record pairs to reduce the uncertainty of model, while the representativeness of a record pairs evaluates how well it reflects the overall input patterns of data. In EM literature, AL utilizes just one of the two query selection criteria listed above, which restrict their performance. We address this issue by proposing a unique AML query strategy, named Combined Score Sampling CSS, for measuring and integrating the two query selection criteria. In light of this, the rest of the paper is structured as follows: In Sect. 2 there are formal definitions of the EM issue. Section 3 describes the entity matching process. Section 4 reviews some related work. In Sect. 5, we highlight the drawbacks of the literature approaches. Section 6 presents our proposed approach, and Sect. 7 discusses the results of a CSS. In Sect. 8, we present our analysis, conclusions, and recommendations for further research.

2 Formal Definitions

This section defines the EM problem and its key concepts, as well as the evaluation criteria for the produced outcomes. These definitions and notations are broad and may encompass domain-specific variants of this problem (Structured_data or Semi_structured_data).

The Profile (Description, Instance or Reference) is the foundational element of the EM data-structure. Each profile includes information about a specific event, venue, organization, or individual in the real world object.

Specifically:

Definition:

p_i: Profile is an element of a database DS ($p_i \subset DS$), offers details of a real_world physical object

e_k: Entity is a collection of p_i ($e_k = \{p_1, \ldots, p_n\}$), every p_i from e_k correspond to the specific physical object.

Depending on the data structure, profiles may be constructed in a variety of ways;

For a relational database, the profile $p_i = R$ $(\mathrm{a}_1, \ldots, \mathrm{a}_k)$, the (R) represents the relation's name and a_k its characteristics.

For semi-structured data, the profiles $p_i = \{(a_l, v_m)\}$ refer to the collection of value_attribute couples. The last approach is more versatile in terms of the model (schema) representing a profile since it permits a diverse collection of feature names, numerous values for the same feature and label-style values.

All pairs of instances in Entity_Collection e_k ((p_i, p_j) s.t $p_i, p_j \in e_k$), is reflexive, commutative, transitive, and symmetric. In the remainder of the review, to simplify notation, like a profiles pair is indicated ($\mathbf{p}_i \equiv \mathbf{p}_j$) and referred to alternatively with two words duplicate and match.

Typically, the data-sets are categorised based on the number of pairs inside every entity. A data-sets with no more than one profile in entity is referred to as Clean Data-sets, while one with several profiles in entity is referred to as Dirty Data-sets. formally:

Dirty Data-sets where $\exists e_k = \{p_i, \ldots, p_k\}$ s.t: $p_i, \ldots, p_k \in DS$.
Clean Data-sets where $\nexists e_k = \{p_i, \ldots, p_k\}$ s.t: $p_i, \ldots, p_k \in DS$.
The formal definition of EM In this context is as follows [11]:
The EM process attempts to create entities from profiles extracted from a collection of data-sets S. The process is named Dirty entity matching D_EM if S contains a single Dirty data sets, Clean-Clean entity matching (CC_EM) if S contains two Clean data-sets, and Multi-source entity matching (MS_EM) if S contains different data-sets.

3 Entity Matching Process

The objective of EM approaches is to turn the input data into a collection or pair of profiles which near as feasible to their matching physical world entity. Traditionally, the input consists of homogeneous structured data or have little schema diversity. In this case, the characteristics may be homogenized manually or automatically. Consequently, the techniques of EM focuses mostly on noise in feature values of pairs and is schema. This job is often referred to as Record Linkage (CleanClean) or Deduplication (Dirty EM), depending on the input [9]. We can divide the deduplication process based on workflow in Fig. 1 into tree steps:

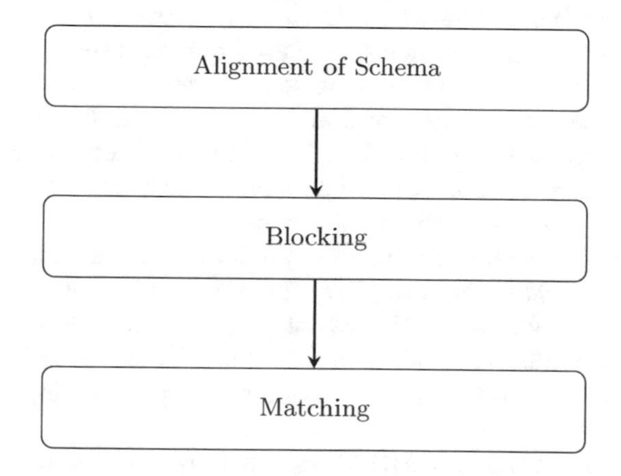

Fig. 1. Illustration of the Entity Matching workflow.

- **Alignment of Schema:** required only for EM across dissimilar Schemes. The purpose is construct connections between the characteristics of input data-sets depending on their relatedness, deduced from the similarity of their architecture, names, and/or contained values [1,20]. Discovering semantically comparable characteristics (such as "profession" and "job") facilitates the schema functioning of subsequent phases.

- **Blocking:** is required to reduce the quadratic complexity operations of the normal EM techniques which compares each p_i to each other p_k, a procedure that cannot scale to huge data sets [9]. It confines potential matches to collection of pairs which have some characteristic in common (e.g., common address). As a consequence, it improves overall time efficiency at the expense of an approximative answer; the more comparisons it eliminates, the more probable it is that duplicates will be overlooked.
- **Matching:** applies a Matching Function to the candidates identified by Blocking step [16]. The resultant grades of similarity and contextual information (e.g., the decision of matching of associated profiles) will be utilized to classify profile pairings into one of two main classes, namely match and non-match.

4 Related Work

Entity matching is described as a problem of classification, The input is two profiles with their similarity and the output is the result (matching or non-matching). They use a labeled data-sets with positive (i.e., matching) and negative (i.e., non-matching) occurrences to overcome this issue. Every profile pairs is defined by a feature vector, each dimension of which generally represents the result of applying a similarity metric to a particular attribute value and returning a score. The labeled data-sets is divided into two separate sets called the training set and the testing set, which are used to create the classification model and evaluate the classification accuracy, respectively.

Classification and regression trees are used in the first supervised methods [12].The feature vector is composed of the Edit_Distance for specific attribute values plus their character length. The taught tree is then pruned to make it simpler, increasing its resilience and generality.

According to Cohen [13], the classifier model trained on collection of boolean characteristics, like Edit_Distance (ED) function, which return True when the ED less than pre-defined value. All possible matches are processed to the model, which generates a similarity graph with pattern confidence-rated edges. To create the final collection of entities, greedy agglomerative clustering is subsequently used.

For the EM job at hand, MARLIN [3] starts by figuring out the best string similarity measurements. In order to determine the appropriate weights for a conventional collection (vector space), it employs the Expectation Maximization method and SVM. Finally,

These measurements are applied to every attribute value for every pair of possible matches. An additional SVM classifier that calculates the chance that the features are duplicates is given the resultant feature vector.

Union_Find information model Monge [22] is used to gradually calculate the transitive closure of found matching pairs, and all probable matches are graded based on the probability of classifier model.

Reyes-Galaviz [26] describe two Multi-layer Perceptron-based models with an only one hidden layer and an one output node. The scoring values of each similarity metric over every attribute, or the scoring values of each attribute over all similarity metrics, are aggregated by each internal node. The relative importance of each similarity measure or attribute value is indicated by the weights learned between internal nodes and out nodes.

In Yan [29], Matching is reframed as a threshold-based ordinal regression job. In reality, the aim is to classify profile pairings into one of five groups: hard-conflict, non-conflict, weak-match, moderate-match, and strong-match. The regression training process teaches it the thresholds that correspond to every category. This allows for the accommodation of various application settings. The reduced cost function may be extended with values for every kind of failure.

The preceding methods usually represent each attribute with the same set of characteristics, in other word the identical similarity algorithms are used to compare each textual feature. Longer training duration, a wider search field, and more complicated learnt models are the results. In De_Carvalho [4] , a technique based on genetic programming is presented to overcome this problem. A tree representation of the matching rules is what it consists of at its heart, with the internal nodes representing operations and the leaves representing attribute values (i.e., similarity functions). These trees are subjected to a sequence of genetic operations until convergence: reproduction preserves a tree in its original form, crossover separates two or multiple young trees from different parent, and genetic change modifies a single tree.

A supervised matching technique may be constructed using any classification algorithm. As a consequence, EM tools have already integrated a number of classifiers, including SVM in Febrl [8] and ID3 decision trees in TAILOR [15]. Magellan [18] also provides Linear & Logistic Regression, Naive_Bayes, xgboost and Random_Forest. A variety metrics similarities, including Jaccard and Jaro similarity, may be utilized to define feature vectors in every tool.

However, none of the suggested categorization models stand out as the winner [19]. In actuality, several classifiers perform better in various settings. Due of this, some research has concentrated on integrating several categorization models to increase accuracy across the board. In Zhao [30], the empirical exploration of a number of ensemble learning strategies,

including four step: bootstraping and boosting steps allows for the mixture of classifiers of the same kind, whereas the cascade and the stack steps allows for a mixture of multiple classifiers, often exceeding all basic models. In Chen [6], more dependable methods that depend on contextual, meta_information level are offered. These characteristics are merged with the underlying classifiers' projected labels to provide a new labeled data-sets that enables the training of context-based meta-classifiers.

Lastly, Chen [7] present a paradigm for risk evaluation that may be learned and interpreted in another area of study. Its objective is to rank all labeled occurrences in order of the risk (probability) of a supervised matching algorithm misclassifying them. It works in three phases to do this. Every labeled profile pairs is first represented using the rules of matching met by most relevant pair of profiles. Second, it aggregates the probability distributions of the different attributes that correspond to each label, modeling each label's uncertainty as a normal distribution. Third, it uses a learning-torank method to fine-tune the parameters of these distributions. As a consequence, experts may adjust how well their supervised matching techniques work using the combinations with high risk profiles.

5 Problematic

Extreme class imbalance in EM negatively impacts the supervised algorithms. The number of positive duplication data rises linearly even as number of input data, but the number of negative duplication data rises quadratically [17]. This shows the majority of record pairings are negative. Typically, undersampling is used to train an effective classification model, which randomly selects as many negative examples as positive ones [24].

The high expense of generating a suitably big and representative labeled data-sets is a further challenge faced by supervised algorithms. Such a data-sets often needs one or more specialists, which is costly and time-consuming [28]. When a large labeled data-sets is needed, the issue is worsened. For instance, a learning-based industrial world for product matching require 1.5 million labeled examples to obtain a F score of 99% [14]. For this reason many proposed methods aim to reduce the human labeling effort by applying AML This is why several suggested solutions try to decrease the amount of human labeling by using AML. Which reduces labeling cost by iteratively selecting the most useful pairs necessary for building a strong and successful classification model and requesting the oracle (e.g., a human annotator) to offer associated labels. Selecting is the most important phase in AL as a result of extreme imbalanced EM classes. For that there have been several ways approaches in literature research. We divide them into the two categories below:

- Heuristic approaches: Choose confusing pairs from the pool based on their feature vectors AdInTDS [10], T3S [2], and LFP/LFN [25].
- Classification-based approaches: They choose confusing unlabeled pairs based on the confidence of classifiers learned on the available training data. In the (QBC) query_by_committee technique, a group of classifiers are trained, and confusing examples are chosen based on the discrepancy between their predictions [21,23]. Random forests provide a natural method, with every single tree serving as the model committee [21] expanded QBC to a classifier-agnostic technique which can be paired with different model. Most uncertain profile pairings are placed at the classifier decision boundary in the margin-based method, which is the 0.5 classifying probabilities for nonlinear models like neural networks or the splitting hyperplane of linear models [21,23] (Table 1).

Table 1. A comparison of the EM active learning strategies

Query Strategy	Type	Advantages	Disadvantages
AdInTDS [10]	Representativeness	solve the problem of class imbalance use blocking to clean the redundant record pairs	Need large number of record pairs
T3S [2]	Informativeness		require more time for select
LFP/LFN [25]	Representativeness		slow and expensive in practice
US/margin-based [21,23]	Informativeness	Easiest method. Pretty quick. Simple to implement. Applicable to any probability model.	myopic. risks getting overconfident about wrong expectations. fail to take use of the abundance of unlabeled record pairs
QBC [5,21]	Informativeness	Comparatively simple and compatible with any basic learning algorithm	challenging to train multiple hypotheses. Still myopic regarding generalization error reduction
Error/variance reduction ALIAS [27]	Informativeness	Effectively achieves the target goal. practically successful	quite computationally complex. hard to implement

However, the above-mentioned AL algorithms only use one of the two criteria (informativeness or representativeness) for query selection, which may considerably restrict their performance.

6 Proposed Approach

The quality and stability of the final classifiers is determined by the representativeness and informativeness of hand labeled record pairs. In real-world situations AL techniques pick either informative or representative unlabeled record pairs to query their labels owing to the fact that they employ a single query approach throughout the whole querying process, which might severely restrict their performance. We address this constraint by using Combined score as query strategy, which aggregate the advantages of different approaches under a single query strategies. We explore the feature space by using the representativeness sampling score to discover something the model has never seen previously and which might completely alter the decision boundary. Utilizing the informativeness sampling score, we utilize specific crucial regions of the feature space, hence enhancing the decision boundary.

The plan is using the both scores concurrently to improve our active learning sample. To do this, we aggregate them into a single measure that we call the Combined sampling score (CSS).

$$CS_{Score}\left(x_i\right) = \begin{cases} R\left(x_i\right) & \text{With probability } \epsilon \\ I\left(x_i\right) & \text{With probability } 1-\epsilon \end{cases} \tag{1}$$

The Combined sampling score $CSS\left(x_i\right)$ is defined as the combination of the representativeness score $R\left(x_i\right)$ and the informativeness score $I\left(x_i\right)$, where ε is a fixed parameter ranging from 0 to 1 that determines the contribution of the two independent scores.

Algorithm 1: Pseudo code

1: $n \Leftarrow NumberOfIterations$
2: $\epsilon \Leftarrow get\_epsilon\_value()$
3: $X \Leftarrow All\_Unabled\_Instances$
4: $model \Leftarrow create\_model()$
5: **while** $i < n$ **do**
6: $\quad p \Leftarrow random(0, 1)$
7: $\quad$ **if** $p < \epsilon$ **then**
8: $\quad\quad utilities \Leftarrow measure\_representativeness(X)$
9: $\quad$ **else**$[p >= \epsilon]$
10: $\quad\quad utilities \Leftarrow measure\_informativeness(X)$
11: $\quad$ **end if**
12: $\quad idx \Leftarrow select\_instances(utilities)$
13: $\quad label \Leftarrow get\_oracl\_label(idx)$
14: $\quad model \Leftarrow update\_model(idx, label)$
15: $\quad i \Leftarrow i + 1$
16: **end while**

Initially the constant ϵ (valued between 0 and 1) is selected by the user, and the number of iteration is fixed (line 1) then the machine learning model is created (line3), in line 4 the active machine learning process started, for each iteration a random value between 0 and 1 was calculated with uniform probability (line 5), and based on this value the utilities of all instances in the pool X is calculated, if $p < \epsilon$ we measure the utility of each instance in pool using representativeness (lines 7–8) otherwise we measure the utility using informativeness (line 9–10). The instance which have high utility value is selected (12), and asking the oracle for its label (line 13), then the model is updated (line 14).

7 Experimental Evaluation

7.1 Experimental Setup

We implemented the CSS algorithm in the Python 3.9 programming language and run the experiments on a machine with Intel i7-7700k CPU @ 4.20 GHz, with 16 GB of RAM, running on Ubuntu 20.04.

7.2 data-Sets

For evaluating the efficacy of CSS described above, we use the benchmark e-commerce pair Abt-Buy from product domains with textual features. A ground truth of 7164 (6067 negative and 1097 positive correspondences) exists between product entities of the two data sets (The common qualities are the product's name, description, and price.)

7.3 Results

We employ CSS as the query strategy sampling in the AL loop and execute it five times with 100 iterations each run. Every iteration is equivalent to human annotation. For the value of ϵ, the strategy of selecting the best value is practically try different values and see which one works best for the test (in our test $\epsilon = 0.3$).

We provide the average F1 scores and accuracy for the final model (Table 2).

Table 2. CSS AML matching results. Comparison of representativeness, informativeness and CSS methods and difference to Supervised Machine Learning.

QUERY	AL		Supervised F1	Δ to Supervised F1
	Accuracy	F1		
representativeness	0.785	0.221	0.818	−0.597
informativeness	0.572	0.421		−0.397
CSS	**0.923**	**0.753**		**−0.065**

8 Conclusion

The proposed approach CSS provides a new active learning query strategy. Which measure and combine the informativeness and representativeness of an unlabeled profile pairs, which make our method guarantees high performance as it produces better models in terms of quality (experimental results showed how our CSS query strategy reached a F1-score (0.92%) of supervised machine learning). On top of the improved high performance, our approach continues showing higher stability and F1 score after 100 iterations.

As a future research direction, we intend to provide a method that permits a dynamic and adaptable trade-off between informativeness and representativeness. In addition, we intend to develop active machine learning approaches with multi-label function.

References

1. Bernstein, P.A., Madhavan, J., Rahm, E.: Generic schema matching, ten years later. Proc. VLDB Endow. **4**(11), 695–701 (2011). https://doi.org/10.14778/3402707.3402710
2. Bianco, G.D., Galante, R., Goncalves, M.A., Canuto, S., Heuser, C.A.: A practical and effective sampling selection strategy for large scale deduplication. IEEE Trans. Knowl. Data Eng. **27**(9), 2305–2319 (2015). https://doi.org/10.1109/tkde.2015.2416734
3. Bilenko, M., Mooney, R.J.: Adaptive duplicate detection using learnable string similarity measures. In: Proceedings of the Ninth ACM SIGKDD International Conference on Knowledge Discovery and Data Mining - KDD 2003. ACM Press (2003). https://doi.org/10.1145/956750.956759
4. de Carvalho, M.G., Laender, A.H.F., Goncalves, M.A., da Silva, A.S.: A genetic programming approach to record deduplication. IEEE Trans. Knowl. Data Eng. **24**(3), 399–412 (2012). https://doi.org/10.1109/tkde.2010.234
5. Chen, X., Xu, Y., Broneske, D., Durand, G.C., Zoun, R., Saake, G.: Heterogeneous committee-based active learning for entity resolution (HeALER). In: Welzer, T., Eder, J., Podgorelec, V., Kamišalić Latifić, A. (eds.) ADBIS 2019. LNCS, vol. 11695, pp. 69–85. Springer, Cham (2019). https://doi.org/10.1007/978-3-030-28730-6_5
6. Chen, Z., Kalashnikov, D.V., Mehrotra, S.: Exploiting context analysis for combining multiple entity resolution systems. In: Proceedings of the 2009 ACM SIGMOD International Conference on Management of data. ACM, June 2009. https://doi.org/10.1145/1559845.1559869
7. Chen, Z., Chen, Q., Hou, B., Li, Z., Li, G.: Towards interpretable and learnable risk analysis for entity resolution. In: Proceedings of the 2020 ACM SIGMOD International Conference on Management of Data. ACM, May 2020. https://doi.org/10.1145/3318464.3380572
8. Christen, P.: Febrl. In: Proceeding of the 14th ACM SIGKDD International Conference on Knowledge Discovery and Data Mining - KDD 2008. ACM Press (2008). https://doi.org/10.1145/1401890.1402020
9. Christen, P.: Data Matching. Springer, Heidelberg (2012). https://doi.org/10.1007/978-3-642-31164-2
10. Christen, P., Vatsalan, D., Wang, Q.: Efficient entity resolution with adaptive and interactive training data selection. In: 2015 IEEE International Conference on Data Mining. IEEE, November 2015. https://doi.org/10.1109/icdm.2015.63
11. Christophides, V., Efthymiou, V., Stefanidis, K.: Entity resolution in the web of data. Synth. Lect. Semant. Web Theory Technol. **5**(3), 1–122 (2015). https://doi.org/10.2200/s00655ed1v01y201507wbe013
12. Cochinwala, M., Kurien, V., Lalk, G., Shasha, D.: Efficient data reconciliation. Inf. Sci. **137**(1-4), 1–15 (2001). https://doi.org/10.1016/s0020-0255(00)00070-0
13. Cohen, W.W., Richman, J.: Learning to match and cluster large high-dimensional data sets for data integration. In: Proceedings of the Eighth ACM SIGKDD International Conference on Knowledge Discovery and Data Mining - KDD 2002. ACM Press (2002). https://doi.org/10.1145/775047.775116
14. Dong, X.L., Rekatsinas, T.: Data integration and machine learning. In: Proceedings of the 2018 International Conference on Management of Data. ACM, May 2018. https://doi.org/10.1145/3183713.3197387

15. Elfeky, M., Verykios, V., Elmagarmid, A.: TAILOR: a record linkage toolbox. In: Proceedings 18th International Conference on Data Engineering (2002). (IEEE Comput. Soc.) https://doi.org/10.1109/icde.2002.994694
16. Elmagarmid, A.K., Ipeirotis, P.G., Verykios, V.S.: Duplicate record detection: a survey. IEEE Trans. Knowl. Data Eng. **19**(1), 1–16 (2007). https://doi.org/10.1109/tkde.2007.250581
17. Getoor, L., Machanavajjhala, A.: Entity resolution. Proc. VLDB Endow. **5**(12), 2018–2019 (2012). https://doi.org/10.14778/2367502.2367564
18. Konda, P., et al.: Magellan. Proc. VLDB Endow. **9**(12), 1197–1208 (2016). https://doi.org/10.14778/2994509.2994535
19. Köpcke, H., Rahm, E.: Frameworks for entity matching: a comparison. Data Knowl. Eng. **69**(2), 197–210 (2010). https://doi.org/10.1016/j.datak.2009.10.003
20. Madhavan, J., Halevy, A.Y.: Composing mappings among data sources. In: Proceedings 2003 VLDB Conference, pp. 572–583. Elsevier (2003). https://doi.org/10.1016/b978-012722442-8/50057-4
21. Meduri, V.V., Popa, L., Sen, P., Sarwat, M.: A comprehensive benchmark framework for active learning methods in entity matching. In: Proceedings of the 2020 ACM SIGMOD International Conference on Management of Data. ACM, May 2020. https://doi.org/10.1145/3318464.3380597
22. Monge, A.E., Elkan, C.P.: An efficient domain-independent algorithm for detecting approximately duplicate database records. In: DMKD (1997)
23. Mozafari, B., Sarkar, P., Franklin, M., Jordan, M., Madden, S.: Scaling up crowdsourcing to very large datasets. Proc. VLDB Endow. **8**(2), 125–136 (2014). https://doi.org/10.14778/2735471.2735474
24. Papadakis, G., Papastefanatos, G., Koutrika, G.: Supervised meta-blocking. Proc. VLDB Endow. **7**(14), 1929–1940 (2014). https://doi.org/10.14778/2733085.2733098
25. Qian, K., Popa, L., Sen, P.: Active learning for large-scale entity resolution. In: Proceedings of the 2017 ACM on Conference on Information and Knowledge Management. ACM, November 2017. https://doi.org/10.1145/3132847.3132949
26. Reyes-Galaviz, O.F., Pedrycz, W., He, Z., Pizzi, N.J.: A supervised gradient-based learning algorithm for optimized entity resolution. Data Knowl. Eng. **112**, 106–129 (2017). https://doi.org/10.1016/j.datak.2017.10.004
27. Sarawagi, S., Bhamidipaty, A.: Interactive deduplication using active learning. In: Proceedings of the Eighth ACM SIGKDD International Conference on Knowledge Discovery and Data Mining - KDD 2002. ACM Press (2002). https://doi.org/10.1145/775047.775087
28. Wu, R., Chaba, S., Sawlani, S., Chu, X., Thirumuruganathan, S.: ZeroER: entity resolution using zero labeled examples. In: Proceedings of the 2020 ACM SIGMOD International Conference on Management of Data. ACM, May 2020. https://doi.org/10.1145/3318464.3389743
29. Yan, L.L., Miller, R.J., Haas, L.M., Fagin, R.: Data-driven understanding and refinement of schema mappings. In: Proceedings of the 2001 ACM SIGMOD International Conference on Management of Data - SIGMOD 2001. ACM Press (2001). https://doi.org/10.1145/375663.375729
30. Zhao, H., Ram, S.: Entity identification for heterogeneous database integration—a multiple classifier system approach and empirical evaluation. Inf. Syst. **30**(2), 119–132 (2005). https://doi.org/10.1016/j.is.2003.11.001

Study of COVID19 Impact on Moroccan Financial Market Based on ARDL

Mohamed Hassan Oukhouya[1(✉)], Nora Angour[2], and Noureddine Aboutabit[1]

[1] Laboratory LIPIM, ENSA Khouribga, University Sultan Moulay Slimane, Khouribga, Morocco
oukhouya.mhassan@gmail.com, n.aboutabit@usms.ma

[2] Faculty of Law, Economics and Social Sciences, Salé, Mohammed V University, Rabat, Morocco
n.angour@um5r.ac.ma

Abstract. The aim of this work is to study the impact of the COVID-19 pandemic new cases on the Moroccan financial market using the Autoregressive Distributed Lag (ARDL) approach. The analysis focuses on the relationship between the natural logarithm of the Moroccan All Shares Index (MASI) price and the natural logarithm of new daily cases of COVID-19 in the short term as well as in the long term. A cointegration test is performed on the daily time series for the period from March 3, 2020 to February 11, 2022. A causality test of Toda-Yamamoto is also applied on the variables. The implementation of the forecast with the ARDL method improves the forecast accuracy by 8% to achieve 26.7%. The implementation of the forecast with the ARDL method shows that the addition of the lag of COVID19, the trend and the seasonality makes it possible to achieve a MAPE of 26.7% by improving it by 8% compared to the forecast with the lag of the price only.

Keywords: ARDL · COVID19 · MASI · Moroccan All Shares Index · Finance · Forecast

1 Introduction

The world has known a number of developments in the health and scientific field throughout history, especially in recent decades, yet a group of diseases and epidemics remain a real threat to humanity.While some epidemics and diseases are confined to specific geographical areas, others are rapidly spreading and reach all parts of the world without exception as with the COVID-19 pandemic. In addition to the human losses that will remain the most important, the rapid spread of this disease has dire effects on the financial situation of citizens, as well as on the economy of the country in relation to several other fields such as health, transport, logistics and agriculture. On the other hand, transactions with other countries can be greatly affected by the fact that the exchange relations in new economic systems pass through supply chains. These facts, along with demographic growth, inter-state travel and climate change make epidemics a global

N. Aboutabit et al. (Eds.): ICMICSA 2022, LNNS 656, pp. 328–339, 2023.
https://doi.org/10.1007/978-3-031-29313-9_29

threat and not tied to a specific region. Therefore, all countries must cooperate and work with each other to develop solutions to confront these diseases.

In this context, the COVID-19 pandemic had a great impact on some sectors such as energy and the petroleum industry which are directly related to the economy of the country and which in turn impacted the financial situation [3].

2 Related Work

This section summarizes some articles that has empirically proven the link between the covid-19 pandemic and financial markets. this part will be the basis for hypothesis evaluated in this article.

[1] This paper contributes to the analysis of how the African financial markets volatility was impacted by the covid-19 pandemic, by obtaining two major results. First result, in the period starting from January 27, 2020 to October 22, 2020 the volatility of these financial markets is significantly and positively affected by the development of the disease and by the fear related to the pandemic. This analysis show that African financial markets was impacted by disease progression more than by panic and fear. The second result is that the mortality rate impacts negatively the volatility, but with no significant coefficient.This may be due to the relatively low mortality rate in the countries covered by this study, as the factors that lead to death from covid-19 are relatively rare in Africa. This suggests that markets have incorporated information on African population structure as a determinant of resistance to COVID-19. Nevertheless, other waves have been spotted in these countries, followed by a virus mutation that appears more dangerous.In times of uncertainty or crisis, the search volume on Google search engine could be a good indicator of the risk in African financial markets.

The authors [2] tested the impact of official COVID-19 publications on financial market volatility, focusing on the pandemic phase of the crisis. For this purpose, they used the realized volatility of the S&P 500 index as an indicator of US financial market volatility and compared the effect of the published data at the global and US levels. The results of the empirical investigation provide evidence that: (i) new infections reported globally and in the US intensify financial volatility, (ii) mortality rate has a positive and significant impact on volatility, (iii) the effect of the COVID-19 data announced globally is stronger than the effect triggered by the data announced in the US, (iv) the effect of the UPR on financial volatility is not too much significant during the pandemic phase of COVID-19. The robust results show that the evolution of the pandemic, and the associated uncertainty, intensifies the volatility of the US financial markets, thus impacting the global financial cycle.

The study [4] mainly focuses on how the financial market was impacted by investment behavior during COVID-19. It was concluded that the variables was affected by both factors financial risk tolerance and general risk tolerance, as well as risk perception, satisfaction and rate of return. COVID-19 has a soft impact on the relationship between them. It is shown that the tolerance of financial risk is considered as an attitude factor in making financial decisions. The study conclude that the return rate is based especially on the assessment of financial risk, also how changes in the return rate change the tolerance of financial risk. The study proves that the previous years have shown that the financial market has harmed the impact of the epidemic, that mainly affects the decisions of a company. It is shown that the effects of this pandemic between general risk tolerance and risk perception can be determined. As the global impacts harmed the business sectors that bring instability for investors. The results of this study show that satisfaction has a positive impact on general risk to tolerance and satisfaction on financial risk to tolerance. In addition, the study proves that the uncertainty of this pandemic moderates between risk perception and general risk at tolerance and moderates between risk perception and financial risk at tolerance. However, the uncertainty of COVID-19 moderates between satisfaction and general risk at tolerance and moderates between rate of return and financial risk at tolerance. Therefore, there is a moderating effect of COVID-19 uncertainty between rate of return and financial risk at tolerance.

This study [5] aims to inspect the relationship between stock market progression, economic development and financial innovation, in Bangladesh on the period from 1980 to 2016. To test the cointegration on the long term, this study applied the autoregressive distributed lag (ARDL) bounds test. Also, the Granger causality test is implemented to detect directional causality between the variables under the term error correction. The results of the ARDL linked test approach study confirm that a long-term association exists between economic growth, stock market development and financial innovation . Furthermore, the results of the Granger causality test confirm a two-way causality between financial innovation, economic growth and stock market development, and economic growth in the long and short run. These results support the theory that market-based financial development and financial innovation in the financial system can stimulate economic development.

In this paper [6], the authors show that the inability to find a significant relationship between the exchange rate and money demand could be due to the assumption of a linear adjustment mechanism between the variables. Once they introduce non-linearity in the short run as well as in the long run through the concept of partial sum, they show that currency appreciation or depreciation could affect money demand asymmetrically, which is demonstrated using data in the Iranian context.

In this paper [7], the authors examine the relationship between oil prices and stock prices in oil-exporting and oil-importing countries in several ways. First, they allow for possible non-linearities in the relationship in order to quantify the asymmetric response of stock prices in these two categories to positive and negative changes in oil prices. Second, in order to capture within-group differences, they allow for the effect of heterogeneity in the cross sections. Third, they assess the relative predictability of linear (symmetric) and nonlinear (asymmetric) Panel ARDL models using the Campbell and Thompson (2008) test. The results show that the stock prices of oil exporting and importing groups respond asymmetrically to changes in the oil price, although the response is stronger in the latter than in the former.

3 Hypothesis Development

The previous section aims to address the current literature about the impact of the this pandemic on financial markets. Which lead to the following assumptions that will build the basement of the current work:

1. Is there a relationship in the short run between COVID-19 pandemic and moroccan financial market.
2. Is there a relationship in the long run between COVID-19 pandemic and moroccan financial market.
3. How the ARDL model can intervene and improve the prediction of future values of the moroccan financial market

4 Methodology

Most studies of causal relationships use VAR modeling to show the relationship between explanatory variables and an explained variable in a financial market. Nevertheless, to implement this method, the series must be integrated of the same order. But,this condition is not satisfied in the majority of macroeconomic series (Plosser and Nelson, 1982). Faced with this shortcoming, Pesaran, Shinet Smith (2001) defined the Auto Regressive Distribution Lag (ARDL) method by taking into account the limitations of the VAR model. This approach has been used in many studies and given the nature of our data and our working hypotheses, this is the reason why we will use this model in our work.

This technique, which is an alternative to cointegration tests, was initiated by Pesaran and Shin (1999) and Pearsan et al. Its objective is to test the long-run on the basis of a number of variables that are not integrated of the same order I (0) or (1) (Senay and Merter, 2010). On the other hand, the application of ARDL allows obtaining unbiased estimates of the long-run relationship (Harris and Sollis, 2003), in addition it is more suitable for small samples (Narayan, 2005). Using the Akaike information criteria (AIC), we can select the optimal number of lags of the dependent variable and the variable of interest. In addition, we will test for the existence of a causal relationship between the two variables.

4.1 Research Model

The ARDL model described by the following equation will allow to estimate the answers to the hypotheses mentioned below.

$$\Delta LogMASI_{(t)} = C + \sum_{i=1}^{p} \alpha_{1i} \Delta LogMASI_{(t-i)} + \sum_{i=0}^{q} \alpha_{2i} \Delta LogCOVID_{(t-i)} + \beta_1 LogMASI_{(t-1)} + \beta_2 LogCOVID_{(t-1)} + \varepsilon_{(t)} \quad (1)$$

The table below describes the different variables of the equation:

Variable	Description
MASI	Price of the Moroccan All Shares Index
COVID	Daily new cases of COVID-19
C	Constant
Log( )	Natural logarithm operator
Δ	First difference operator
$\alpha_1; \alpha_2$	Short-run coefficients
$\beta_1; \beta_2$	Long-run dynamics
$\varepsilon(t)$	Error term

Note that we must start by a cointegration test before the estimation of the ARDL model, since for the variables that are not cointegrated, it's not possible to make an estimation of an error correction model, nor the estimation of the short and long term effects. For the long-run case, we compute the "cointegration limit test" based on the Fisher statistic on the hypothesis:

$$H_0 : \beta_1 = \beta_2 = 0$$

The rejection of H_0 implies the presence of cointegration. H_0 is rejected when the F-statistic exceeds the value of the upper bounds I(1), while it is accepted if the F-statistic is less than the value of the lower critical bounds I(0). Otherwise, we cannot conclude. In the presence of cointegration, the long-run relationship is obtained by the cancellation of the variables in first difference (Morley, 2006 and Antoniou et al., 2013). Based on Eq. (1), we deduce that it is represented by the following equation:

$$LogMASI_{(t)} = -\left(\frac{C}{\beta_1}\right) - \left(\frac{\beta_2}{\beta_1}\right) LogCOVID_{(t)} \tag{2}$$

An error correction model (ECM) for Eq. 2 can help confirm the existence or not of cointegration between the variables as follows:

$$\Delta LogMASI_{(t)} = \sum_{i=1}^{p} \alpha_{1i} \Delta LogMASI_{(t-i)} + \sum_{i=0}^{q} \alpha_{2i} \Delta LogCOVID_{(t-i)} + \beta_1 ECM_{(t-1)} + \varepsilon_t \tag{3}$$

$$ECM_{(t)} = LogMASI_{(t)} - \left[-\left(\frac{C}{\beta_1}\right) - \left(\frac{\beta_2}{\beta_1}\right) LogCOVID_{(t)}\right] \tag{4}$$

4.2 Data and Description

The data, subject of our study are the daily closing price of the Moroccan All Shares Index (MASI) downloaded from the website www.investing.com. To study the impact of the COVID19 pandemic, the daily number of confirmed cases is downloaded from the official website of the Moroccan Ministry of Health. The data are from March 03, 2020 (the day a new case of covid is reported in Morocco) to February 11, 2022. Both variables are log-transformed before being processed.

5 Results and Discussion

5.1 Descriptive Statistics

Table 1 presents the descriptive statistics for the series of the two variables. MASI refers to the closing price of the MASI index. NEW_CASES refers to the number of new confirmed daily cases of COVID-19 in Morocco (Figs. 1, 2 and 3).

Table 1. Descriptive statistics of MASI and COVID-19 new cases (2 March 2020 to 11 February 2022.)

	MASI	NEW_CASES
Mean	11524.53	1576.031
Median	11517.92	579.0000
Maximum	13991.47	12039.00
Minimum	8987.890	0.000000
Std. Dev	1347.250	2219.878
Skewness	0.018541	2.233916
Kurtosis	1.856586	8.076603
Jarque-Bera	26.44808	924.1962
Observations	485	485

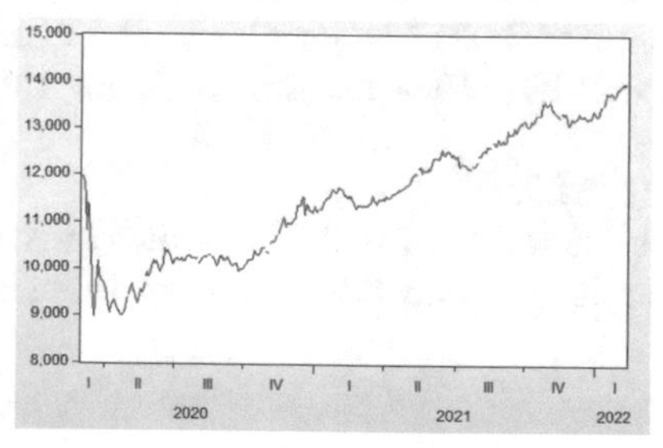

Fig. 1. Movement of MASI price in the concerned period

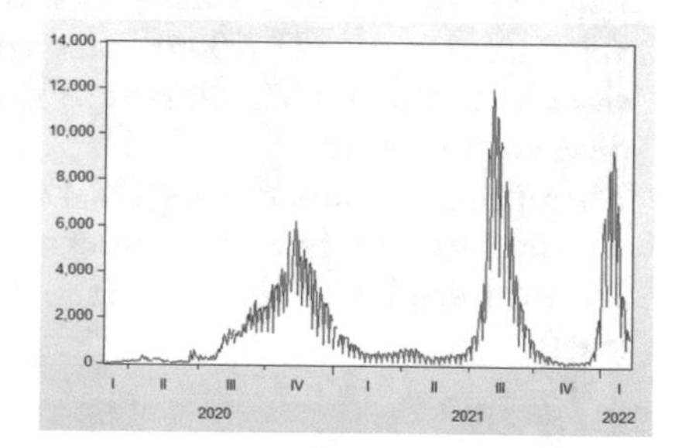

Fig. 2. COVID19 new cases in the concerned period

5.2 Stationnarity (Unit Root Tests)

A time series whose the moving average and/or variance varies over time is concidered as non-stationary, this non-stationarity (of the deterministic or stochastic type), if it is not treated (stationarization), can lead to "spurious" regressions.

Several tests help to verify the stationary character or not (existence of a unit root) of a series: augmented Dickey-Fuller/ADF test, Phillippe-Perron/PP test, Andrews and Zivot/AZ test, Ng test -Perron, KPSS, etc. Of all these tests, the first two are easy to apply and commonly used. In fact, the ADF test is effective in the event of autocorrelation of errors and the PP test is suitable in the presence of heteroscedasticity. In this study, we used the ADF and PP tests with test critical values Mackinnon (1996), the results are given as follows:

Table 2. ADF Unit Root test onthe log level of variables

Variables	Level					Integration order
	T-statistic	1%	5%	10%	P-value	
Log MASI	−0.829160	−3.444128	−2.867509	−2.570012	0.8095	I(1)
Log MASI 1st Difference	−5.306658	−3.444128	−2.867509	−2.570012	0.0000	
Log COVID	−3.044266	−3.444280	−2.867576	−2.570048	0.0317	I(0)

The augmented Dickey-Fuller (ADF) statistic, used in the test, is a negative number. The more negative it is, the stronger the rejection of the hypothesis that there is a unit root at some level of confidence.

- Log MASI Level : test statistics value = −0.829160 is greater than the critical value (5%) = −2.867509 and the p-value=0.8095 is greater than 0.05, thus the data is not stationary.
- Log MASI 1st difference : test statistics value = −5.306658 is lower than the critical value (5%) = −2.867509 and also the p-value = 0 is very much less than the significant value 0.05. Thus rejecting the null hypothesis and considering the data as stationary.
- Log COVID19 : test statistics value = −3.044266 is lower than the critical value (5%) = −2.867576 and also the p-value = 0.0317 is very lower than the significant value 0.05. Thus rejecting the null hypothesis and considering the data as stationary.
- The optimal number of lags is 17 based on AIC.
- All the variables have a significance at the 1% level since all the values at the 1% level are lower than 3.5 (the 1st value lower than 3.5 is that of the 1% level)

Table 3. PP Unit Root test on the log level of variables

Variables	Level					Integration order
	T-statistic	1%	5%	10%	P-value	
Log MASI	−0.369520	−3.443635	−2.867292	−2.569896	0.9114	I(1)
Log MASI 1st Difference	−18.19505	−3.443663	−2.867304	−2.569902	0.0000	
Log COVID	−5.637943	−3.443776	−2.867354	−2.569929	0.0000	I(0)

5.3 ARDL Estimation

We use the Akaike Information Criterion (AIC) to select the optimal ARDL model, the one that offers statistically significant results with the least parameters. The model is estimated with the "constant & trend" option since it's very significant (Prob < 1%). Below are the estimation results of the optimal ARDL model retained (Tables 2, 3, 4, 5, 6, 7, 8 and 9.

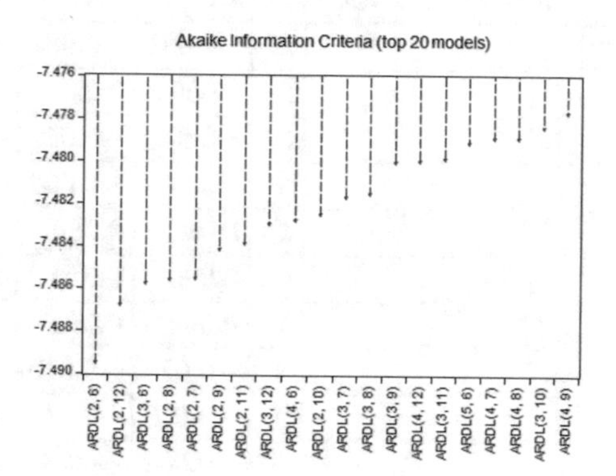

Fig. 3. Akaike information criteria

The model that offers the smallest AIC value is the best, which in our case is the ARDL(2,6), also it is globally significant with Prob (F-statistic) = 0.0000.

Table 4. ARDL estimation

Variable	Coefficient	Std. Error	t-Statistic	Prob.*
MASI_LOG(−1)	1.196800	0.043909	27.25662	0.0000
MASI_LOG(−2)	−0.257500	0.043916	−5.863430	0.0000
COVID19_LOG	0.001027	0.000702	1.461975	0.1444
COVID19_LOG(−1)	−0.000982	0.000711	−1.382608	0.1675
COVID19_LOG(−2)	0.000570	0.000559	1.020708	0.3079
COVID19_LOG(−3)	−0.001067	0.000566	−1.885020	0.0601
COVID19_LOG(−4)	−0.000443	0.000557	−0.795811	0.4265
COVID19_LOG(−5)	−0.000417	0.000710	−0.586752	0.5577
COVID19_LOG(−6)	0.001191	0.000681	1.749257	0.0809
C	0.555984	0.128045	4.342094	0.0000
@TREND	$5.14E-05$	$1.22E-05$	4.212603	0.0000
R-squared	0.997024	Mean dependent var		9.346391
Adjusted R-squared	0.996960	S.D. dependent var		0.118471
S.E. of regression	0.006532	Akaike info criterion		−7.201220
Sum squared resid	0.019756	Schwarz criterion		−7.104652
Log likelihood	1717.689	Hannan-Quinn criter.		−7.163241
F-statistic	15512.26	Durbin-Watson stat		2.022289
Prob(F-statistic)	0.000000			

*Note: p-values and any subsequent tests do not account for model selection.

Most of the coefficients are significant and also the model it is globally significant. However, it is important to check if it is valid by doing validity tests including autocorrelation tests.

Table 5. Autocorrelation of residuals.

Autocorrelation	Partial Correlation		AC	PAC	Q-Stat	Prob*
.|. |	.|. |	1	−0.031	−0.031	0.4672	0.494
.|. |	.|. |	2	0.030	0.029	0.8915	0.640
.|. |	.|. |	3	−0.059	−0.057	2.5576	0.465
.|. |	.|. |	4	−0.055	−0.060	4.0298	0.402
.|* |	.|* |	5	0.082	0.082	7.2305	0.204
.|. |	.|. |	6	0.020	0.026	7.4317	0.283
.|* |	.|* |	7	0.099	0.090	12.155	0.096
.|. |	.|. |	8	−0.029	−0.019	12.565	0.128
.|. |	.|. |	9	−0.019	−0.015	12.741	0.175
.|. |	.|. |	10	−0.038	−0.032	13.453	0.199

*Probabilities may not be valid for this equation specification.

The Ljung-Box test shows that the probability of the Q-statistic is above the 5% and 10% thresholds for all results, this suggest the absence of autocorrelation in the model errors, which is important for further estimation. The presence of autocorrelation of residuals leads to an estimation of parameters that are inconsistent, due to the presence of a lagged dependent variable that appears as an exogenous variable in the model.

Table 6. Bound test to cointegration results

F-Bounds Test		Null Hypothesis: No levels relationship		
Test Statistic	Value	Signif.	Lower bound I(0)	Upper bound I(1)
			Asymptotic: n = 1000	
F-statistic	9.433081	10%	5.59	6.26
k	1	5%	6.56	7.3
		2.5%	7.46	8.27
		1%	8.74	9.63

Here we can see that the F test value of 9.43 is bigger than most of values of the I(1) bound hence there is a **cointegration** between the variables, with a significance at the level 2.5%. Which gives the possibility of estimating the long-term effects on Log_COVID19 on Log_MASI.

Table 7. Dynamics of the short-run and long-run

Conditional Error Correction Regression				
Variable	Coefficient	Std. Error	t-Statistic	Prob.
MASI_LOG(−1)*	−0.060700	0.013991	−4.338372	0.0000
COVID19_LOG(−1)	−0.000121	0.000223	−0.545517	0.5857
D(MASI_LOG(−1))	0.257500	0.043916	5.863430	0.0000
D(COVID19_LOG)	0.001027	0.000702	1.461975	0.1444
D(COVID19_LOG(−1))	0.000166	0.000680	0.243598	0.8077
D(COVID19_LOG(−2))	0.000736	0.000706	1.042008	0.2980
D(COVID19_LOG(−3))	−0.000331	0.000706	−0.469430	0.6390
D(COVID19_LOG(−4))	−0.000774	0.000677	−1.144776	0.2529
D(COVID19_LOG(−5))	−0.001191	0.000681	−1.749257	0.0809
EC = Log_MASI - (-0.0020*COVID19_LOG)				
Long-Run Coefficients				
Variable	Coefficient	Std. Error	t-Statistic	Prob.
COVID19_LOG	−0.002001	0.003616	−0.553284	0.5803
C	0.555984	0.128045	4.342094	0.0000
@TREND	$5.14E-05$	$1.22E-05$	4.212603	0.0000

The long term relationship is described as follows :

- Log_MASI = -0.0020*Log_COVID.

The results show that there is a significant negative long-term relationship between COVID 19 and the stock market in Morocco: a 100% increase in the daily number of confirmed cases of COVID-19 resulted in a 0.2% decrease in the MASI price. In the short-term relationship, it appears that there is no significance between all variables, but Log_COVID delayed by 5 days (t-5) has a positive impact on Log_MASI in day (t) at the 10% level.

Since correlation does not necessarily imply causality, we must test the causality that may exist between the variables, we use the Toda Yamamoto causality test.

Table 8. Toda-Yamamoto causality test.

Dependent variable: MASI_LOG			
Excluded	Chi-sq	df	Prob.
COVID19_LOG	15.32437	2	0.0005
All	15.32437	2	0.0005
Dependent variable: COVID19_LOG			
Excluded	Chi-sq	df	Prob.
MASI_LOG	4.914155	2	0.0857
All	4.914155	2	0.0857

A causal relationship from Log_COVID to Log_MASI is confirmed by the Toda-Yamamoto causality test (Prob = 0.0005, the null hypothesis is rejected). However, there is no causality between Log_MASI and Log_COVID (Pob = 0.0857).

5.4 Forecasting Results

The forecast performance measure is the MAPE, which is the average of the deviations in absolute value from the observed values. It is a percentage and therefore a practical indicator of comparison, describe with the following formula:

$$MAPE = 100\% - \frac{100\%}{n} \sum_{t=1}^{n} \left| \frac{Forecasted\_value_t}{Real\_value_t} \right| \tag{5}$$

With **n** the number of forcasted values.

Based on the ARDL results in the previous subsection, a MASI time series forecast was performed to assess the impact of each element on the accuracy of the forecast. In iteration (a), the MASI prices of the last 2 days are used to predict the day in question, iteration (b) consists of the forecast with the addition of the values of the new confirmed cases of the last 6 days to forecast the value of the price of the day in question. While iteration (c) aims to predict the value of the price of the day in question by adding the trend and seasonality to iteration (b).

Table 9. Forecasting results.

	Iteration		
	(a) MASI lag only	(b) (a)+COVID19 lag	(c) (b) + Trend & Seasonality
MAPE	34.7%	32.2%	26.7%
Improvement	–	2.5%	8%

As shown in the table, iteration (a) predicted MASI prices with a MAPE of 34.7%, after which it improved in iteration (b) by 2.5% with the addition of new lags confirmed cases of covid19, eventually arriving at a MAPE of 26.7% in iteration (c) with trend and seasonality.

6 Conclusion

The study deals with the modeling of the effects of coronavirus on the stock market in Morocco during the period from March 3, 2020 to February 11, 2022 with the ARDL estimation approach. The natural logarithm of the Moroccan All Share Index (MASI) price is the endogenous variable while the number of daily confirmed cases is the prevalence measure of Covid. Previous research on the effects of the Covid-19 epidemic on the Arab stock market is not numerous, while this study is a contribution to this recently emerging literature. Furthermore, it analyzes not only the short-term relationship between the number of new confirmed cases of coronavirus and the Moroccan stock market, but also the long-term and causality. The results indicate the presence of a negative long-term relationship between Log MASI and Log Covid. Causality between the variables is unidirectional from Log Covid to Log MASI.

References

1. Del Lo, G., Basséne, T., Séne, B.: COVID-19 and the African financial markets: less infection, less economic impact? Finan. Res. Lett. **45**, 102148 (2022)
2. Albulescu, C.T.: COVID-19 and the United States financial markets' volatility. Finan. Res. Lett. **38**, 101699 (2021)
3. Norouzi, N., de Rubens, G.Z., Choupanpiesheh, S., Enevoldsen, P.: When pandemics impact economies and climate change: exploring the impacts of COVID-19 on oil and electricity demand in China. Energy Res. Soc. Sci. **68**, 101654 (2020)
4. Wang, F., Zhang, R., Ahmed, F., Shah, S.M.M.: Impact of investment behaviour on financial markets during COVID-19: a case of UK. Econ. Res.-Ekonomska Istraživanja **35**, 2273–2291 (2022)
5. Qamruzzaman, M., Wei, J.: Financial innovation, stock market development, and economic growth: an application of ARDL model. Int. J. Finan. Stud. **6**, 69 (2018)
6. Bahmani-Oskooee, M., Bahmani, S.: Nonlinear ARDL approach and the demand for money in Iran. Econ. Bullet. **35**, 381–391 (2015)
7. Salisu, A.A., Isah, K.O.: Revisiting the oil price and stock market nexus: a nonlinear panel ARDL approach. Economic Modell. **66**, 258–271 (2017)

Augmented Analytics Big Data Warehouse Based on Big Data Architecture and LOD System

Abdelghafour Benoualy(✉), Nassima Soussi, and Imad Hafidi

LIPIM, National School of Applied Sciences, Sultan Moulay Slimane University, Khouribga, Morocco
abdelbenoualy@gmail.com

Abstract. With the development of new technologies, the Internet and social networks in the last twenty years, the production of digital data has been increasing. This has motivated different companies and organizations to adapt and update their information systems, technology and data architectures in order to improve the extraction of added value for decision making by varying the integration of multiple data sources for advanced analysis and to meet analytical and reporting needs.

Today with the explosion of data in the world of big data and the challenges and pain point faced by businesses to fulfill strategic requirements and to make the right decision The ability to analyze available of this data is a valuable asset for any successful business, especially when the analysis is based on data from multiple data sources and enriched by a flexible integration to answer business KPIs based on predefined objectives.

To address this issue of analysis complexity and sometimes lack of data we provide a vision and implementation of a new architecture that covers the entire data life-cycle and leverages new big data technologies with classic DW capabilities in order to increase the Big DW based on the combination of data from internal DW and external DW fed by an external data model on demand.

Keywords: Big Data · MapReduce · Spark · Sqoop Data warehouse · ETL.Big Data · Business KPI · LOD · OpenData

1 Introduction

Compared to traditional data warehouses (DW) the arrival of Big Data technologies presents a significant advantage in the new Data-Driving systems, by enabling the management of extremely large volumes of data. It serves as the foundation for collecting and analyzing structured, semi-structured and unstructured data in its native format to gain new insights, better predictions and improved optimization. In contrast to the classic and traditional data warehouse (DW), DataLake can process video, audio, logs, text, social media, sensor

N. Aboutabit et al. (Eds.): ICMICSA 2022, LNNS 656, pp. 340–348, 2023.
https://doi.org/10.1007/978-3-031-29313-9_30

data and documents to power applications, analytics and AI. DataLake can be designed as a component of a data matrix architecture to deliver the right data, at the right time, regardless of its location.

To get the best results for effective decision making and take advantage of these big data technologies, companies today tend to increase their DataLake by combining a variety of external and internal data sources and take advantage of the modeling capability and other benefits of internal DW including.

To feed a Big Data warehouse and as mentioned above it is necessary to take advantage of the internal DW and Their importance for strategic decision making within companies [3,6].

The usage of a Data Warehouse offers the possibility to improve the management and decision making. Through different and flexible analytical processing tools by a simple and graphical and automated mode, in order to facilitate the analysis and decision making. It brings together all the functional data of an organization.

The flexibility of BI systems provides business management with the ability to make the right analytical decisions based on all available data, allowing for a potential boost in productivity, due to the intuitive reports generated by Business Intelligence systems.

Today with the explosion of data and to meet several business needs the necessity of integrating and invest on other data sources including that of big data source systems, The Traditional ETLs prove difficult and slow to store and process large data and Siloed [4].

On the modern data architecture Leveraging a target Big Data source system, Linked Open Data (LOD) refers to all external data that are published on the web platforms that can provide a new value to increase and augmented a Big Data Warehouse we can find as a popular LOD [5] YAGO and DBpedia which is qualified by two main features that are relevant to Data Warehouses it is open and generally available to organizations, which can find useful insights by supplementing their sources.

The advantage of these open data systems (Linked Open Data) is that they are accessible and simple to reuse in different environments and the data is based on Semantic technologies.

To answer these different pain points and to respond to several business issues when making decisions in this paper, we propose a global architecture concept that covers the entire life cycle of a Big Data warehouse, from the definition of the objectives that require the basic internal and external data sources (LOD) to the integration of all sources that satisfy the objectives defined by the business deciders.

2 Related Work

The ability to analyze available data is an essential asset for any company, especially when the analysis produces knowledge and indicators for decision support.

For this reason various researches proposed to integrate semantic data, social data, NoSQL data, DataLake data and LOD, and been made to enrich the internal data Waterhouse by external data [2] based on the added value of open data on the different business domains and processes.

Other works have focused on building multi-dimensional templates from LOD or log data sets and have operated the LOD integration workflow using the traditional ETL process or an ETL adaptation of extracting, processing and querying (ETQ) the requested on demand data.

Some work has proposed new architecture based on the scalability of cloud technology and involving the business throughout the construction phase of the logical data model feeding the data warehouse. Based on a large volume and heterogeneity of data and on big data technologies and an on-demand cluster for the storage and processing, transformation of these data to meet the needs of Analytics and reporting of the company [10–12].

In this article, we propose a complete architecture approach that covers the entire life cycle of building an Analytics Big Data warehouse based on the definition of objectives and Business KPI [4,7] requiring internal data warehouse and external master data sources Waterhouse based on the business objective and KPIs [3,4,7,10] to the integration of all sources that meet the objectives defined by the business and decisional maker.

2.1 Proposed Architecture

Globally in the standard company's architectures, and to satisfy decisional and analytical needs we find several types of data sources with different formats (structured or semi-structured) and with different integration methods (batch mode or Near Real Time) which are involved in the construction of the enterprise data warehouse.

To process these types of data companies usually use traditional and classical data integration tools. In our experience and based on several case studies, we have used several Big Data technologies covering the entire life cycle of the data with and benefit from all the new technologies and capabilities of the Big Data mode.

2.2 Data Sourcing

The presence and the choice of an external data sources can be seen as a key element and a real opportunity to build a target big data warehouse for a reporting and analysts usage on the different companies.

Proposed Architecture:

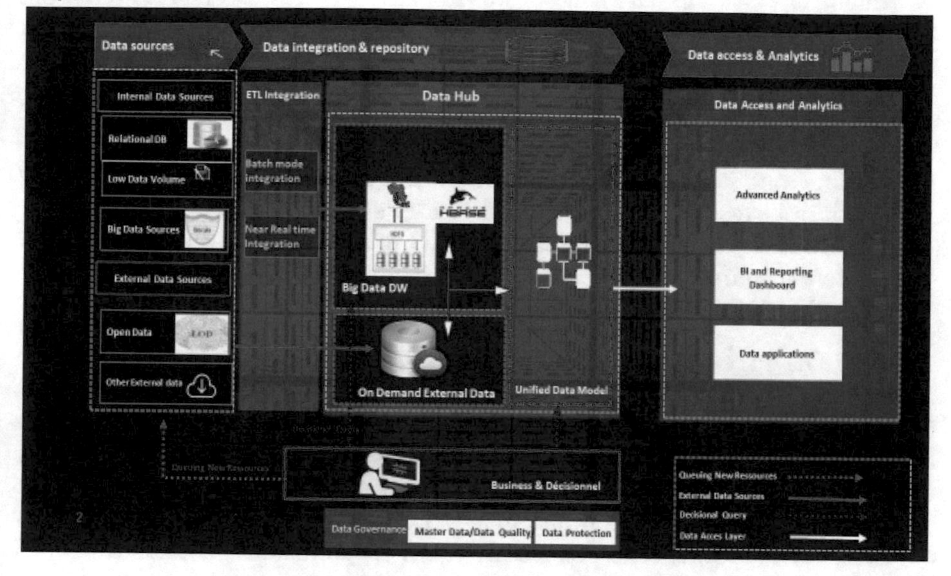

Fig. 1. Architecture of the hybrid data integration and the proposed system technologies

For this integration of external + internal data sources to have more value for decision making it must be driven by the business requirements like as the key performance indicators (KPIs) [3] that need to be identified early in the conception process to be fully useful to the company's strategies. Also Data should be collect from a variety of different sources. Often, it is a mix of structured, Semi- structured and unstructured data. While each organization will use different data streams, some common sources include:

- web server logs;
- cloud applications;
- mobile applications;
- social media content;
- text from customer emails and survey responses;
- mobile phone records;
- Machine data captured by sensors connected to the internet of things (IoT);
- External and open data sources LOD;
- Internal Data Warehouse.

Among the internal data sources we also used the internal DW to build our target Big Data warehouse with a combined integration that we will detail later on the integration phase.

To identify the added value of internal and external sources in the target Big DW we suggest two approaches:

- The first one is source-driven. It provides quantitative indications of DW growth in terms of multidimensional concepts and instances.
- The second approach based on the objective orientation. It estimates the value of external sources of LOD based on their ability to assess the performance of the objectives that the company identified for the DW defined [4].

The LOD augmented our on demand External DWs with free available web open data [1,2] Many studies have focused on building a multidimensional business design from LOD datasets or LOD queries. Other studies have managed the LOD integration process using the traditional extract, process and load process of required data on demand. Our approach offers a new vision by estimating the value of an augmented LOD from our Big Data DW, following a new business objective oriented integration approach based on predefined KPIs as explained above (Fig. 1).

Based on the maturity of Big Data technologies we attempt to model the different phases of the design and construction of our target architecture schema with the unified model that will be guided by the KPI defined by the business objectives.

This phase of modeling will contribute to complete with external and internal data sources. These modeling efforts are then presented in the integration sections.

2.3 Data Integration

Multiple data integration technologies are available on the legacy of Big Data Solution and platform we suggest using apache Sqoop open source and SparkQL. To integration data as-is from the LOD to the on demand DW then processing to the target Big DW.

In the proposed architecture we have also implemented a logical DW based on end-user on-demand design the integration using LOD data in the DW must be driven by business needs to be fully useful to business strategies.

In our case we are using an integration scenario based on queries defined by decisional maker objectives and KPIs [3].

This scenario of integration corresponds to build an on demand Data Warehouse to feed and augmented the target big Data Warehouse, which the uses data is retrieved incrementally from the existing DW and LOD (if needed) and loaded into the DW only when needed to answer some OLAP cube queries (Fig. 3).

To detail more this phase of integration scenario and define the design phase to feed the data warehouse with external data.

As mentioned this phase is based on Query-data-Driven Design, and corresponds to the on-demand data warehouse to feed the target Big DW.

Two scenarios have been identified to raise a big data warehouse that meets the business needs according to the objectives and the Kips established on the first architecture Vision step:

Scenario 2

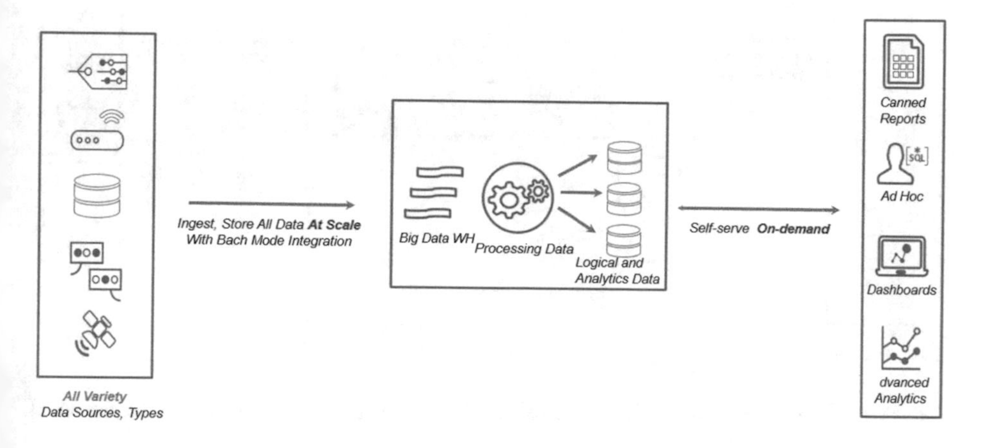

Fig. 2. Building Big Data WH only from internal DW and Other Data Sources

In this scenario the first step is to validate if only internal data can meet the business needs, in this case we will collect the data as is from the internal DW and the different internal legacy data sources in order to feed the big data warehouse.

The data will be structured and stored on several zone on HDFS the (Hadoop Data storage system) [8] :

- **Raw Zone Build**: Data is simply loaded from all sources unaltered as-is, never deleted and only appended.
- **Conformed Zone Build**:
  - Initially, some common norms are applied to the Data like formatting to the standard pattern, enrichment with technical metadata like point of origin id and timestamp, clean-up of data formats, this layer is the cornerstone or traceability, lineage, audit and must be compliance.
  - Secondly, common business logic is applied to the data to drive new value from the data available on the internal DW and all other sources.

- **Analytics Zone Build**: Here, Data is optimized and based from business requirement and objectives defined before by a set of KPI.

Scenario 2

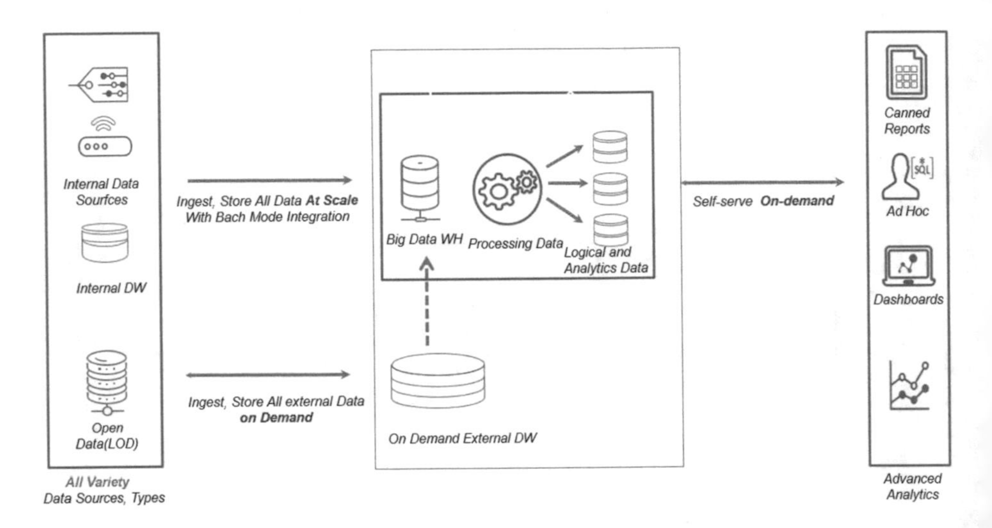

Fig. 3. Building Big Data WH from internal DW Other Data Sources with external Data

The data integration process is designed on the level and purpose of query and LOD data fragments. The value and result estimation of the KPIs are built in the same method as the integration process. The result query is stored directly on the LOD and also on the DW on demand, and the final query results (on the LOD) are merged with the query results executed on the internal DW to perform the matching KPI value:

- If Internal Data can't meet the business needs an integration of external data will be necessary on demand to enrich the target big DW.
- The choice of data will be made according to the business objectives and their integration will be stored as is after transformation in column-oriented formats (Hbase) and Hive Database [8,9] on the DW on demand as shown on the proposed architecture.
- The phase of modeling and mapping between internal and external data will be done and stored logically on the target big DW.

In both scenarios applied we find that the business value is guaranteed by increasing the big data warehouse based on the business objectives and the associated metrics.

2.4 Processing and Modeling

Processing

For the processing of massive data (in our case given in anticipation of external LOD systems) and given the limitation of classical ETL tools [4] that do not perfectly meet the needs we suggest integrating adapted and performing tools for the management of Big Data flows can be collected and processed efficient.

Modeling

This part of modeling consists in respecting the integration of different sources identified to meet the business objectives predefined in order to solve the problem of the unavailability of KPIs in the internal sources of the company.

The modeling is done mainly based on the variety of heterogeneous data sources. When the Big DW takes into account external sources of LOD and internal models on the DW, this variety intensifies and is solved by choosing the appropriate model. It can unify the sources with a logical unified data model this model must be unified at the schema and the and column-oriented database format (Hbase).

Conclusion

In this work we have merged the advantages of new big data technology with the capability of warehousing the KPI modeling functionality of classic data warehouse.

Adding external data based on metrics and business objectives brings a lot of flexibility and advantage in terms of decision making and improves the analytics and dashboarding part.

We proposed then to exploit the external data based on metrics and business objectives brings a lot of flexibility and advantage in terms of decision making and improves the analytics . Another approach that has attracted less attention from research on DW functionality is related to the results of the value added of the built DW and its ability to satisfy the strategic business needs. We proposed to renovate the traditional DW design methodology, proposing a new data architecture based on the web open data of Data context that provides decision makers with metrics (KPIs) and business objectives to assess the value of the built DW.

References

1. Gallinucci, E., Golfarelli, M., Rizzi, S., Abell'o, A., Romero, O.: Interactive multidimensional modeling of linked data for exploratory OLAP. Inf. Syst. **77**, 86–104 (2018)
2. Rizzi, S., Gallinucci, E., Golfarelli, M., Abell'o, A., Romero, O.: Towards exploratory OLAP on linked data. In: SEBD, pp. 86-93 (2016)
3. Berkani, N., Bellatreche, L., Khouri, S., Ordonez, C.: Value-driven approach for designing extended data warehouses. In: DOLAP (2019)

4. Barone, D., Jiang, L., Amyot, D., Mylopoulos, J.: Reasoning with key performance indicators. In: Johannesson, P., Krogstie, J., Opdahl, A.L. (eds.) PoEM 2011. LNBIP, vol. 92, pp. 82–96. Springer, Heidelberg (2011). https://doi.org/10.1007/978-3-642-24849-8_7
5. Berkani, N., Bellatreche, L., Khouri, S., Ordonez, C.: The contribution of linked open data to augment a traditional data warehouse. J. Intell. Inf. Syst. **55**(3), 397–421 (2020). https://doi.org/10.1007/s10844-020-00594-w
6. Berkani, N., Bellatreche, L., Guittet, L.: ETL processes in the era of variety. In: Hameurlain, A., Wagner, R., Benslimane, D., Damiani, E., Grosky, W.I. (eds.) Transactions on Large-Scale Data- and Knowledge-Centered Systems XXXIX. LNCS, vol. 11310, pp. 98–129. Springer, Heidelberg (2018). https://doi.org/10.1007/978-3-662-58415-6_4
7. Khouri, S., Berkani, N., Bellatreche, L., Lanasri, D.: Data cube is dead, long life to data cube in the age of web data. In: Madria, S., Fournier-Viger, P., Chaudhary, S., Reddy, P.K. (eds.) BDA 2019. LNCS, vol. 11932, pp. 44–64. Springer, Cham (2019). https://doi.org/10.1007/978-3-030-37188-3_4
8. Zdravevski, E., Lameski, P., Kulakov, A.: Row key designs of NoSQL database tables and their impact on write performance. In: Proceedings - 24th Euromicro International Conference on Parallel, Distributed, and Network-Based Processing, PDP 2016, pp. 10-17 (2016). https://doi.org/10.1109/PDP.2016.84
9. Thusoo, A., et al.: Hive - a petabyte scale data warehouse using hadoop (2010). https://doi.org/10.1109/ICDE.2010.5447738
10. Herodotou, H., Dong, F., Babu, S.: No one (cluster) size fits all: automatic cluster sizing for data-intensive analytics. In: Proceedings of the 2nd ACM Symposium on Cloud Computing, p. 18. ACM (2011)
11. Ranjan, R.: Streaming big data processing in datacenter clouds. IEEE Cloud Comput. **1**(1), 78–83 (2014). https://doi.org/10.1109/MCC.2014.22
12. Hu, H., Wen, Y., Chua, T.S., Li, X.: Toward scalable systems for big data analytics: a technology tutorial. IEEE Access **2**, 652–687 (2014). https://doi.org/10.1109/ACCESS.2014.2332453

Analyzing Instagram Images to Predict Personality Traits

Siham El Bahy(✉), Noureddine Aboutabit, and Imad Hafidi

Laboratory of Process Engineering, Computer Science and Mathematics, National School of Applied Sciences Khouribga, University Sultan Moulay Slimane, Beni Mellal, Morocco

siham.elbahy@gmail.com, n.aboutabit@usms.ma

Abstract. Our posts on social media are a way of expressing ourselves and our inclinations. In this research, we focused on Instagram (a social media platform for sharing images) to study the relationship between the personality of users and the content of photos they post. We also predicted their personalities based on the relationship between them and their images. This study is the first work that focuses on the Moroccan population. We collected a database of 316 Instagram Moroccan users, larger than the other databases used in previous works. We asked the users to fill out a NEO PI form to extract a Big Five that we chose to express their personality. And then we downloaded the images of users and extracted from them three categories of features; visual features (9 colors and lighting), image content (objects, animals, faces, people, light sources, dark, buildings), and emotional features (anger, disgust, fear, happy, neutral, sad, surprise). A root mean square error (RMSE) was used to indicate prediction accuracy, which relates to the [1,5] score scale. We succeeded in outperforming the best results obtained in the previous works, especially in predicting extraversion and neuroticism traits. We got an RMSE value of 0.83 versus 0.90 for extraversion and 0.87 versus 0.89 for neuroticism.

Keywords: Instagram · Personality · Big Five · Emotions · Content · Images

1 Introduction

Our social media posts have become a way of expressing ourselves and presenting ourselves the way we want to appear. Everything we post reflects our inclinations and describes a part of our personality.

A body of research has been interested in studying the personality reflection of social media users on their posts. Some of these studies extracted personality traits from the profile images of Facebook users [1], and others extracted them from a single profile image containing a face [2]. Additionally, another study

Supported by LIPIM laboratory.

N. Aboutabit et al. (Eds.): ICMICSA 2022, LNNS 656, pp. 349–360, 2023.
https://doi.org/10.1007/978-3-031-29313-9_31

investigated the relationship between the traits of Instagram users and the content of shared images [3]. In addition, others have predicted personality from content [4] and visual features [5].

Our study examines the Instagram application to determine how personality affects content, emotions, and colors in images posted on the platform. Moreover, we used the features extracted from their photos to predict their personalities. The steps of our work are in "Fig. 1".

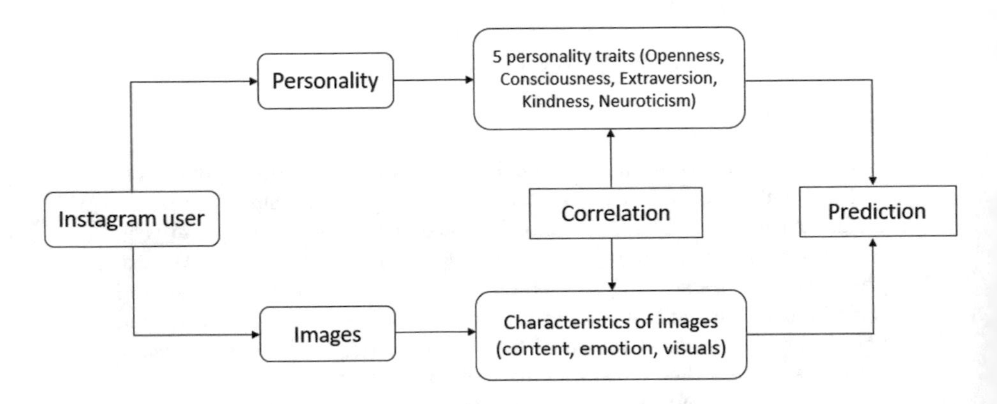

Fig. 1. Our work process.

We collected a database of 316 Instagram users aged 15 to 38. To express their personalities, we used the Big Five model, which describes users' personalities with five traits (Openness, Conscientiousness, Extraversion, Agreeableness, and Neuroticism). First, we asked them to complete the NEO FI survey; as a result, the value of each trait was between 0 and 100. Next, we divided the features of images into three groups: the first group for visual features, the second group for the components of image content, and the third for the emotions that the images express.

The remainder of this article is structured as follows: In the second section, we present related work, while in the third, we present the methods used. The fourth section will study the relationship between personality traits and image characteristics. Finally, in the last two sections, we predict personality traits and discuss the results obtained compared to previous works.

2 Related Work

Researchers have shown that published images reflect the personality traits of social media users [1,6]. Moreover, they have shown a correlation between the photos posted on social media and personality traits, which opens new ways to extract personality traits from social media and facilitate personalized systems. For example, a study on the Facebook platform [7] found that Facebook profile pictures convey thorough information about users. The same applies to Twitter profile pictures [8]. Furthermore, the results of [7] showed that the information encoded in profile pictures could be exploited to classify the personality traits and interaction styles of Facebook users. Nevertheless, there are differences between the traits, for example, openness and conscientiousness, which have few examples and require more experiments. However, agreeableness and extraversion scored the highest among the personality traits. While dominance and affect reach slightly higher performances. However, neuroticism is the most difficult trait to predict. This is consistent with the results that human raters accurately estimate extraversion and agreeableness and that extraversion is related to the expressiveness of the pictures. Another study [2] was conducted on personality trait prediction on the same platform using profile pictures. They proposed an application of convolutional neural networks to model the personality prediction of individuals by analyzing their Facebook profile pictures. They showed that personality prediction models could predict users' personalities from a single profile picture on Facebook [2].

Furthermore, on Twitter [8], this research was conducted on 1290 Twitter users, based on profile pictures and tweets, to predict Twitter users' personalities automatically. They extracted facial features using a face detection model combined with smile detection and other features like hue, saturation, and value. The results show that image features can predict personality better than text features and the combination of text and image features.

[1] analyzed the user's choice of profile image to extract the users' personality traits. The semantics of the image can be classified into person identification, event semantics, concept semantics, and location semantics. They used facial features and found that nervous and open-minded users post images containing fewer people and express mostly negative emotions. On the other hand, conscientious users post what is needed for a profile picture and express positive emotions. Therefore, extraverts and agreeable users show aspects of the good life like pets, travel, and food. They like colorful images but not blurred and bright images. Facial expressions are smiling and display mostly positive mood expressions.

On the Instagram platform, studies have been conducted to extract users' personality traits through the images posted on their accounts. For example, in [4], when they conducted a study on a sample of 113 users in the United States, they could predict personality using visual characteristics (hue, value, saturation) of Instagram images. The results show similar trends to previous work on personality extraction from social media [9]. However, they outperformed it in predicting most of the personality traits. The most successful prediction was openness, conscientiousness, and agreeableness. However, the most complex traits were extraversion and neuroticism.

Based on the hue of the user's image, this study [10] aimed to infer the user's personality traits since applying filters to images before publishing them is considered a method of expressing personal characteristics. They explored the relationship between image hue and personality traits. In addition, they entered it into the SVM classifier and the CNN model. They succeeded in classifying the personality traits.

[5] examined personality prediction from Instagram image features. They used a database of 193 Instagram users. They exploited the images' visual features (e.g., hue, valence, saturation) and content characteristics (architecture, body parts, clothing, musical instruments, art, entertainment, botany, cartoons, animals, food, sports, vehicles, electronics, babies, hobbies, jewelry, weapons). As a result, they obtained a better prediction of certain personality traits using visual and content features than [9,11]. Furthermore, they showed that both visual and content features can be used to predict personality and that they generally perform well. However, combining the two does not lead to an increase in predictive power. Therefore, they do not add more value than they already have independently. Nevertheless, [13] found features that could improve prediction when combined.

According to [14], the prediction of self-reported and attributed personality traits can be achieved by using the images' features marked as favorites on Flickr. Nevertheless, they noticed a clear covariation for attributes but not self-reported traits. The authors explained that participants may have used information such as their personal history and life experiences to determine their traits [15], although their favorite pictures did not contain all of these things. In addition, they differentiated between two activities that have fundamentally different goals. The first is production, the act of posting an image, and the second is the consumption of social media content, which is the act of liking an image [12].

This research [3] studied the reflection of personality traits of Instagram users on the content of shared images, they worked on a database of 193 Instagram users, and the number of images in the database was 54,962. Using the Google Vision API, they retrieved 4090 unique tags from the Instagram images, and using the doc2vec approach, they gathered them into 400 groups, manually resulting in 17 categories afterward. Their results showed that each user with one of the personality traits posts a specific type of image. For example, those with a high score for openness to experience generally posted more images composed of musical instruments, while conscientious participants more frequently shared images containing clothing or sports, unlike extroverts who tended to post pictures composed of electronic devices. In addition, positive correlations were found between the agreeableness trait and the categories (clothing) and (hobbies), meaning that agreeable participants' Instagram photo collections consisted of photos containing clothing or hobbies. On the other hand, those with high neuroticism scores tend to have less content with clothing and more with jewelry in their photos.

3 Methodology

3.1 Database

To carry out this study, we asked a sample of Moroccan volunteers (133 males and 183 females.) who have an Instagram account to participate in this research, aged between 15 to 38. We asked the volunteers to allow us to download their images and collected 17,847 images.

3.2 Features

User Features

We chose the Big Five model [16] to describe users' personalities; it is a descriptive model of the personality in five central traits (Openness, Conscientiousness. Extraversion, Agreeableness, and Neuroticism).

We asked participants to take an online personality test at the link: https://truity.com/test/big-fivepersonality-test. It is a survey that includes 60 questions about personality. The results are values between 0 and 100 of each of the five personality traits of the volunteers.

Image Features

To extract image features, we measured these features in three classes: content category, emotion, and visual features (Table 1).

Table 1. Features used by category

Content category	Emotion category	Visual characteristic category	Other features
abstract, buildings, dark, object, people, cat, dog, sports ball, Tv, laptop, cup, bottle, horse, handbag percentage of Percentage of images contain faces, face number.	angry, disgust, fear, happy, neutral, sad, surprise.	brightness, red, orange, yellow, green_yellow, green, cyan, blue, violet, rose, warm color, cold color.	images number.

Content category

We used different methods to extract 16 contents from the images, which are: Abstract, Buildings, Dark, Objects, People, Cats, Dogs, Sports balls, TV, Laptops, Cup, Bottle, Horse, Purse, percentage of the images contains faces, number of faces, Light sources.

First, we used "InceptionV3" to detect the occupancy of each category (Abstract, Buildings, Dark, Object, People) in the image with percentage. In addition, we calculated the average of each of these features for all images.

Second, we used the MTCNN model to detect faces in an image; it is a model that has three convolutional networks (P-Net, R-Net, and O-Net). Moreover, it can outperform many detection benchmarks of the face while maintaining real-time performance. We count the number of faces in each image for each user and consider it a feature. Furthermore, the average was calculated for all the images to determine how often the images contained faces.

Third, we used the YOLO method to detect the following contents: Cat, Dog, Sports ball, TV, Laptop, Cup, Bottle, Horse, and Purse. Then, for each user, we calculate the average of the images that contain the content.

To detect he light sources in images, we transformed them into HSV before calculating the average of pixels having a Value parameter greater than 240. These values will be averaged over all the images of the user.

Emotion Category
We used CNN to detect seven emotions (Anger, Disgust, Fear, Joy, Neutral, Sadness, and Surprise). We calculated the percentage of each emotion in the images. Then, these values will be averaged over all a user's images.

Visual Feature Category
We extracted nine colors (Red, Orange, Yellow, Yellow-green, Green, Cyan, Blue, Violet, and Pink) and two features: one group, the warm colors (Red, Orange, Yellow, Yellow-green, and Pink), and the other groups the cold (Green, Cyan, Blue, Violet) ones.

We transformed the images into HSV, and we divided the range of the parameter hue into intervals that correspond to the hues. For each interval, we divided the total number of pixels by the total number of pixels in the image. Then we calculated the average of these values for all images.

4 Result and Discussion

4.1 Relationship Between Personality and Image Features

Scatter Plots Between the Features Extracted from the Images and the Personality Traits
This step aims to draw the relationship between the five personality traits and the characteristics of Instagram users' images. The relationship is defined by the influence of the independent variables (content category, emotion, pixels, number of images) on the dependent variables (the value of each trait), which can be accomplished using machine learning techniques. To find out the types

of dependencies between the different parameters, we plotted the scatter plot of each feature against each parameter. The "Fig. 2" shows examples of these curves.

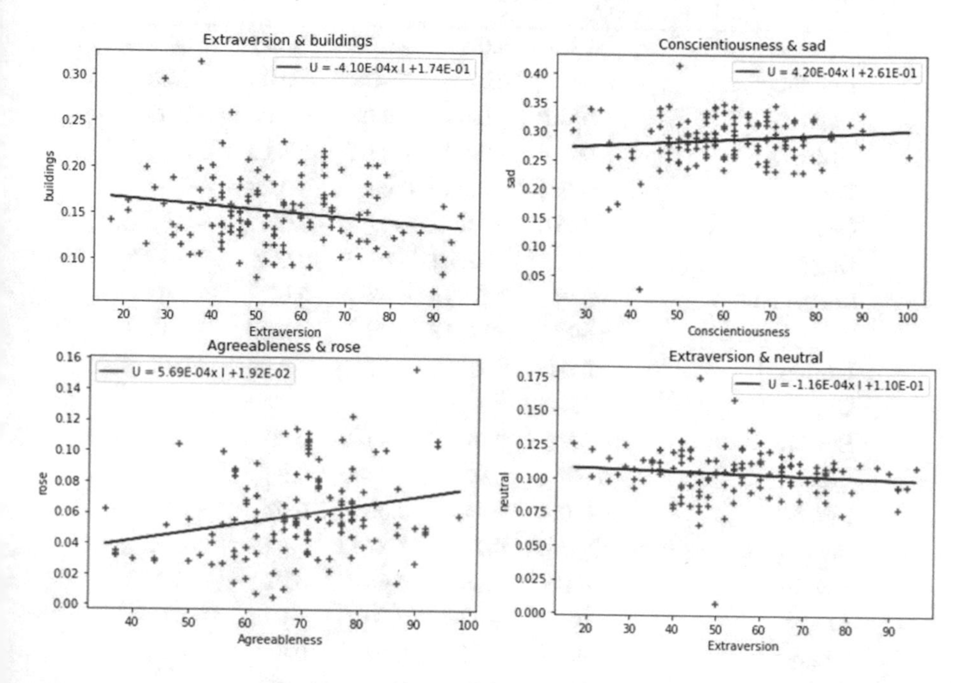

Fig. 2. Examples of scatter plots between the characteristics extracted from the images and the personality traits.

We noticed that most of these curves represent linear dependencies. For this reason, we used the Pearson correlation method.

Pearson Correlation

A Pearson correlation determines the relationship between two variables. This method returns a value between +1 and −1. The value 0 indicates that the two variables do not have any association. Positive associations consist of values greater than 0. When the value is less than 0, the association is negative.

We notice that there are correlations between the characteristics and the personality traits. We also see that openness to experience and conscientiousness are the traits with the most significant correlations with image characteristics, followed by agreeableness (Table 2).

Table 2. Pearson correlation matrix between image characteristics and personality traits ((O) Openness, (C) conscientiousness, (E) extraversion, (A) agreeableness, (N) neuroticism).

	O	C	E	A	N
Abstract	***0.16**	0.02	−0.02	***0.17**	0.08
Building	0.09	0.08	^^**-0.11**	−0.03	−0.05
Dark	****0.19**	−0.05	−0.03	−0.02	−0.02
Object	0.09	−0.005	−0.02	***0.14**	0.08
People	****-0.24**	−0.02	0.09	−0.09	−0.02
Cat	^^**0.10**	0.01	−0.04	0.02	−0.05
Dog	0.07	^^**0.12**	0.07	0.007	^^**-0.13**
Sports ball	−0.06	****0.18**	0.03	0.07	−0.03
Tv	0.09	****0.18**	0.08	−0.05	^^**-0.12**
Laptop	−0.04	^^**0.11**	0.02	−0.09	−0.07
Cup	0.03	***0.16**	0.09	0.07	−0.07
Bottle	−0.07	0.02	−0.02	0.02	−0.04
Horse	0.05	0.05	−0.03	0.06	−0.02
Purse	−0.006	0.11	0.02	0.07	−0.06
Images contains faces	^^**-0.14**	0.1	^^**0.12**	−0.03	−0.02
Number of faces	−0.05	0.06	0.03	0.03	−0.04
Angry	0.01	−0.07	0.001	−0.02	0.08
Disgust	0.03	0.014	0.02	0.08	0.02
Fear	0.02	0.05	0.06	−0.03	−0.04
Happy	−0.03	0.02	−0.04	0.08	0.05
Neutral	0.04	−0.04	−0.06	−0.004	0.03
Sad	0.07	−0.002	0.06	−0.07	−0.03
Surprise	−0.09	0.01	−0.03	0.017	−0.06
Red	^^**0.11**	−0.06	−0.008	−0.01	0.01
Orange	−0.02	0.04	−0.04	0.01	^^**-0.13**
Yellow	0.007	^^**0.14**	−0.07	^^**-0.14**	−0.07
Yellow-green	−0.02	0.02	−0.02	−0.07	−0.06
Green	0.001	0.02	−0.06	^^**-0.11**	−0.06
Cyan	0.002	0.04	−0.02	−0.05	^^**-0.11**
Blue	−0.09	−0.06	0.05	0.05	^^**0.11**
Violet	−0.06	0.03	0.08	***0.14**	0.0.05
Rose	−0.01	^^**0.11**	0.01	0.06	0.001
Warm	^^**0.10**	0.03	−0.05	−0.04	−0.06
Cold	^^**-0.10**	−0.03	0.05	0.04	0.06
Light sources	0.06	−0.07	−0.02	0.10	^^**0.12**
Number of images	0.04	0.09	0.05	0.06	−0.04

Note.^^p<.05, *p<.01, **p<.001

In general, we found that personality type influences the content of images shared on Instagram. We will explain the influence of each trait on the content of images:

Openness
Those with high openness to experience usually post pictures that do not contain faces [1,4]. This type of user prefers to post images containing abstract things and animals, such as cats. In addition, they publish fewer images containing people [1]. Also, users with high aperture tend to have darker [4] and more colorful images in warm colors, especially red. The pictures are rich in negative emotions [1], especially sadness and disgust, and weaker in positive emotions, such as joy.

Conscientiousness
Conscientious users more frequently share images containing faces; their pictures are colored yellow and pink. This type of user tends to post images of animals, especially dogs, equipment such as televisions and laptops, and objects such as cups, sports balls, and bags.

Extraversion
The percentage of images containing faces in the extroverted user's account is high. Their pictures contain more people and fewer buildings. Moreover, this type of user publishes colored images in cold colors [4].

Agreeableness
Agreeable users usually post images containing abstract objects and things, rich in purple and lacking in green and yellow. They tend to have images full of positive emotions [1].

Neuroticism
Users with a high neuroticism score usually share bright [4], colored in blue, less orange, and cyan, and do not contain animals.

4.2 Personality Prediction from Shared Images

The prediction of the value of each personality trait will be performed separately, assuming that there is no overlap between the traits. We predicted each trait from the features of the users' images. We trained the predictive models with linear regression, Random Forest, Decision Tree, Linear Regression and SVM. We periodically selected 10% of the data for testing and 90% for training and returned the experiment for ten iterations. We used the reported root-mean-square error (RMSE) ($r \in [1,5]$) to indicate the prediction performance of the personality traits. The RMSE is calculated as the average of the accuracies obtained in each iteration.

Furthermore, we compare our results with previous work [4,5] on personality prediction from images posted on Instagram. Finally, for each classifier we used, we report the RMSE in Table 3 to show the RMSE difference between the predicted and observed values.

Table 3. Comparison of different classifiers to predict personality prediction compared to previous work [4,5] according to the RMSE. Numbers in bold represent results that surpassed previous work.

	[4]			[5]	Our results			
	Radial base function network	Random forest	M5 Roles	RBF Network	Random forest	Decision tree	Linear regression	SVR
O	0.68	0.71	0.77	0.71	0.74	0.77	0.80	0.76
C	0.66	0.67	0.73	0.62	0.79	0.81	0.84	0.80
E	0.90	0.95	0.96	0.98	**0.85**	**0.89**	**0.88**	**0.83**
A	0.69	0.71	0.78	0.61	0.72	0.77	0.73	0.77
N	0.95	1.01	0.97	0.89	**0.88**	0.90	0.90	**0.87**

The results showed that personality traits influence the content shared in Instagram images. In general, we obtained a good result for all characteristics, but the remaining ones that were difficult to predict were extraversion and neuroticism, knowing these traits are the hardest to predict [4,9].

We show an improvement in predicting some personality traits compared to previous work. For example, we outperformed previous [4,5] work with random forests, decision trees, linear regression, and SVR in predicting the extraversion trait. Moreover, we could also outperform them with random forests and SVRs in predicting neuroticism traits.

5 Conclusion

In this study, we showed the influence of personality traits on the image features shared on Instagram by analyzing the relationship between these traits and the contents of images, the emotions expressed, and the dominant color in the images. The result indicates that the images users post convey much information about their personalities. Furthermore, we outperformed previous work [4,5] in predicting traits that are the hardest to predict: neuroticism and extraversion traits [4,9].

Some of the results obtained in our study about the influence of personality traits on the content of published images are similar to those obtained in previous works 1.4, but others are not similar. The reason can be that the image content and the meaning of colors may be influenced by cultural factors, which limits the generalizability of results.

We propose, in future work, to train two models, one for women and the other for men, to predict the personality of each gender separately due to the difference in nature and preferences between the sexes. It can give accurate results and be useful for referral, diagnostic, and recruitment services. In addition, use a more diverse sample covering different age groups, occupations, and countries.

References

1. Naik, A., Asnani, K.: Personality prediction using profile pictures. Int. J. Eng. Res. Comput. Sci. Eng. (2018)
2. Akshat, D.: Application of convolutional neural network models for personality prediction from social media images and citation prediction for academic papers (2016)
3. Ferwerda, B., Tkalcic, M.: You are what you post: what the content of instagram pictures tells about users' personality. In: The 23rd International on Intelligent User Interfaces (2018)
4. Ferwerda, B., Schedl, M., Tkalcic, M.: Using instagram picture features to predict users' personality. In: International Conference on Multimedia Modeling (2016). https://doi.org/10.1007/978-3-319-27671-771
5. Ferwerda, B., Tkalcic, M.: Predicting users' personality from instagram pictures: using visual and/or content features? In: Proceedings of the 26th Conference on User Modeling, Adaptation and Personalization, pp. 157–161 (2018). https://doi.org/10.1145/3209219.3209248
6. Celli, F., Bruni, E., Lepri, B.: Automatic personality and interaction style recognition from Facebook profile pictures. ACMMM (2014). https://doi.org/10.1145/2647868.2654977
7. Hall, J.A., Pennington, N., Lueders, A.: Impression management and formation on Facebook: a lens model approach (2013). https://doi.org/10.1177/1461444813495166
8. Christian, G., Suhartono, D., Kamal, M.F., Suryaningrum, K.M.: Automatic personality prediction using deep learning based on social media profile picture and posts. In: 2021 4th International Seminar on Research of Information Technology and Intelligent Systems (ISRITI) (2021). https://doi.org/10.1109/ISRITI54043.2021.9702873
9. Quercia, D., Kosinski, M., Stillwell, D., Crowcroft, J.: Our twitter profiles, our selves: predicting personality with twitter. In: Conference: PASSAT/SocialCom 2011, Privacy, Security, Risk and Trust (PASSAT), 2011 IEEE Third International Conference on and 2011 IEEE Third International Confernece on Social Computing (SocialCom), Boston, MA, USA, 9–11 Oct., 2011 (2011). https://doi.org/10.1109/PASSAT/SocialCom.2011.26
10. Chu-Chien Wu, N.X.M.-S.C. Ping-Yu Hsu, Chen, Y.-Y.: Prediction of personality traits through instagram photo HSV. In: Kurosu, M. (ed.) Human-Computing Interaction Technology Innovation, vol. 13303, pp. 279–287 (2022) https://doi.org/10.1007/978-3-031-05409-9_21
11. Ferwerda, B., Schell, S., Yang, E., Tkalcic, M.: Personality traits predict music taxonomy preferences. In: Proceedings of the 33rd Annual ACM Conference Extended Abstracts on Human Factors in Computing Systems, pp. 2241–2246 (2015). https://doi.org/10.1145/2702613.2732754
12. Lay, A., Ferwerda, B.: Predicting users' personality based on their 'liked' images on instagram. IEEE Trans. Affect. Comput. 268–285 (2018)
13. Skowron, M., Ferwerda, B., Tkalcic, M., Schedl, M.: Fusing social media cues: personality prediction from twitter and instagram. In: Proceedings of the 25th International Conference Companion on World Wide Web. International World Wide Web Conferences Steering Committee, pp. 107–108 (2016) https://doi.org/10.1145/2872518.2889368

14. Segalin, C., Perina, A., Cristani, M., Vinciarelli, A.: The pictures we like are our image: continuous mapping of favorite pictures into self-assessed and attributed personality traits. IEEE Trans. Affect. Comput. 268–285 (2017). https://doi.org/10.1109/TAFFC.2016.2516994
15. Wright, A.G.C.: Current directions in personality science and the potential for advances through computing. IEEE Trans. Affect. Comput. 292–296 (2014). https://doi.org/10.1109/TAFFC.2014.2332331
16. Costa, P., McCrae, R.R.: Revised neo personality inventory (neo pi-r): professional manual (2008). https://doi.org/10.4135/9781849200479.n9

Heuristic Algorithm for Extracting Frequent Patterns in Transactional Databases

Meryem Barik(✉), Imad Hafidi, and Yassir Rochd

Process Engineering, Computer Science and Mathematics Laboratory, ENSA Khouribga, Khouribga, Morocco
meryem.barik@usms.ac.ma, {i.hafidi,y.rochd}@usms.ma

Abstract. Frequent Itemsets Mining (FIM), the extraction of frequent patterns from a transactional database, is regarded as one of the most effective data mining strategies. The FIM issue may often be treated using either exact or metaheuristic-based approaches. For small to medium datasets, exact approaches, such as the Apriori algorithm, are very successful. In the case of huge datasets, however, these approaches suffer from temporal complexity. The must of metaheuristic-based approaches are still inefficient, despite the fact that their speed is increasing. Several experiments were done to enhance metaheuristics-based techniques by integrating the Apriori algorithm with the Genetic Algorithm (GA), Particle Swarm Optimization (PSO), and Bees Swarm Optimization (BSO). This combination produced three distinct methods: GA-GD, PSO-GD, and BSO-GD. This research compares the performance of different algorithms in terms of two metrics, including execution time and the number of frequently occurring itemsets created. In terms of execution time, the findings indicate that the techniques outperform the Apriori algorithm. In addition, the found solutions converge to 100%.

Keywords: Frequent Itemset Mining · Apriori algorithm · Heuristics · Genetic algorithm · Swarm Intelligence algorithm

1 Introduction

Data Mining is a science used to extract important information from large datasets. It is regarded as a multi-disciplinary process in repute for more than two decades, involving specialists in various fields (algorithms, machine learning, mathematics, database, etc.) [23]. This is a very important field of research that has attracted the interest of scientists, industrialists, commercial companies, and

Process Engineering, Computer Science and Mathematics Laboratory, ENSA Khouribga, Morocco.

N. Aboutabit et al. (Eds.): ICMICSA 2022, LNNS 656, pp. 361–371, 2023.
https://doi.org/10.1007/978-3-031-29313-9_32

all the players in various fields and organizations in view of its significant scientific and socio-economic impacts. The large datasets can be a valuable asset for predictive analytics [10].

The creation of Data Mining algorithms on large datasets in the academic field presents an inspiring research challenge and opportunity. In fact, the application of traditional data mining techniques on large datasets is very difficult. Moreover, as the number of data increases, the proportion of people who can interpret it increases [1]. Frequent Itemset Mining (FIM) is one of the popular techniques.

Frequent Itemset Mining consists of extracting frequent patterns from the transactional database. Extraction of frequent patterns in a transactional database is an NP-Hard problem. The number of itemsets that can be generated from a database of n items is $2^n - 1$. The time complexity of extracting frequent patterns using traditional algorithms, such as Apriori-based algorithms, is $O(2^n)$. This indicates that as the number of objects increases, the running time grows exponentially, due to the multiple scans of an entire transactional database. In response to this, the researchers proposed metaheuristics-based methods, including genetic algorithms [8,18] or swarm intelligence [7,17].

In a short amount of time, metaheuristic-based algorithms extract a subset of all frequent item sets, but they cannot locate all conceivable frequent item sets in a database. In other words, the efficiency of the solution acquired through the use of metaheuristic-based approaches is inferior than the optimal quality achieved through the use of conventional methods, which identify all feasible frequent item sets. How the randomized search of the itemsets space is conducted determines the efficacy of the solutions produced by FIM metaheuristic-based approaches. Existing FIM metaheuristic-based approaches do not take into consideration the fundamental qualities of the FIM solution space in order to improve a search, according to our argument. The most essential of these features is that frequent item sets are recursive; if an item set of size k is frequent, then all of its sub-item sets of sizes s = {1,..., k−1} are likewise frequent. This function is crucial to the Apriori algorithm, although it is seldom used by FIM metaheuristic-based approaches. In the literature, we uncover FIM-based metaheuristic algorithms that use the recursive characteristic of frequently recurring item sets in a database. Among these methods are GA-GD [6], PSO-GD [6], and BSO-GD [6], which use a genetic algorithm, particle swarm optimization, and bees swarm optimization, respectively. The approaches GA-GD [6] and PSO-GD [6] are an improvement of the existing algorithms GA-FIM [8] and PSO-FIM [17] by defining new search space intensification and diversification operators, such as crossover and mutation for GA-FIM and particle positioning and velocity for PSO-FIM, that take into account the recursive property of frequent item sets.

This paper provides information on FIM and presents the Apriori algorithm and different types of metaheuristic-based methods in Sect. 2. Section 3 introduces some performance metrics for comparing metaheuristic-based methods. Finally, a conclusion can be found in Sect. 4.

2 Background and Literature Review

2.1 Frequent Itemset Mining

Frequent Itemset Mining (FIM) is an essential part of Data Mining. FIM's goal is to discover all the frequent itemset from a massive dataset, i.e., if an itemset or several itemsets occurs/seem several times in a Data, they are interesting, this is according to a minimum frequency threshold given by a user. This activity was developed in the early 1990s by Agrawal et al. [2] for identifying frequently co-occurring items in market basket analysis. In this section, formal definitions associated with this task are explained.

Definition 1 (Pattern). *A pattern is a collection of elements, events, or things that appear often in a database. Formally, a pattern P in a database D is defined as a subset of elements $P \subseteq \{i_1, \ldots, i_n\} \in D$ that explain important data characteristics.*

Definition 2 (Support). *In a database D, the set of transactions that contain a pattern x is called its coverage. The support of a pattern x, denoted sprt(x, D) is the number of transactions that contain x, i.e., the cardinal of its coverage.*

$$sprt(x, D)_{abs} = |t_k \in D/x \subseteq t_k| \tag{1}$$

The equation gives the absolute value of the support (a positive integer less than or equal to the size of the database), which can also be expressed in relative terms, i.e., by a real number between zero and one, representing a percentage by dividing it by the size of the data set as shown in the formula below.

$$sprt(x, D)_{abs} = \frac{|t_k \in D/x \subseteq t_k|}{|D|} \tag{2}$$

In order to lighten the writing, we simply note it sprt(x) without any abs submention or reference to a transaction base that will be understood from the context.

Definition 3 (Frequent Pattern). *A pattern X is frequent if and only if its support exceeds a minimal threshold s specified in advance. Formally, X is frequent in D relative to the threshold s if and only if: sprt(X) > s.*
The problem of pattern mining consists in enumerating the set $F_D(s)$ of all frequent patterns derived from A present in the transaction base D with respect to the minimal threshold of support s. To simplify the notation, this last set set will, from now on, be noted F for short.

$$F_D(s) = x \in /sprt(x) \geq s \tag{3}$$

2.2 Apriori Algorithm

The Apriori method [2] is based on the iterative and recursive generation of all candidates for itemsets of size k from frequent itemsets of size k−1. This process is performed until no candidates for itemsets are produced during an iteration.

Let us consider a database has 9 transactions: {{a, b, e}, {b, d}, {b, c}, {a, b, d}, {a, c}, {b, c}, {a, c}, {a, b, c, e}, {a, b, c}}. The Apriori heuristic is applied to this database. The steps involved in using the Apriori heuristic with a MinSup of 30% are shown in 2. First, the database is explored to determine the support for each candidate of a single item. The frequent 1-itemsets are then selected (C1). In this instance, the most common itemsets are {a, b, c} (F1), since their supports are more than 0.3. In the second iteration, the candidate itemsets of size 2 (C2) are formed by merging the frequent itemsets of size 1, and then the support of each candidate itemset of size 2 is computed (F2). The most frequent itemsets are {ab, ac, bc} (F2). By concatenating these three frequent itemsets, we get the candidate itemsets of size 3 {abc} (C3)that procedure is terminated since its support is less than 0.3. As an output, we acquire the collection of all frequent itemsets F by concatenating the frequent itemsets of sizes 1 and 2. Consequently, F = {a, b, c, ab, ac, bc}. The search of the complete database at each iteration to assess the support of each potential itemset is the most significant drawback of the Apriori heuristic (Fig. 1).

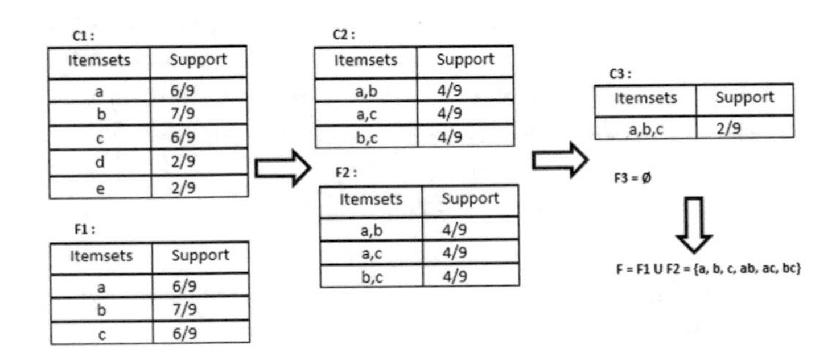

Fig. 1. Apriori algorithm

2.3 Related Work on Metaheuristic-Based FIM Methods

Exact Methods

In addition to the Apriori algorithm, we have numerous exact algorithms, including AIS [2], DHP (Dynamic Hashing and Pruning) [22], DIC (Dynamic Itemset Counting) [4], and FPGrowth [11]. AIS is the first introduced algorithm before Apriori. The Apriori method inspired DHP (Dynamic Hashing and Pruning) and DIC (Dynamic Itemset Counting). The FPGrowth method has been proposed by Han et al. [11]. As a representation of the database, the latter utilizes a compact data structure called FPTree (Frequent Pattern tree), which is a kind of prefix tree or sort [20] enriched with support information. Consequently, each node of the tree has an item and an integer representing the support of the pattern constructed from the objects discovered along the route from the root to this node. Large database instances render exact techniques wasteful in terms of time and space complexity. Numerous causes, including the exponential expansion of self-generated data from several sources, have led to the emergence of incredibly huge databases as contemporary signs [5]. To extract frequent itemsets, more efficient techniques are required for this challenge.

In [24], the PrePost algorithm based on Hadoop is proposed. PrePost is a fresh method for mining frequent itemsets based on N-lists, has been proposed. PrePost typically performs better than other state-of-the-art algorithms at the moment. A Hadoop-based improvement on PrePost that makes greater use of a HashMap to efficiently traverse the PPC tree (Pre-order Post-order Code tree), which is an FP-tree type structure, and accelerates the creation of N-lists linked to 1-itemsets. The authors integrate Hadoop's features with our own in order to process enormous amounts of data. Experience has shown that, in terms of performance and scalability, the HPrePostPlus algorithm outperforms the most recent techniques.

Metaheuristic-Based Methods

In metaheuristic-based intelligence approaches, we discovered genetic algorithms and swarm-based methodologies. For genetic algorithm-based methods, we have GAR [19]. It is the first genetic algorithm-based method for mining association rules and frequent itemsets. Individuals are represented inefficiently in GAR, which is a shortcoming of the algorithm. Individuals are represented based on their size, or the number of items that they contain. Individuals may vary in size across populations, which impairs the efficacy of both crossover and mutation operators. Then, we have AGA [3] and ARMGA [27]. The mutation and crossover operators are the primary distinction between ARMGA and AGA. ARMGA utilizes a two-points crossover, while AGA favors a simple crossover. G3PARM [25] is an alternative approach that use G3P (Grammar Guided Genetic Programming) to prevent locating incorrect people. GAFIM algorithm was developed by [8]. The primary innovation of GAFIM is a deletion and decomposition method that divides infrequent item sets into pairs of frequent item sets. Recently, in [6], the authors suggest an effective method known as GA-GD. It combines the recursive characteristic of the Apriori algorithm with the genetic algorithm's operators to enhance the space exploration of item sets. Similar to GAFIM, GA-GD applies the basic method typical of FIM techniques affected by genetic algorithms. However, GA-initialization, GD's crossover, mutation, and selection operators are specified differently from GAFIM's.

Recent research indicates that swarm intelligence approaches may be effectively used to data mining issues such as feature collection, clustering, and periodic itemsets mining [9]. In [21], the first swarm intelligence-based method using ACO (Ant Colony Optimization) is proposed [15]. It integrates clustering with ACO. This approach is extended for continuous domains by the [16] group. HUIM-ACS is a more contemporary ACO-based approach that modifies the TWU heuristic to guide ants in exploring the solution space. The greatest drawback of ACO-based FIM techniques is runtime efficiency. The authors of [14] recommend particle swarm optimization for the FIM issue. By shifting the front and rear locations of each particle, the neighborhood space is determined. This method beats AGA, although the search based on front and back points produces more neighbourhoods, preferring the intensification search over the diversity search. This algorithm has been updated by PSOFIM to establish a balance between intensity and diversity. In [7], the FIM issue is handled utilizing swarm optimization by bees. The search area of the bees is determined initially, and then one bee searches each region for typical item sets. Using a dance table, the bees coordinate their movements to overlap at each iteration. At each iteration, the bees use a dance table to choose the strongest collection of item sets. According to [12] and [13], genetic algorithms and particle swarm optimization surpass other existing bio-inspired approaches in terms of runtime efficiency and solution consistency when tackling the FIM issue. However, owing to the design of the utilized randomized search mechanism, which does not account for the FIM problem's complexity, the accuracy of the solutions provided by these algorithms remains poor. BATFIM has lately exploited the bat metaheuristic to address the FIM conundrum [28]. A colony of bats explores the same region of the solution space for relevant item sets. There are other ways given for selecting the most repeated bats among the total bats. The results demonstrate that BATFIM outperforms the existing evolutionary and swarm intelligence-based FIM methods. More recently, in [6], the authors describe a technique called as PSO-GD and BSO-GD, which combines the recursive aspect of the Apriori algorithm with the PSO and BSO algorithms. The recursive feature of frequent itemsets is employed to update the particle locations by updating the operators and the areas of bees for the PSO and BSO algorithms, respectively.

Feature selection is one of the most important data preparation procedures for classification problems. In [26], the effective data mining technique of association rules mining based on fuzzy grids is used for feature selection in an application for detecting network abuse. The fundamental purpose of this method is to detect the relationships between items in massive datasets so that it may uncover correlations between system inputs and then decrease duplicate system inputs. A fuzzy ARTMAP neural network with gravitational search algorithm-optimized training parameters is utilized to recognize the attempts. In the same application, the performance of the proposed system and current machine learning approaches are compared. The best "feature subset size-adjustment" parameter improves the performance of the proposed system in terms of detection rate, false alarm rate, and cost per example in classification tasks, as shown by experimental results. In addition, selecting the smaller feature set decreases computational complexity by more than 8.4%.

3 Comparison Performance Metrics

In this section, we are going to present comparative studies between the latest metaheuristic-based methods that exist in literature and the first traditional method, the Apriori algorithm. In order to evaluate the performance of the GA-GD, PSO-GD, and BSO-GD methods, several studies on medium, high, and large databases were done. The algorithms have been implemented in Java and the experiments were done under windows 10 using a laptop equipped with an Intel I5 processor and 8 GB memory.

3.1 Description of Datasets

In these experiments, we have been used well-known data instances. These instances are obtained from an open-source data mining library[1,2]

The various data instances, as well as the number of transactions, number of items, and average number of items per transaction, are shown in Table 1.

Table 1. Data instances description

Instance name	N. of transactions	N. of items	Avg. size of transactions
Zoo	102	17	17
Segment	2310	19	19
Chess	3196	75	37
Mushroom	8124	119	23
Pumbs_star	40385	7116	50
BMS-WebView-1	59602	497	2.5
BMS-WebView-2	77512	3340	5

3.2 Performance Metrics

Any FIM algorithm must have certain performance metrics. FIM techniques may encounter difficulties if certain performance parameters are not reached. In this part, we are going to present two metrics including execution time and the number of frequent itemsets generated.

Execution Time

Without a doubt, the most significant purpose for many FIM algorithms was to provide a quicker algorithm. Its significance is projected to grow in the future, particularly when attempting to examine enormous data sets. Lately, many metaheuristic-based techniques, such as GA-GD, and PSO-GD, attempted to enhance existing FIM algorithms, like Apriori, by extracting frequent patterns in a shorter amount of time.

[1] https://archive.ics.uci.edu/ml/datasets.html, http://fimi.ua.ac.be/data/, and https://sourceforge.net/projects/ibmquestdatagen/.

[2] http://fimi.ua.ac.be/data/, and https://sourceforge.net/projects/ibmquestdatagen/.

In Fig. 2, the results revealed that PSO-GD and GA-GD outperform the Apriori algorithm in computing time. When dealing with small instances (such as Zoo, Segment, Chess, and Mushroom), the execution time is almost the same with all small instances. Moreover, with the largest instances, like, Pumbs_star, BMS_WebView-1, and BMS_WebView-2, the runtime of metaheuristic-based approaches is lower than the Apriori algorithm.

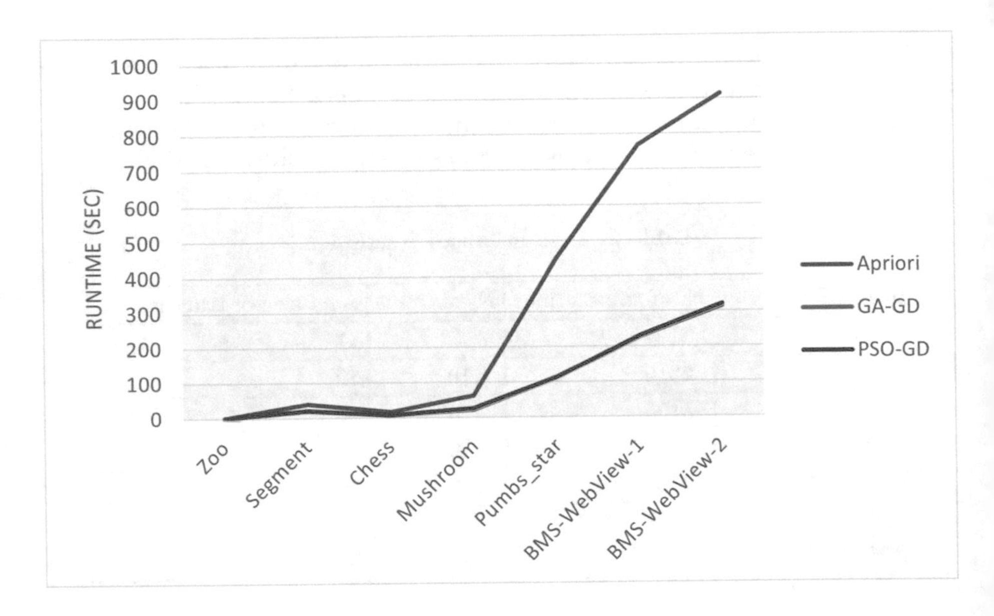

Fig. 2. Comparison of execution time

Number of Frequent Itemsets Generated

Another crucial performance indicator for FIM algorithms is the number of frequently created itemsets. The objective of metaheuristic-based approaches is to enhance the quality solution, i.e., the proportion of frequent itemsets identified to be closer to that of conventional methods, in a short amount of time.

Figure 3 compares the solution quality of the GA-GD, PSO-GD, and BSO-GD algorithms using the data from Table 1. The minimum support for all findings is set at 10%. Notably, the Apriori algorithm identifies every common item collection or 100%. The findings demonstrate that the approaches converge on 100% of identified answers.

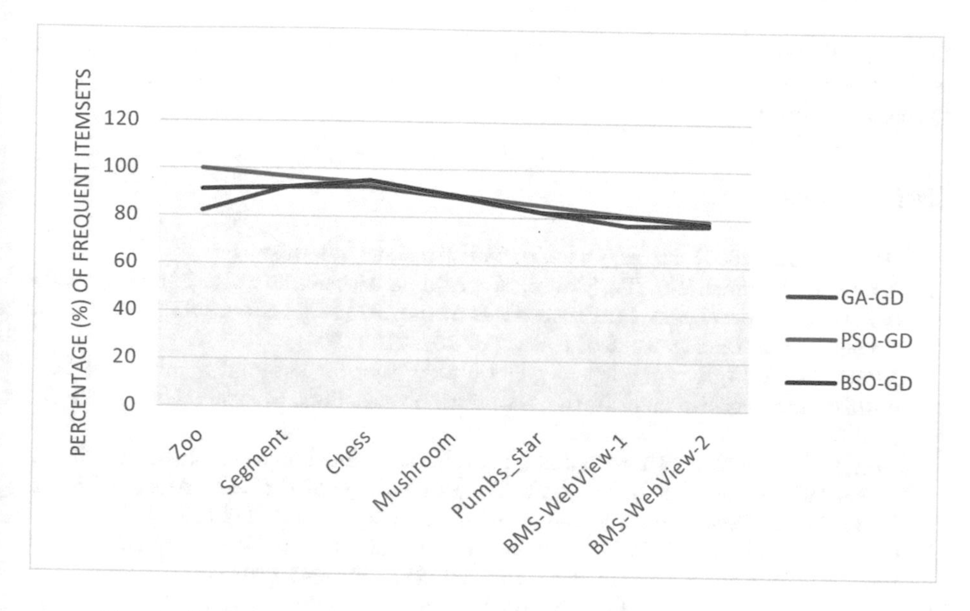

Fig. 3. Comparison in terms of number of frequent itemsets generated

4 Conclusion

This work proposes intriguing ideas for future research to enhance algorithms for extracting frequent items. As we have seen in this work, a number of methods have been developed for discovering frequently occurring things. These include algorithms based on metaheuristics. Recently, three metaheuristic-based solutions to the FIM issue have been offered. The itemset space is explored by combining the recursive feature of frequent itemsets with the stochastic search mechanism of metaheuristics. Using this strategy, three novel metaheuristic-based FIM algorithms have been created. In the first GA-GD, the crossover operator creates itemsets of size k from frequent itemsets of size k−1 for each iteration k, while the mutation operator identifies frequent itemsets from the itemsets formed by the crossover. In the second PSO-GD, the direction and velocity of particles exploring the solution space of frequent itemsets are determined by the recursive property of those itemsets. In the third BSO-GD, the common itemsets detected in the $k-1^{th}$ iteration are utilized to determine the bees' locations in the kth iteration.

Several research on medium, high, and large database instances were conducted to determine the effectiveness of various methods. In terms of runtime complexity, the findings demonstrated that metaheuristic-based techniques beat the Apriori algorithm. In the future, we want to improve the GA-GD method by adjusting the crossover operator or the mutation operator in order to achieve a decent runtime and number of frequently formed item sets. In addition, as we are aware, several metaheuristic-based techniques have been created to operate with distributed frameworks like as Hadoop, MapReduce, and Spark in recent years, and we expect to distribute the metaheuristic-based methods utilizing Apache Spark in future work.

References

1. Data mining (2005). https://www.cs.waikato.ac.nz/ml/weka
2. Agrawal, R., Imieliński, T., Swami, A.: Mining association rules between sets of items in large databases. In: Proceedings of the 1993 ACM SIGMOD International Conference on Management of Data, pp. 207–216 (1993)
3. Alataş, B., Akin, E.: An efficient genetic algorithm for automated mining of both positive and negative quantitative association rules. Soft. Comput. **10**(3), 230–237 (2006)
4. Brin, S., Motwani, R., Ullman, J.D., Tsur, S.: Dynamic itemset counting and implication rules for market basket data. In: Proceedings of the 1997 ACM SIGMOD International Conference on Management of Data, pp. 255–264 (1997)
5. Chen, C.P., Zhang, C.Y.: Data-intensive applications, challenges, techniques and technologies: a survey on big data. Inf. Sci. **275**, 314–347 (2014)
6. Djenouri, Y., Djenouri, D., Belhadi, A., Fournier-Viger, P., Lin, J.C.W.: A new framework for metaheuristic-based frequent itemset mining. Appl. Intell. **48**(12), 4775–4791 (2018)
7. Djenouri, Y., Drias, H., Habbas, Z.: Bees swarm optimisation using multiple strategies for association rule mining. Int. J. Bio-Inspired Comput. **6**(4), 239–249 (2014)
8. Djenouri, Y., Nouali-Taboudjemat, N., Bendjoudi, A.: Association rules mining using evolutionary algorithms. In: The 9th International Conference on Bio-inspired Computing: Theories and Applications (BIC-TA 2014). LNCS. Springer, Cham (2014)
9. Fong, S., Wong, R., Vasilakos, A.V.: Accelerated PSO swarm search feature selection for data stream mining big data. IEEE Trans. Serv. Comput. **9**(1), 33–45 (2015)
10. Junqué de Fortuny, E., Martens, D., Provost, F.: Predictive modeling with big data: is bigger really better? Big Data **1**(4), 215–226 (2013)
11. Han, J., Pei, J., Yin, Y.: Mining frequent patterns without candidate generation. ACM SIGMOD Rec. **29**(2), 1–12 (2000)
12. del Jesus, M.J., Gamez, J.A., Gonzalez, P., Puerta, J.M.: On the discovery of association rules by means of evolutionary algorithms. Wiley Interdiscipl. Rev. Data Min. Knowl. Discov. **1**(5), 397–415 (2011)
13. Krishna, G.J., Ravi, V.: Evolutionary computing applied to customer relationship management: a survey. Eng. Appl. Artif. Intell. **56**, 30–59 (2016)
14. Kuo, R.J., Chao, C.M., Chiu, Y.: Application of particle swarm optimization to association rule mining. Appl. Soft Comput. **11**(1), 326–336 (2011)
15. Kuo, R., Lin, S., Shih, C.: Mining association rules through integration of clustering analysis and ant colony system for health insurance database in Taiwan. Expert Syst. Appl. **33**(3), 794–808 (2007)
16. Kuo, R., Shih, C.: Association rule mining through the ant colony system for national health insurance research database in Taiwan. Comput. Math. Appl. **54**(11–12), 1303–1318 (2007)
17. Lin, J.C.W., et al.: Mining high-utility itemsets based on particle swarm optimization. Eng. Appl. Artif. Intell. **55**, 320–330 (2016)

18. Martín, D., Alcalá-Fdez, J., Rosete, A., Herrera, F.: NICGAR: a niching genetic algorithm to mine a diverse set of interesting quantitative association rules. Inf. Sci. **355**, 208–228 (2016)
19. Mata, J., Alvarez, J.L., Riquelme, J.C.: An evolutionary algorithm to discover numeric association rules. In: Proceedings of the 2002 ACM Symposium on Applied Computing, pp. 590–594 (2002)
20. Mehta, D.P., Sahni, S.: Handbook of Data Structures and Applications. Chapman and Hall/CRC, Boca Raton (2004)
21. Olmo, J.L., Luna, J.M., Romero, J.R., Ventura, S.: Mining association rules with single and multi-objective grammar guided ant programming. Integr. Comput.-Aided Eng. **20**(3), 217–234 (2013)
22. Park, J.S., Chen, M.S., Yu, P.S.: An effective hash-based algorithm for mining association rules. ACM SIGMOD Rec. **24**(2), 175–186 (1995)
23. Piateski, G., Frawley, W.: Knowledge Discovery in Databases. MIT Press, Cambridge (1991)
24. Rochd, Y., Hafidi, I.: Performance improvement of prepost algorithm based on Hadoop for big data. Int. J. Intell. Eng. Syst. **11**(5), 226–235 (2018)
25. Romero, C., Zafra, A., Luna, J.M., Ventura, S.: Association rule mining using genetic programming to provide feedback to instructors from multiple-choice quiz data. Expert. Syst. **30**(2), 162–172 (2013)
26. Sheikhan, M., Sharifi Rad, M.: Gravitational search algorithm-optimized neural misuse detector with selected features by fuzzy grids-based association rules mining. Neural Comput. Appl. **23**(7), 2451–2463 (2013)
27. Yan, X., Zhang, C., Zhang, S.: Genetic algorithm-based strategy for identifying association rules without specifying actual minimum support. Expert Syst. Appl. **36**(2), 3066–3076 (2009)
28. Yang, M.H., et al.: The efficacy of individual-donation and minipool testing to detect low-level hepatitis b virus DNA in Taiwan. Transfusion **50**(1), 65–74 (2010)

Author Index

N. Aboutabit et al. (Eds.): ICMICSA 2022, LNNS 656, pp. 373–374, 2023.
https://doi.org/10.1007/978-3-031-29313-9

Printed in the United States
by Baker & Taylor Publisher Services